Undergraduate Lecture Notes in Physics

Undergraduate Lecture Notes in Physics (ULNP) publishes authoritative texts covering topics throughout pure and applied physics. Each title in the series is suitable as a basis for undergraduate instruction, typically containing practice problems, worked examples, chapter summaries, and suggestions for further reading.

ULNP titles must provide at least one of the following:

- An exceptionally clear and concise treatment of a standard undergraduate subject.
- A solid undergraduate-level introduction to a graduate, advanced, or non-standard subject.
- A novel perspective or an unusual approach to teaching a subject.

ULNP especially encourages new, original, and idiosyncratic approaches to physics teaching at the undergraduate level.

The purpose of ULNP is to provide intriguing, absorbing books that will continue to be the reader's preferred reference throughout their academic career.

More information about this series at http://www.springer.com/series/8917

Victor Ilisie

Lectures in Classical Mechanics

With Solved Problems and Exercises

 Springer

Victor Ilisie
Detectors for Molecular Imaging
Instituto de Instrumentacion para Imagen
Molecular
Valencia, Spain

ISSN 2192-4791   ISSN 2192-4805   (electronic)
Undergraduate Lecture Notes in Physics
ISBN 978-3-030-38584-2   ISBN 978-3-030-38585-9   (eBook)
https://doi.org/10.1007/978-3-030-38585-9

This Springer imprint is published by the registered company Springer Nature Switzerland AG
The registered company address is: Gewerbestrasse 11, 6330 Cham, Switzerland

To Carmen, Alicia and the new one ...

Preface

Practical exercises can be easily found in many classical books in the literature. This book aims to be rather more conceptual than practical. Special emphasis is made not only on learning some formulae and applying them to somewhat random exercises but on understanding the underlying concepts of Classical Mechanics. This is why many exercises that are proposed (and solved) in this book are more conceptual rather than practical. Part of the theory is sometimes presented or being taught also in the form of exercises. This way the reader is challenged to first solve some theoretical exercises that follow immediately (or not so immediately) from the minimum set of explained theory. This is also why, during the presentation of the theoretical part in each chapter, no exercise or problem is proposed. Exercises are left exclusively for the end. This way one is able to follow the line of reasoning of the theoretical concepts with no interruption. Thus the reader is invited to discover concepts, instead of directly being shown all the theory from the beginning. More practical exercises are also considered, but always trying to teach something conceptually new, this is why they try to be non-repetitive.

A second aim of the book is clarity and organization. All related concepts are grouped together in a somewhat rigid manner, in their corresponding chapters and not scattered around, as it is unfortunately done sometimes when teaching or writing. Special emphasis is made on the correct notion of vector, symmetries and invariance, especially when introducing vector analysis in Cartesian and curvilinear coordinates. These concepts are not at all trivial and are oftenly missing from the literature. As any advanced subject, Classical Mechanics is full of subtleties that, as a professor, one assumes as trivial. However, for a student, not being able to infer the answer to these subtle questions might result very frustrating. Thus, all along this book I try, as much as possible, to analyse and give an answer to most of them.

Appendices are added for completeness to describe basic notions of tensor analysis in Cartesian and curvilinear coordinates and for a better understanding of Special Relativity. Other appendices also contain the description of active and passive transformations, a step-by-step deduction of vector operators in curvilinear coordinates, an introduction to Euler's angles, etc. Somewhat difficult concepts which remain mysterious for many students, such as non-inertial reference frames, rigid body motion, variable mass systems, basic tensor algebra and calculus, are extensively described. The equations for the motion in non-inertial reference systems are derived in two independent ways, providing the student with a more complete understanding of the underlying phenomena. Also alternative deductions of the equations of motion for variable mass problems are presented. Lagrangian and Hamiltonian formulations of mechanics are studied for non-relativistic cases, with relevant examples and comments on their limitations also made. Other concepts such as inertial reference frames and the equivalence principle are introduced and commented upon properly. This way the reader is not left with somewhat vague ideas on the problem.

Valencia, Spain   Victor Ilisie

Acknowledgements

I would like to thank Prof. Pepe Peñarrocha, with whom I shared the teaching of this subject, for many interesting conversations and helpful material, that I used for my classes. Some of the solved exercises are inspired in the problem definitions I used during my time at the University of Valencia.

Many thanks to Prof. José María Benlloch, for all his efforts for making my now-a-days home institute (I3M), a warm place to work, think and innovate. I would also like to thank him for reviewing this manuscript.

Thank also, to my always supporting family and specially my parents. Finally, I will always be grateful to my bellowed wife Carmen for always enduring me, and being there for me. Without you, this would not have been possible. And Alicia, thank you for being such an extraordinary and special kid.

Contents

Chapter 1
Vector Analysis in Cartesian Coordinates

Abstract A review of the main aspects of vector analysis in $\mathbb{R}^3$ in Cartesian coordinates is presented. This chapter is intended to serve as a reminder of the basic operations with vectors and the reader is supposed to be at least familiar with the notions presented here. Along this chapter, as it is a very important topic that is often neglected, we introduce and insist upon the notion of basis-invariance as well as the correct definition of the terms scalar, scalar field, vector and vector field and their properties; these properties are also analysed for vector operators. Important conclusions and reflections regarding invariance are presented at the end of the chapter. We briefly mention the concept of active and passive transformations, concept that is explained in greater detail in Appendix C.

1.1 Introduction

In $\mathbb{R}^3$, an orthonormal right-handed reference frame will by given by the set $\{\mathcal{O}, \mathbf{i}, \mathbf{j}, \mathbf{k}\} \equiv \{\mathcal{O}, \mathbf{e}_i\}$ (we shall use both notations indistinctly), as shown in Fig. 1.1. Here $\mathcal{O}$ stands for the origin of the coordinate system. The basis is said to be orthonormal when its components have the *norm/modulus/magnitude* equal to 1 and they are orthogonal under the *inner/scalar/dot product* "·". This means that

$$\mathbf{i} \cdot \mathbf{i} = 1, \quad \mathbf{j} \cdot \mathbf{j} = 1, \quad \mathbf{k} \cdot \mathbf{k} = 1,$$
$$\mathbf{i} \cdot \mathbf{j} = \mathbf{j} \cdot \mathbf{i} = 0, \quad \mathbf{i} \cdot \mathbf{k} = \mathbf{k} \cdot \mathbf{i} = 0, \quad \mathbf{j} \cdot \mathbf{k} = \mathbf{k} \cdot \mathbf{j} = 0. \tag{1.1}$$

The basis is said to be right-handed because it satisfies the following properties under the *vector product* "×"

$$\mathbf{i} \times \mathbf{j} = \mathbf{k}, \quad \mathbf{j} \times \mathbf{k} = \mathbf{i}, \quad \mathbf{k} \times \mathbf{i} = \mathbf{j}. \tag{1.2}$$

Using the second notation, the properties (1.1) and (1.2) can be compactly written as

$$\mathbf{e}_i \cdot \mathbf{e}_j = \delta_{ij}, \quad \mathbf{e}_i \times \mathbf{e}_j = \sum_k \epsilon_{ijk} \mathbf{e}_k, \tag{1.3}$$

© Springer Nature Switzerland AG 2020
V. Ilisie, *Lectures in Classical Mechanics*, Undergraduate Lecture
Notes in Physics, https://doi.org/10.1007/978-3-030-38585-9_1

Fig. 1.1 Cartesian reference
frame with an orthonormal
right-handed basis

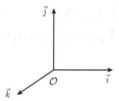

where δ_{ij} is the Kronecker delta defined the usual way ($\delta_{ij} = 0$ for $i \neq j$ and $\delta_{ij} = 1$ for $i = j$) and where ϵ_{ijk} is the totally antisymmetric Levi-Civita symbol, also defined the usual way ($\epsilon_{ijk} \neq 0$ if $i \neq j \neq k \neq i$ with $\epsilon_{123} = 1$ and $\epsilon_{ijk} = \epsilon_{jki} = \epsilon_{kij} = -\epsilon_{jik} = -\epsilon_{ikj} = -\epsilon_{kji}$, otherwise $\epsilon_{ijk} = 0$, i.e., if $i = j$ or $j = k$ or $i = k$).

One can define reference systems that are neither orthonormal nor right-handed but we have no special interest in doing that. We will always use orthonormal right-handed bases (unless stated otherwise) in order to simplify our calculations. Thus, given a basis a vector is written as

$$\mathbf{v} = v_x \mathbf{i} + v_y \mathbf{z} + v_z \mathbf{k} \equiv \sum_i v_i \mathbf{e}_i , \qquad (1.4)$$

where v_i are the components of the vector in the given basis.

1.2   Operations with Vectors

Given a vector $\mathbf{v}$ one can define its squared modulus/magnitude/norm as the inner/scalar product

$$\mathbf{v}^2 \equiv v^2 = \mathbf{v} \cdot \mathbf{v} = v_x^2 + v_y^2 + v_z^2 = \sum_i v_i^2 . \qquad (1.5)$$

Given two vectors $\mathbf{u}$ and $\mathbf{v}$ one can define their inner/dot/scalar product as

$$\mathbf{u} \cdot \mathbf{v} = \mathbf{v} \cdot \mathbf{u} = u_x v_x + u_y v_y + u_z v_z = \sum_i u_i v_i = u v \cos \theta , \qquad (1.6)$$

where u and v are the moduli of the vectors and θ the angle between them. This angle lays within the interval $[0, \pi]$. Schematically this is shown in Fig. 1.2.[1]

Another operation that one can define is *parity* P. Under parity the components of a vector change their sign, i.e., $\mathbf{v}$ transforms as

---

[1]The reader should remember that, in order to give the correct geometric interpretation of the angle θ, both vectors must be placed at a common origin and their arrows must point outwards. More details about the origin, direction, sense, for a given vector will be discussed later on in this chapter.

Fig. 1.2 Schematic representation of the angle *theta* formed by **u** and **v** in two cases, for $\theta < \pi/2$ and for $\theta > \pi/2$

$$P\,\mathbf{v} = -\mathbf{v} = -v_x\,\mathbf{i} - v_y\,\mathbf{j} - v_z\,\mathbf{k} = -\sum_i v_i\,\mathbf{e}_i\,. \tag{1.7}$$

Also, one can define the vector product as the matrix determinant

$$\mathbf{w} = \mathbf{u} \times \mathbf{v} = \begin{vmatrix} \mathbf{i} & \mathbf{j} & \mathbf{k} \\ u_x & u_y & u_z \\ v_x & v_y & v_z \end{vmatrix} = -\mathbf{v} \times \mathbf{u}$$

$$= \mathbf{i}(u_y v_z - v_y u_z) - \mathbf{j}(u_x v_z - v_x u_z) + \mathbf{k}(u_x v_y - v_x u_y)\,, \tag{1.8}$$

or, in terms of $\mathbf{e}_i$ and the Levi-Civita symbol

$$\mathbf{w} = \sum_i w_i\,\mathbf{e}_i\,, \quad \text{with} \quad w_i = \sum_{j,k} \epsilon_{ijk}\,u_i\,v_j\,. \tag{1.9}$$

The element **w** is a *pseudo-vector*, orthogonal to **u** and **v**. It is called pseudo-vector because under parity it does not satisfy the property (1.7) i.e.,

$$P\,\mathbf{w} = (P\,\mathbf{u}) \times (P\,\mathbf{v}) = (-\mathbf{u}) \times (-\mathbf{v}) = \mathbf{w}\,. \tag{1.10}$$

One can explicitly check that it is orthogonal to **u** and **v** (that we shall denote as $\mathbf{w} \perp \mathbf{u}$ and $\mathbf{w} \perp \mathbf{v}$) by checking that the inner products are null

$$\mathbf{w} \cdot \mathbf{u} = 0 = \mathbf{w} \cdot \mathbf{v}\,. \tag{1.11}$$

The modulus of **w** is the area of the parallelogram with its sides given by **u** and **v** and it can be written as

$$w = u\,v\,\sin\theta\,, \tag{1.12}$$

where, again, u and v are the moduli of the vectors and θ the angle between them.

Finally, consider the coordinates of a point-like particle in a given reference frame

$$\mathbf{r} = x\,\mathbf{i} + y\,\mathbf{j} + z\,\mathbf{k}\,. \tag{1.13}$$

An infinitesimal displacement in Cartesian coordinates will be given by

$$d\mathbf{r} = dx\,\mathbf{i} + dy\,\mathbf{j} + dz\,\mathbf{k}. \qquad (1.14)$$

An infinitesimal surface and volume elements will take the form

$$
\begin{aligned}
d\mathbf{s} &= dy\,dz\,\mathbf{i} + dx\,dz\,\mathbf{j} + dx\,dy\,\mathbf{k} \\
&\equiv ds_{yz}\,\mathbf{i} + ds_{xz}\,\mathbf{j} + ds_{xy}\,\mathbf{k}, \\
dV &= dx\,dy\,dz.
\end{aligned}
\qquad (1.15)
$$

1.3 Vector Operators

The *nabla* operator is defined as the following vector operator

$$\nabla \equiv \mathbf{i}\frac{\partial}{\partial x} + \mathbf{j}\frac{\partial}{\partial y} + \mathbf{k}\frac{\partial}{\partial z} \equiv \sum_i \mathbf{e}_i \frac{\partial}{\partial x_i}, \qquad (1.16)$$

that can act over scalar and vector functions as shown next. Given a scalar function (scalar field) $\phi(x, y, z)$ one can define its gradient

$$\nabla\phi = \frac{\partial\phi}{\partial x}\mathbf{i} + \frac{\partial\phi}{\partial y}\mathbf{j} + \frac{\partial\phi}{\partial z}\mathbf{k}. \qquad (1.17)$$

Given a vector function (more correctly called a vector field)

$$\mathbf{F}(x, y, z) = F_x(x, y, z)\mathbf{i} + F_y(x, y, z)\mathbf{j} + F_z(x, y, z)\mathbf{k}, \qquad (1.18)$$

we can define its divergence as

$$\nabla \cdot \mathbf{F} = \frac{\partial F_x}{\partial x} + \frac{\partial F_y}{\partial y} + \frac{\partial F_z}{\partial z}. \qquad (1.19)$$

For a vector field $\mathbf{F}$ one can also define the curl (or rotational) as

$$\nabla \times \mathbf{F} = \begin{vmatrix} \mathbf{i} & \mathbf{j} & \mathbf{k} \\ \dfrac{\partial}{\partial x} & \dfrac{\partial}{\partial y} & \dfrac{\partial}{\partial z} \\ F_x & F_y & F_z \end{vmatrix}. \qquad (1.20)$$

Explicitly the previous equation reads

$$\nabla \times \mathbf{F} = \mathbf{i}\,(\partial_y F_z - \partial_z F_y) - \mathbf{j}\,(\partial_x F_z - \partial_z F_x) + \mathbf{k}\,(\partial_x F_y - \partial_y F_x), \qquad (1.21)$$

where we have introduced the shorthand notation $\partial/\partial x \equiv \partial_x$, etc., that will often come in handy.

1.4   Change of Basis

Before considering the transformations of the components of a vector due to a change of basis we must clarify the **correct** notion and, intrinsic defining properties of what we call vector. A vector in $\mathbb{R}^3$ is an *arrow* which is characterised by its direction (the infinite line along which the vector will be placed), its modulus (the length of the arrow), its sense (towards where the arrow points) and its origin. These are the four defining characteristics of a vector, as schematically shown in Fig. 1.3.

Note that we have described a vector without making use of any coordinate system (or reference frame). Given a reference frame, the vector can be described by explicitly using its components in the given basis. By changing the reference system the **orientation, sense, magnitude and origin remain invariant**, however its components will be different when written in terms of the new basis. Thus we simply *place* a coordinate system or another *next* to the vector in order to describe it. Its intrinsic characteristics do not get modified in any way if we use one reference system or another.

If we do not specify its origin, and consider parallel directions as equivalent, then we are talking about **free** vectors. This is normally the case that we deal with in Cartesian coordinates in Classical Mechanics.

Having made the previous considerations let's consider again the coordinates of a point-like particle in a given reference frame

$$\mathbf{r} = x\,\mathbf{i} + y\,\mathbf{j} + z\,\mathbf{k}. \qquad (1.22)$$

The squared modulus is simply given by $r^2 = \mathbf{r}^2 = x^2 + y^2 + z^2$. Under a translation of the reference frame (as shown in Fig. 1.4) given by the *segment* $\mathbf{c}$ with $\mathbf{c} = \overline{OO'} =$

Fig. 1.3 Intrinsic properties of a vector

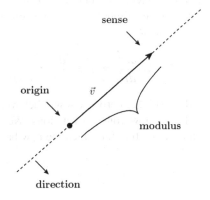

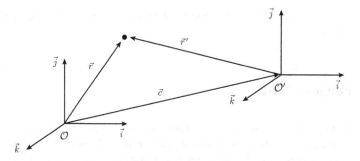

Fig. 1.4 Transformation of the coordinates of a given point originated by a translation of the reference frame

(c_x, c_y, c_z), the coordinates will transform as

$$\mathbf{r} \rightarrow \mathbf{r}' = x'\mathbf{i} + y'\mathbf{j} + z'\mathbf{k} = (x - c_x)\mathbf{i} + (y - c_y)\mathbf{j} + (z - c_z)\mathbf{k}. \qquad (1.23)$$

We can immediately observe that neither one of the intrinsic properties of a vector hold for $\mathbf{r}$ (its direction, sense, etc., have changed). This is why we will avoid using the term *vector* for $\mathbf{r}$ and we shall correctly call its components (as we are already doing), the **coordinates** of a point.

Consider now that the coordinates of a point vary with time t, thus

$$\mathbf{r}(t) = x(t)\mathbf{i} + y(t)\mathbf{j} + z(t)\mathbf{k}, \qquad (1.24)$$

is the trajectory of a moving point-like particle in a the given reference frame. The time derivative of $\mathbf{r}$

$$\mathbf{v} = \frac{d\mathbf{r}}{dt} \equiv \dot{\mathbf{r}} = \frac{dx}{dt}\mathbf{i} + \frac{dy}{dt}\mathbf{j} + \frac{dz}{dt}\mathbf{k} \equiv \dot{x}\mathbf{i} + \dot{y}\mathbf{j} + \dot{z}\mathbf{k}, \qquad (1.25)$$

will therefore describe the velocity of the particle. The velocity of a particle is indeed a well-defined, well-behaved vector. Under translations (due to the fact that $\mathbf{c}$ is constant) its components simply remain invariant. Similarly, the acceleration $\mathbf{a}$ is also a vector

$$\mathbf{a} = \frac{d^2\mathbf{r}}{dt^2} \equiv \ddot{\mathbf{r}} = \frac{d^2x}{dt^2}\mathbf{i} + \frac{d^2y}{dt^2}\mathbf{j} + \frac{d^2z}{dt^2}\mathbf{k} \equiv \ddot{x}\mathbf{i} + \ddot{y}\mathbf{j} + \ddot{z}\mathbf{k}, \qquad (1.26)$$

and also higher derivatives will behave as vectors.

Let us now consider rotations. We shall call a proper rotation of the reference frame, a rotation for which the new basis is also orthonormal and right-handed just as the initial basis.[2] This transformation will be given by an orthogonal matrix with determinant $+1$, i.e., a matrix R with the following two properties

1. $\det(R) = +1$,
2. $R R^T = R^T R = I$ or equivalently $R^T = R^{-1}$,   (1.27)

where I stands for the identity matrix, R^T for the transposed matrix and R^{-1} for the inverse matrix. The relation between the rotated basis (the primed basis) and the original one can be then written in matrix form as

$$\begin{pmatrix} \mathbf{i}' \\ \mathbf{j}' \\ \mathbf{k}' \end{pmatrix} = \begin{pmatrix} R_{11} & R_{12} & R_{13} \\ R_{21} & R_{22} & R_{23} \\ R_{31} & R_{32} & R_{33} \end{pmatrix} \begin{pmatrix} \mathbf{i} \\ \mathbf{j} \\ \mathbf{k} \end{pmatrix} \equiv R \begin{pmatrix} \mathbf{i} \\ \mathbf{j} \\ \mathbf{k} \end{pmatrix}.$$ (1.28)

In terms of $\mathbf{e}_i$ we can write the previous relation as $\mathbf{e}'_i = \sum_j R_{ij}\, \mathbf{e}_j$, where i stands for the row and j for the column. Let's now deduce how the components of a vector change under the previous transformation. Using matrix notation, an arbitrary vector $\mathbf{v}$ is written as

$$\mathbf{v} = (\mathbf{i}, \mathbf{j}, \mathbf{k}) \begin{pmatrix} v_x \\ v_y \\ v_z \end{pmatrix} = (\mathbf{i}', \mathbf{j}', \mathbf{k}') \begin{pmatrix} v'_x \\ v'_y \\ v'_z \end{pmatrix} = (\mathbf{i}, \mathbf{j}, \mathbf{k})\, R^T \begin{pmatrix} v'_x \\ v'_y \\ v'_z \end{pmatrix}.$$ (1.29)

As the components of the basis are independent we find the following

$$\begin{pmatrix} v_x \\ v_y \\ v_z \end{pmatrix} = R^T \begin{pmatrix} v'_x \\ v'_y \\ v'_z \end{pmatrix}.$$ (1.30)

Multiplying both sides of the previous equation by R and using the fact that $R R^T = I$, we finally find

$$\begin{pmatrix} v'_x \\ v'_y \\ v'_z \end{pmatrix} = R \begin{pmatrix} v_x \\ v_y \\ v_z \end{pmatrix},$$ (1.31)

which is the same transformation as (1.28). We conclude, that when dealing with proper orthogonal matrices, the components of a vector (written as a 1-column matrix) obey the same transformation law as the basis.

---

[2]We could work with generic linear transformations of the components of the basis, but we have no special interest in doing that. We are interested in maintaining the orthonormality and right-handiness properties of the bases, thus we will always work with proper rotations unless stated otherwise.

If we consider $\mathbf{v}$ to be the velocity, as R is a constant matrix, taking the time derivative on both sides of the previous equation, one obtains that the acceleration also transforms as a vector i.e.,

$$\begin{pmatrix} a'_x \\ a'_y \\ a'_z \end{pmatrix} = R \begin{pmatrix} a_x \\ a_y \\ a_z \end{pmatrix} . \tag{1.32}$$

If, in (1.29), instead of $\mathbf{v}$ one writes the coordinates $\mathbf{r}$, one obtains that under rotations

$$\begin{pmatrix} x' \\ y' \\ z' \end{pmatrix} = R \begin{pmatrix} x \\ y \\ z \end{pmatrix} . \tag{1.33}$$

We can therefore conclude that $\mathbf{r}$ also behaves like a vector under rotations. However, we must insist that **this is only true in Cartesian coordinates and under rotations**. The previous relation no longer holds true in curvilinear coordinates as we shall see in the next chapter.

So far we have considered transformations of coordinates and of the components of vectors due to a change of the reference system. This type of transformations are called *passive* transformations and it is what we normally use in physics. They correspond to observers from different reference frames that describe the same phenomena. There is also another type of transformation which is called *active*. In this case the reference frame remains unchanged and the (observed) object is being displaced. Both approaches are related. Further details and examples are given in Appendix C.

1.5   Proposed Exercises

1.1 Show that the matrix elements R_{ij} from (1.28) are given by the scalar product:

$$R_{ij} = \mathbf{e}'_i \cdot \mathbf{e}_j . \tag{1.34}$$

Solution: As we have already mentioned, the transformation (1.28) can be written in the simplified form $\mathbf{e}'_i = \sum_k R_{ik} \mathbf{e}_k$. Multiplying both sides of the previous equation by $\mathbf{e}_j$ we obtain

$$\mathbf{e}'_i \cdot \mathbf{e}_j = \sum_k R_{ik} \mathbf{e}_k \cdot \mathbf{e}_j = \sum_k R_{ik} \delta_{kj} = R_{ij} , \tag{1.35}$$

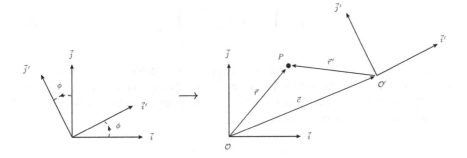

Fig. 1.5 Change of coordinates given by a rotation about the **k** axis and a translation in the (x, y) plane

where we have used the orthonormality property of the basis $\mathbf{e}_i \cdot \mathbf{e}_j = \delta_{ij}$.

1.2 Find the generic expression for the coordinates of a point P in a basis $\{\mathcal{O}', \mathbf{i}', \mathbf{j}', \mathbf{k}'\}$ that is obtained from a rotation of the basis $\{\mathcal{O}, \mathbf{i}, \mathbf{j}, \mathbf{k}\}$ about the **k** axis and a translation in the (x, y) plane, as shown in Fig. 1.5.

Solution: The relation between the coordinates in the original basis and the primed basis is given by

$$\mathbf{r} = \mathbf{c} + \mathbf{r}'. \tag{1.36}$$

Writing $\mathbf{r}$ and $\mathbf{c}$ in the non-primed basis and $\mathbf{r}'$ in the primed basis we obtain

$$(\mathbf{i}, \mathbf{j}, \mathbf{k}) \begin{pmatrix} x \\ y \\ z \end{pmatrix} = (\mathbf{i}, \mathbf{j}, \mathbf{k}) \begin{pmatrix} c_x \\ c_y \\ c_z \end{pmatrix} + (\mathbf{i}', \mathbf{j}', \mathbf{k}') \begin{pmatrix} x' \\ y' \\ z' \end{pmatrix}. \tag{1.37}$$

Introducing the rotation matrix we can write the primed basis in terms of the original basis, as we have already done previously

$$(\mathbf{i}', \mathbf{j}', \mathbf{k}') = (\mathbf{i}, \mathbf{j}, \mathbf{k})\, R^T. \tag{1.38}$$

Plugging this expression back into (1.37) and simplifying we obtain

$$\begin{pmatrix} x \\ y \\ z \end{pmatrix} = \begin{pmatrix} c_x \\ c_y \\ c_z \end{pmatrix} + R^T \begin{pmatrix} x' \\ y' \\ z' \end{pmatrix}. \tag{1.39}$$

Thus, the expression for the coordinates $\mathbf{r}'$ int the primed basis is simply given by

$$\begin{pmatrix} x' \\ y' \\ z' \end{pmatrix} = R \begin{pmatrix} x - c_x \\ y - c_y \\ z - c_z \end{pmatrix}. \tag{1.40}$$

This equation is generic and it holds for any proper orthogonal rotation and a translation, and it will turn out to be useful many times later on. The only thing left now is to particularize it for our case. From Fig. 1.5 we observe that we can write $\mathbf{i}'$ and $\mathbf{j}'$ as a linear combination of $\mathbf{i}$ and $\mathbf{j}$ as

$$\begin{aligned} \mathbf{i}' &= \cos\phi\,\mathbf{i} + \sin\phi\,\mathbf{j}, \\ \mathbf{j}' &= -\sin\phi\,\mathbf{i} + \cos\phi\,\mathbf{j}. \end{aligned} \tag{1.41}$$

The translation is performed in the (x, y) plane, thus $c_z = 0$. Therefore (1.40) particularized to our case simply reads

$$\begin{pmatrix} x' \\ y' \end{pmatrix} = \begin{pmatrix} \cos\phi & \sin\phi \\ -\sin\phi & \cos\phi \end{pmatrix} \begin{pmatrix} x - c_x \\ y - c_y \end{pmatrix}, \tag{1.42}$$

and $z = z'$.

1.3 Show that the following identities hold true for arbitrary scalar and vector fields $\phi = \phi(x, y, z)$ and $\mathbf{F} = F_x(x, y, z)\mathbf{i} + F_y(x, y, z)\mathbf{j} + F_z(x, y, z)\mathbf{k}$:

$$\begin{aligned} \nabla \times (\nabla\phi) &= \mathbf{0}, \\ \nabla \cdot (\nabla \times \mathbf{F}) &= 0. \end{aligned} \tag{1.43}$$

Solution: Introducing the explicit expression of $\nabla\phi$ into the generic expression of the curl one simply obtains

$$\nabla \times (\nabla\phi) = \begin{vmatrix} \mathbf{i} & \mathbf{j} & \mathbf{k} \\ \dfrac{\partial}{\partial x} & \dfrac{\partial}{\partial y} & \dfrac{\partial}{\partial z} \\ \dfrac{\partial\phi}{\partial x} & \dfrac{\partial\phi}{\partial y} & \dfrac{\partial\phi}{\partial z} \end{vmatrix}. \tag{1.44}$$

Expanding the previous determinant, we get

$$\begin{aligned} \nabla \times (\nabla\phi) &= \mathbf{i}\,(\partial_y\partial_z\phi - \partial_z\partial_y\phi) - \mathbf{j}\,(\partial_x\partial_z\phi - \partial_z\partial_x\phi) + \mathbf{k}\,(\partial_x\partial_y\phi - \partial_y\partial_x\phi) \\ &= \mathbf{0}, \end{aligned} \tag{1.45}$$

where we have assumed that ϕ is a (well-behaved) differentiable function i.e., $\partial_x\partial_y\,\phi = \partial_y\partial_x\,\phi$, etc. For the second identity, using the expression for $\nabla \times \mathbf{F}$ from (1.21), we obtain the following

$$\mathbf{\nabla} \cdot (\mathbf{\nabla} \times \mathbf{F}) = \partial_x (\partial_y F_z - \partial_z F_y) - \partial_y (\partial_x F_z - \partial_z F_x) + \partial_z (\partial_x F_y - \partial_y F_x)$$
$$= 0, \tag{1.46}$$

where, again, we have supposed that F_x, F_y and F_z are differentiable functions.

1.4 Show that the following identities hold true for arbitrary vectors **a**, **b**, **c** and for $\mathbf{r} = x\mathbf{i} + y\mathbf{j} + z\mathbf{k}$:

1. $\mathbf{a} \times (\mathbf{b} \times \mathbf{c}) = \mathbf{b}(\mathbf{a} \cdot \mathbf{c}) - \mathbf{c}(\mathbf{a} \cdot \mathbf{b})$, \hfill (1.47)
2. $(\mathbf{a} \times \mathbf{b}) \times \mathbf{c} = \mathbf{b}(\mathbf{a} \cdot \mathbf{c}) - \mathbf{a}(\mathbf{b} \cdot \mathbf{c})$, \hfill (1.48)
3. $\mathbf{\nabla} \times \mathbf{r} = 0$, \hfill (1.49)

4. $\mathbf{a} \cdot \mathbf{b} \times \mathbf{c} = \mathbf{b} \cdot \mathbf{c} \times \mathbf{a} = \mathbf{c} \cdot \mathbf{a} \times \mathbf{b} = \begin{vmatrix} a_1 & a_2 & a_3 \\ b_1 & b_2 & b_3 \\ c_1 & c_2 & c_3 \end{vmatrix}$. \hfill (1.50)

Solution: The demonstrations are straightforward by introducing the explicit expressions in each case as done in the previous exercise.

1.5 Prove that the scalar product of two vectors $\mathbf{u} \cdot \mathbf{v}$ is invariant under rotations.

Solution: Transposing (1.31) we obtain the following transformations of the components of a vector written as a 1-row matrix

$$(v_x', \ v_y', \ v_z') = (v_x, \ v_y, \ v_z) \, R^T. \tag{1.51}$$

Writing the product $\mathbf{u} \cdot \mathbf{v}$ as a matrix product in the primed basis and manipulating the expression we obtain

$$\mathbf{u} \cdot \mathbf{v} = (u_x', \ u_y', \ u_z') \begin{pmatrix} v_x' \\ v_y' \\ v_z' \end{pmatrix} = (u_x, \ u_y, \ u_z) \, R^T R \begin{pmatrix} v_x \\ v_y \\ v_z \end{pmatrix}$$

$$= (u_x, \ u_y, \ u_z) \begin{pmatrix} v_x \\ v_y \\ v_z \end{pmatrix} = u_x v_x + u_y v_y + u_z v_z. \tag{1.52}$$

It is worth mentioning that this is the reason why the inner product is called scalar product, because the result is a scalar, and **a scalar is a basis-independent quantity**.[3]

> **1.6** A scalar function in $\mathbb{R}^3$ by definition is a function that remains invariant under a change of coordinates i.e.,
>
> $$\phi(x', y', z') = \phi(x, y, z). \tag{1.53}$$
>
> Prove that in Cartesian coordinates, the components of $\nabla\phi$ in a given basis are actually the components of a vector.

Solution: We have to show that $\nabla\phi$ transforms as a vector under translations, parity and rotations. In the case of a translation, $\mathbf{i}$, $\mathbf{j}$ and $\mathbf{k}$ remain invariant and, the derivatives with respect to x, y and z do not change as they are only shifted by a constant. In the case of parity, the proof is trivial. The remaining task is to show that

$$\nabla\phi = \frac{\partial\phi}{\partial x}\mathbf{i} + \frac{\partial\phi}{\partial y}\mathbf{j} + \frac{\partial\phi}{\partial z}\mathbf{k} = \frac{\partial\phi}{\partial x'}\mathbf{i'} + \frac{\partial\phi}{\partial y'}\mathbf{j'} + \frac{\partial\phi}{\partial z'}\mathbf{k'}, \tag{1.54}$$

under rotations (1.28). Let's write down $\nabla\phi$ in the primed basis and operate

$$\nabla\phi = \left(\mathbf{i'}, \mathbf{j'}, \mathbf{k'}\right) \begin{pmatrix} \dfrac{\partial\phi}{\partial x'} \\[2mm] \dfrac{\partial\phi}{\partial y'} \\[2mm] \dfrac{\partial\phi}{\partial z'} \end{pmatrix} = (\mathbf{i}, \mathbf{j}, \mathbf{k})\, R^T \begin{pmatrix} \dfrac{\partial\phi}{\partial x'} \\[2mm] \dfrac{\partial\phi}{\partial y'} \\[2mm] \dfrac{\partial\phi}{\partial z'} \end{pmatrix}. \tag{1.55}$$

Thus, we have to prove that

$$R^T \begin{pmatrix} \dfrac{\partial\phi}{\partial x'} \\[2mm] \dfrac{\partial\phi}{\partial y'} \\[2mm] \dfrac{\partial\phi}{\partial z'} \end{pmatrix} = \begin{pmatrix} \dfrac{\partial\phi}{\partial x} \\[2mm] \dfrac{\partial\phi}{\partial y} \\[2mm] \dfrac{\partial\phi}{\partial z} \end{pmatrix}. \tag{1.56}$$

Transposing we get an equivalent relation

---

[3]If one takes $\mathbf{u} = \mathbf{v}$, then the result of the scalar product is the modulus of a vector, which is by definition, basis-invariant.

$$\left(\frac{\partial \phi}{\partial x'}, \frac{\partial \phi}{\partial y'}, \frac{\partial \phi}{\partial z'}\right) R = \left(\frac{\partial \phi}{\partial x}, \frac{\partial \phi}{\partial y}, \frac{\partial \phi}{\partial z}\right). \tag{1.57}$$

We can now conveniently introduce the transformation (1.33) that we have previously analysed for Cartesian coordinates under rotations. This transformation reads

$$\begin{aligned} x' &= R_{11} x + R_{12} y + R_{13} z, \\ y' &= R_{21} x + R_{22} y + R_{23} z, \\ z' &= R_{31} x + R_{32} y + R_{33} z. \end{aligned} \tag{1.58}$$

As R is a constant matrix, one can take derivatives with respect to x, y and z on both sides of the previous expressions and thus re-express the components of R as

$$R_{11} = \frac{\partial x'}{\partial x}, \qquad R_{12} = \frac{\partial x'}{\partial y}, \qquad R_{13} = \frac{\partial x'}{\partial z}, \tag{1.59}$$

and so on. Calculating the rest of the matrix components, and inserting these expression into (1.57) and using the chain rule i.e.,

$$\frac{\partial \phi}{\partial x'}\frac{\partial x'}{\partial x} + \frac{\partial \phi}{\partial y'}\frac{\partial y'}{\partial x} + \frac{\partial \phi}{\partial z'}\frac{\partial z'}{\partial x} = \frac{\partial \phi}{\partial x}, \tag{1.60}$$

one obtains the desired result.

The statement we have just proven for Cartesian coordinates is actually much more general. It holds for any generic transformations of coordinates. We will check that this statement is true for curvilinear coordinates in the next chapter. The next exercise is also related to the concept of basis-invariance and some further comments will be made at the end of the chapter.

> **1.7** Show that in Cartesian coordinates, the quantity $\mathbf{V} \cdot \mathbf{A}$ is basis independent, where $\mathbf{A} = A_x(x, y, z)\mathbf{i} + A_y(x, y, z)\mathbf{j} + A_z(x, y, z)\mathbf{k}$ is a vector field.

Solution: Again we have three cases, as in the previous exercise. Let us first clarify the notion of vector field. A vector field is a function (map/application) that assigns a vector to each point in space. Thus, for a given point in space, with coordinates (x_0, y_0, z_0) in a certain reference frame, there will be a vector associated to it and it will be given by $\mathbf{A}_0 \equiv \mathbf{A}(x_0, y_0, z_0)$. Under a translation of the system of coordinates, by definition a vector does not suffer any transformations and thus, we can conclude that the vector field remains invariant. For parity, the demonstration is also trivial. Again, the remaining task is to show that

$$\mathbf{V} \cdot \mathbf{A} = \frac{\partial A_x}{\partial x} + \frac{\partial A_y}{\partial y} + \frac{\partial A_z}{\partial z} = \frac{\partial A'_x}{\partial x'} + \frac{\partial A'_y}{\partial y'} + \frac{\partial A'_z}{\partial z'}, \qquad (1.61)$$

under rotations. Just as in the previous exercise, writing $\mathbf{V} \cdot \mathbf{A}$ in the primed basis and introducing the components of the matrix R as in (1.59) when transforming the components of $\mathbf{A}$ from the primed basis to the non-primed one, we obtain

$$\frac{\partial A'_x}{\partial x'} + \frac{\partial A'_y}{\partial y'} + \frac{\partial A'_z}{\partial z'} = \frac{\partial}{\partial x'} \left(\frac{\partial x'}{\partial x} A_x + \frac{\partial x'}{\partial y} A_y + \frac{\partial x'}{\partial z} A_z \right) + \cdots . \qquad (1.62)$$

Keeping in mind that the components of the matrix R are constants, i.e.,

$$\frac{\partial}{\partial x'} \left(\frac{\partial x'}{\partial x} \right) = 0, \qquad \frac{\partial}{\partial x'} \left(\frac{\partial x'}{\partial y} \right) = 0, \qquad \text{etc.,} \qquad (1.63)$$

and rearranging terms, one finally obtains

$$\left(\frac{\partial x'}{\partial x} \frac{\partial}{\partial x'} + \frac{\partial y'}{\partial x} \frac{\partial}{\partial y'} + \frac{\partial z'}{\partial x} \frac{\partial}{\partial z'} \right) A_x + \cdots = \frac{\partial A_x}{\partial x} + \frac{\partial A_y}{\partial y} + \frac{\partial A_z}{\partial z}. \qquad (1.64)$$

Again, this result is more general and we shall check explicitly that it also holds for the case of curvilinear coordinates, in the next chapter.

1.8 Check explicitly that in Cartesian coordinates, the quantity $\mathbf{V}^2 \phi$ is a scalar function (where obviously ϕ is a scalar function) and that $\mathbf{a} \times \mathbf{b}$ and $\mathbf{V} \times \mathbf{A}$ (with $\mathbf{a}$ an $\mathbf{b}$ vectors and $\mathbf{A}$ a vector field) are pseudo-vectors.

Solution: Using the tools given in the previous exercises together with the theoretical introduction, one can easily check the veracity of the previous affirmations.

Using tensor formalism all these results can be easily generalized for **any transformation of coordinates**. However tensor algebra and tensor calculus are topics that are too advanced for this course and shall only be treated here in some detail. One can check Appendices A and B for a brief introduction to tensors in Cartesian coordinates, Minkowsky space (Special Relativity) and curvilinear coordinates. We must insist however, that Appendix A is only recommended for Special Relativity and that Appendix B is only written for completeness and for the interested reader, and its lecture its optional.

We have insisted upon the notion of reference-frame invariance and the notion of basis-independent quantities for one reason. All physics is built upon such concepts. If this weren't true, observers from different reference frames would describe the same physical phenomena using different equations. Could we, in that case, even talk about physical laws? All equations that describe Nature (at least the ones known up till now) obey such symmetries. We will see this in more detail later on. As one goes deeper into the study of physics, one realises that it would had been impossible to come so far without relying on, or having previously analysed symmetries. All the way from the realm of particle physics to solid-state physics, to General Relativity and cosmology, everything is related to some symmetry group or some type of invariance under certain transformations. Symmetries are of course present, perhaps in a more intuitive way, in Classical Mechanics, as we shall discover along this book.

Further Reading

1. J.V. José, E.J. Saletan, *Classical Dynamics: A Contemporary Approach*. Cambridge University Press
2. S.T. Thornton, J.B. Marion, *Classical Dynamics of Particles and Systems*
3. H. Goldstein, C. Poole, J. Safko, *Classical Mechanics*, 3rd edn. Addison Wesley
4. J.R. Taylor, *Classical Mechanics*
5. D.T. Greenwood, *Classical Dynamics*. Prentice-Hall Inc.
6. D. Kleppner, R. Kolenkow, *An Introduction to Mechanics*
7. C. Lanczos, *The Variational Principles of Mechanics*. Dover Publications Inc.
8. W. Greiner, *Classical Mechanics: Systems of Particles and Hamiltonian Dynamics*. Springer
9. H.C. Corben, P. Stehle, *Classical Mechanics*, 2nd edn. Dover Publications Inc.
10. T.W.B. Kibble, F.H. Berkshire, *Classical Mechanics*. Imperial College Press
11. M.G. Calkin, *Lagrangian and Hamiltonian Mechanics*
12. A.J. French, M.G. Ebison, *Introduction to Classical Mechanics*

Chapter 2
Vector Analysis in Curvilinear Coordinates

Abstract In this chapter, we introduce the generic concept of curvilinear coordinates and study the three relevant cases, i.e., cylindrical, polar, and spherical. We analyse the properties of the new bases and explain the peculiarities of vectors when compared to Cartesian coordinates. Again, we insist upon the notion of basis invariant quantities, which are not at all obvious, related to vectors and vector operators, always comparing them to the Cartesian case. We derive some results on vector operators using curvilinear coordinates and deduce the rest in one of the appendices.

2.1 Introduction

Consider the Cartesian coordinates of a point-like particle

$$\mathbf{r} = x\,\mathbf{i} + y\,\mathbf{j} + z\,\mathbf{k}. \tag{2.1}$$

If x, y, z are differentiable functions of some independent parameters α, β, γ, i.e., $x = x(\alpha, \beta, \gamma)$, $y = y(\alpha, \beta, \gamma)$ and $z = z(\alpha, \beta, \gamma)$, then we can define some new coordinate basis $\{\mathbf{u}_\alpha, \mathbf{u}_\beta, \mathbf{u}_\gamma\}$ given by the partial derivatives of $\mathbf{r}$ with respect to $\alpha,\ \beta$ and γ:

$$
\begin{aligned}
\mathbf{u}_\alpha &= \frac{\partial \mathbf{r}/\partial \alpha}{|\partial \mathbf{r}/\partial \alpha|} \equiv \frac{1}{h_\alpha}\partial_\alpha \mathbf{r} = \frac{1}{h_\alpha}\left(\frac{\partial x}{\partial \alpha}\mathbf{i} + \frac{\partial y}{\partial \alpha}\mathbf{j} + \frac{\partial z}{\partial \alpha}\mathbf{k}\right), \\
\mathbf{u}_\beta &= \frac{\partial \mathbf{r}/\partial \beta}{|\partial \mathbf{r}/\partial \beta|} \equiv \frac{1}{h_\beta}\partial_\beta \mathbf{r} = \frac{1}{h_\beta}\left(\frac{\partial x}{\partial \beta}\mathbf{i} + \frac{\partial y}{\partial \beta}\mathbf{j} + \frac{\partial z}{\partial \beta}\mathbf{k}\right), \\
\mathbf{u}_\gamma &= \frac{\partial \mathbf{r}/\partial \gamma}{|\partial \mathbf{r}/\partial \gamma|} \equiv \frac{1}{h_\gamma}\partial_\gamma \mathbf{r} = \frac{1}{h_\gamma}\left(\frac{\partial x}{\partial \gamma}\mathbf{i} + \frac{\partial y}{\partial \gamma}\mathbf{j} + \frac{\partial z}{\partial \gamma}\mathbf{k}\right).
\end{aligned}
\tag{2.2}
$$

In each case, we have divided each partial derivative by its magnitude

$$h_\alpha = |\partial_\alpha \mathbf{r}| = (\partial_\alpha \mathbf{r} \cdot \partial_\alpha \mathbf{r})^{1/2}, \tag{2.3}$$

© Springer Nature Switzerland AG 2020
V. Ilisie, *Lectures in Classical Mechanics*, Undergraduate Lecture
Notes in Physics, https://doi.org/10.1007/978-3-030-38585-9_2

and similar for β and γ, explicitly,

$$h_\alpha = \sqrt{\left(\frac{\partial x}{\partial \alpha}\right)^2 + \left(\frac{\partial y}{\partial \alpha}\right)^2 + \left(\frac{\partial z}{\partial \alpha}\right)^2}, \tag{2.4}$$

etc., in order to ensure that the new basis is unitary.

For an arbitrary set of curvilinear coordinates (α, β, γ) the unitary basis we have just defined is not necessarily orthogonal i.e., $\mathbf{u}_\alpha \cdot \mathbf{u}_\beta \neq 0$ for some α and β. However as we will only be working with **cylindrical, polar** and **spherical** coordinates, these bases **will turn out to be orthogonal and also right handed**:

$$\mathbf{u}_\alpha \cdot \mathbf{u}_\beta = 0, \qquad \mathbf{u}_\alpha \cdot \mathbf{u}_\gamma = 0, \qquad \mathbf{u}_\beta \cdot \mathbf{u}_\gamma = 0,$$
$$\mathbf{u}_\alpha \times \mathbf{u}_\beta = \mathbf{u}_\gamma, \qquad \mathbf{u}_\beta \times \mathbf{u}_\gamma = \mathbf{u}_\alpha, \qquad \mathbf{u}_\gamma \times \mathbf{u}_\alpha = \mathbf{u}_\beta. \tag{2.5}$$

This means that the matrix R that accounts for the change of basis (2.2) given by

$$\begin{pmatrix} \mathbf{u}_\alpha \\ \mathbf{u}_\beta \\ \mathbf{u}_\gamma \end{pmatrix} = \begin{pmatrix} \dfrac{1}{h_\alpha}\dfrac{\partial x}{\partial \alpha} & \dfrac{1}{h_\alpha}\dfrac{\partial y}{\partial \alpha} & \dfrac{1}{h_\alpha}\dfrac{\partial z}{\partial \alpha} \\ \dfrac{1}{h_\beta}\dfrac{\partial x}{\partial \beta} & \dfrac{1}{h_\beta}\dfrac{\partial y}{\partial \beta} & \dfrac{1}{h_\beta}\dfrac{\partial z}{\partial \beta} \\ \dfrac{1}{h_\gamma}\dfrac{\partial x}{\partial \gamma} & \dfrac{1}{h_\gamma}\dfrac{\partial y}{\partial \gamma} & \dfrac{1}{h_\gamma}\dfrac{\partial z}{\partial \gamma} \end{pmatrix} \begin{pmatrix} \mathbf{i} \\ \mathbf{j} \\ \mathbf{k} \end{pmatrix} \equiv R \begin{pmatrix} \mathbf{i} \\ \mathbf{j} \\ \mathbf{k} \end{pmatrix}, \tag{2.6}$$

will be orthogonal: $R^T R = R R^T = I$.

Therefore, given an arbitrary vector $\mathbf{v}$ in Cartesian coordinates

$$\mathbf{v} = v_x \mathbf{i} + v_y \mathbf{j} + v_z \mathbf{k}, \tag{2.7}$$

in curvilinear coordinates it will read

$$\mathbf{v} = v_\alpha \mathbf{u}_\alpha + v_\beta \mathbf{u}_\beta + v_\gamma \mathbf{u}_\gamma, \tag{2.8}$$

with its components in the new basis given by the transformation

$$\begin{pmatrix} v_\alpha \\ v_\beta \\ v_\gamma \end{pmatrix} = R \begin{pmatrix} v_x \\ v_y \\ v_z \end{pmatrix}. \tag{2.9}$$

In order to deduce this transformation we may use same arguments as in (1.29). An infinitesimal displacement, in terms of the coordinates in the new basis can be written as

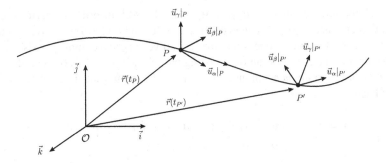

Fig. 2.1 Local bases in curvilinear coordinates at two different points of the trajectory of a point-like particle

$$d\mathbf{r} = \frac{\partial \mathbf{r}}{\partial \alpha} d\alpha + \frac{\partial \mathbf{r}}{\partial \beta} d\beta + \frac{\partial \mathbf{r}}{\partial \gamma} d\gamma = h_\alpha \, d\alpha \, \mathbf{u}_\alpha + h_\alpha \, d\beta \, \mathbf{u}_\beta + h_\gamma \, d\gamma \, \mathbf{u}_\gamma. \quad (2.10)$$

The velocity vector will be thus given by

$$\mathbf{v} = \dot{\mathbf{r}} = h_\alpha \, \dot{\alpha} \, \mathbf{u}_\alpha + h_\beta \, \dot{\beta} \, \mathbf{u}_\beta + h_\gamma \, \dot{\gamma} \, \mathbf{u}_\gamma \equiv v_\alpha \, \mathbf{u}_\alpha + v_\beta \, \mathbf{u}_\beta + v_\gamma \, \mathbf{u}_\gamma, \quad (2.11)$$

where we have introduced the "dot" notation for time derivative as in the previous chapter ($\dot{\mathbf{r}} \equiv d\mathbf{r}/dt$, $\dot{\alpha} \equiv d\alpha/dt$, etc.). At first sight, checking that the components of the velocity vector transform as in (2.9) might not be straightforward. Obviously they have to, otherwise this new basis would be ill defined. It will be thus a nice exercise to explicitly check that this holds true for cylindrical and spherical coordinates at the end of this chapter.

A peculiar property of the newly defined basis is that it is **local**. The components of the basis are not fixed as in Cartesian coordinates, but vary for each point in space i.e., they are functions of the coordinates (α, β, γ)

$$\mathbf{u}_\alpha = \mathbf{u}_\alpha(\alpha, \beta, \gamma), \qquad \mathbf{u}_\beta = \mathbf{u}_\beta(\alpha, \beta, \gamma), \qquad \mathbf{u}_\gamma = \mathbf{u}_\gamma(\alpha, \beta, \gamma). \quad (2.12)$$

This is schematically shown in Fig. 2.1, where we have considered the trajectory of a point-like particle $\mathbf{r}(t)$.

Thus, in order to completely specify a reference system in curvilinear coordinates we will need an origin P, with coordinates given by $\mathbf{r}(t_P)$, and the components of the basis evaluated in P. The reference system for every point P of the trajectory $\mathbf{r}(t)$ will be given by

$$\{P, \mathbf{u}_\alpha|_P, \mathbf{u}_\beta|_P, \mathbf{u}_\gamma|_P\}. \quad (2.13)$$

Note that in curvilinear coordinates, as the basis is local (depends on the point P), **it does not make any sense to operate with vectors with different origins** i.e., a vector product such as $\mathbf{u}_\alpha|_P \times \mathbf{u}_\gamma|_{P'}$ is absurd.[1] In order to perform such operations consistently with vectors defined in different bases (with different origins) in curvilinear coordinates, one has to express all vectors in a common basis.

Moving on, the acceleration of a moving particle will by given by both the time derivative of the components of the velocity ($\dot{v}_\alpha$, $\dot{v}_\beta$, and $\dot{v}_\gamma$) and the time derivatives of the basis

$$
\begin{aligned}
\mathbf{a} = \dot{\mathbf{v}} &= \dot{v}_\alpha\,\mathbf{u}_\alpha + v_\alpha\,\dot{\mathbf{u}}_\alpha + \dot{v}_\beta\,\mathbf{u}_\beta + v_\beta\,\dot{\mathbf{u}}_\beta + \dot{v}_\gamma\,\mathbf{u}_\gamma + v_\gamma\,\dot{\mathbf{u}}_\gamma \\
&\equiv a_\alpha\,\mathbf{u}_\alpha + a_\beta\,\mathbf{u}_\beta + a_\gamma\,\mathbf{u}_\gamma .
\end{aligned}
\tag{2.14}
$$

We are able to write down $\mathbf{a}$ as in the last line of the previous equation because the time derivatives of the components of the basis can be written as a linear combination of the basis components themselves i.e.,

$$
\dot{\mathbf{u}}_\alpha = a\,\mathbf{u}_\alpha + b\,\mathbf{u}_\beta + c\,\mathbf{u}_\gamma ,
\tag{2.15}
$$

and so on. This obviously has to hold because our basis is complete and thus any vector can be written as a linear combination of the components of the given basis. The components of the acceleration (a_α, a_β and a_γ) as defined previously in (2.14) **will also obey the transformation** (2.9). Again this is not obvious at first sight, but this transformation has to hold as $\mathbf{a}$ is a well defined vector. Just as in the case of $\mathbf{v}$ we shall check as an exercise that this holds true in the case of cylindrical and spherical coordinates.

One source of confusion might be (2.9) itself. By taking time derivatives on both sides of this equation, one will obtain an additional term proportional to dR/dt (which is absent in Cartesian coordinates because R is constant) and therefore this transformation might seem not to hold for the acceleration. This approach is wrong, as $\dot{v}_\alpha \neq a_\alpha$, $\dot{v}_\beta \neq a_\beta$ and $\dot{v}_\gamma \neq a_\gamma$ in curvilinear coordinates, which is clear from (2.14).

Let us now continue and deduce the expression for a differential volume and surface in curvilinear coordinates. Introducing the Jacobian matrix, a differential volume simply takes the form

$$
dV = dx\,dy\,dz = J\,d\alpha\,d\beta\,d\gamma ,
\tag{2.16}
$$

where J as usual, is given by the determinant

$$
J = \left| \frac{\partial(x, y, z)}{\partial(\alpha, \beta, \gamma)} \right| =
\begin{vmatrix}
\dfrac{\partial x}{\partial \alpha} & \dfrac{\partial x}{\partial \beta} & \dfrac{\partial x}{\partial \gamma} \\[6pt]
\dfrac{\partial y}{\partial \alpha} & \dfrac{\partial y}{\partial \beta} & \dfrac{\partial y}{\partial \gamma} \\[6pt]
\dfrac{\partial z}{\partial \alpha} & \dfrac{\partial z}{\partial \beta} & \dfrac{\partial z}{\partial \gamma}
\end{vmatrix}
= \left| \frac{\partial \mathbf{r}}{\partial \alpha} \cdot \left(\frac{\partial \mathbf{r}}{\partial \beta} \times \frac{\partial \mathbf{r}}{\partial \gamma} \right) \right| .
\tag{2.17}
$$

---

[1] We are accustomed to do that in Cartesian coordinates because we normally work with free vectors (as the basis does not depend on P).

In order to check the last equality, one can write $\mathbf{r}$ in Cartesian coordinates as

$$\mathbf{r} = x(\alpha, \beta, \gamma)\mathbf{i} + y(\alpha, \beta, \gamma)\mathbf{j} + z(\alpha, \beta, \gamma)\mathbf{k}, \tag{2.18}$$

and perform the vector and scalar products in Cartesian coordinates. Introducing (2.2), the previous expression takes the form

$$J = h_\alpha h_\beta h_\gamma \left| \mathbf{u}_\alpha \cdot (\mathbf{u}_\beta \times \mathbf{u}_\gamma) \right| = h_\alpha h_\beta h_\gamma. \tag{2.19}$$

For the last equality we have used (2.5). We conclude that in orthogonal curvilinear coordinates dV reads

$$dV = h_\alpha h_\beta h_\gamma \, d\alpha \, d\beta \, d\gamma. \tag{2.20}$$

In a similar way one can also deduce the expression for $d\mathbf{s}$

$$
\begin{aligned}
d\mathbf{s} &= h_\alpha h_\beta \, d\alpha \, d\beta \, \mathbf{u}_\gamma + h_\alpha h_\gamma \, d\alpha \, d\gamma \, \mathbf{u}_\beta + h_\beta h_\gamma \, d\beta \, d\gamma \, \mathbf{u}_\alpha \\
&\equiv ds_{\alpha\beta} \, \mathbf{u}_\gamma + ds_{\alpha\gamma} \, \mathbf{u}_\beta + ds_{\gamma\beta} \, \mathbf{u}_\alpha.
\end{aligned} \tag{2.21}
$$

2.2 Vector Operators

We will now deduce the generic expression for the gradient in curvilinear coordinates. Consider the infinitesimal quantity $d\phi$ for a scalar function $\phi(x, y, z)$ in Cartesian coordinates. It can be written in terms of its gradient $\nabla\phi$ and the differential displacement $d\mathbf{r}$ as follows

$$d\phi = \partial_x \phi \, dx + \partial_y \phi \, dy + \partial_z \phi \, dz = \nabla\phi \cdot d\mathbf{r}. \tag{2.22}$$

If we wish to re-express $d\phi$ in terms of curvilinear coordinates, the previous expression turns into

$$
\begin{aligned}
d\phi &= \partial_\alpha \phi \, d\alpha + \partial_\beta \phi \, d\beta + \partial_\gamma \phi \, d\gamma \\
&= \nabla\phi \cdot \frac{\partial \mathbf{r}}{\partial \alpha} \, d\alpha + \nabla\phi \cdot \frac{\partial \mathbf{r}}{\partial \beta} \, d\beta + \nabla\phi \cdot \frac{\partial \mathbf{r}}{\partial \gamma} \, d\gamma \\
&= h_\alpha \, d\alpha \, \nabla\phi \cdot \mathbf{u}_\alpha + h_\beta \, d\beta \, \nabla\phi \cdot \mathbf{u}_\beta + h_\gamma \, d\gamma \, \nabla\phi \cdot \mathbf{u}_\gamma
\end{aligned} \tag{2.23}
$$

Because $d\phi$ is a basis-independent quantity[2] both expressions (in curvilinear and Cartesian coordinates) must be equal. One can finally conclude that

---

[2] $d\phi$ is the infinitesimal increment of the function ϕ, and just as in the case of vectors, by definition, it cannot depend on any particular choice of coordinates!

$$\nabla\phi \;=\; \frac{1}{h_\alpha}\frac{\partial\phi}{\partial\alpha}\,\mathbf{u}_\alpha + \frac{1}{h_\beta}\frac{\partial\phi}{\partial\beta}\,\mathbf{u}_\beta + \frac{1}{h_\gamma}\frac{\partial\phi}{\partial\gamma}\,\mathbf{u}_\gamma\,. \tag{2.24}$$

For the expression of the divergence $\nabla\cdot\mathbf{A}$ of a vector field $\mathbf{A}$ in curvilinear coordinates, looking at (2.24) one could be tempted to generically define the ∇ operator as $\nabla \equiv h_\alpha^{-1}\,\partial/\partial\alpha\,\mathbf{u}_\alpha + \cdots$, however this would be the wrong approach. First, it would lead to an ambiguity with respect to when the derivation must take place. Is it either before or after performing the scalar product of $\mathbf{A}$ with the components of the basis $\mathbf{u}_\alpha$, $\mathbf{u}_\beta$, and $\mathbf{u}_\gamma$ from ∇? Remember that now the basis depends on α, β, γ and performing the scalar product before or after the derivation would lead to different results. Either way the operation would not lead to the correct answer.

There are a few ways of deducing the correct result and none of them is trivial (if one does not use tensor formalism). Here we prefer to use the geometric approach as it seems to be the most intuitive. The deduction for the expressions of $\nabla\cdot\mathbf{A}$ and $\nabla\times\mathbf{A}$ can be found in Appendix D. Here we shall directly present the results:

$$\nabla\cdot\mathbf{A} \;=\; \frac{1}{h_\alpha h_\beta h_\gamma}\left[\frac{\partial}{\partial\alpha}\big(A_\alpha\,h_\beta\,h_\gamma\big) + \frac{\partial}{\partial\beta}\big(A_\beta\,h_\alpha\,h_\gamma\big)\right.$$
$$\left. + \frac{\partial}{\partial\gamma}\big(A_\gamma\,h_\alpha\,h_\beta\big)\right], \tag{2.25}$$

and

$$\nabla\times\mathbf{A} \;=\; \frac{1}{h_\alpha h_\beta h_\gamma}\begin{vmatrix} h_\alpha\,\mathbf{u}_\alpha & h_\beta\,\mathbf{u}_\beta & h_\gamma\,\mathbf{u}_\gamma \\[4pt] \dfrac{\partial}{\partial\alpha} & \dfrac{\partial}{\partial\beta} & \dfrac{\partial}{\partial\gamma} \\[8pt] h_\alpha\,A_\alpha & h_\beta\,A_\beta & h_\gamma\,A_\gamma \end{vmatrix}. \tag{2.26}$$

As it was already mentioned in the previous chapter, the components of $\nabla\phi$ and $\nabla\times\mathbf{A}$ will obey (2.9), as they are the components of a vector and a pseudo-vector. Also $\nabla\cdot\mathbf{A}$ will be basis invariant. We shall verify these properties later on with a few exercises.

2.3 Cylindrical and Polar Coordinates

Let us now particularize the previously obtained results to cylindrical coordinates where $(\alpha,\,\beta,\,\gamma) = (\rho,\,\phi,\,z)$. The relation between Cartesian and cylindrical coordinates is given by

$$x \;=\; \rho\cos\phi\,, \qquad y \;=\; \rho\sin\phi\,, \qquad z \;=\; z\,, \tag{2.27}$$

with $0 \leqslant \phi < 2\pi$ and $\rho = \sqrt{x^2 + y^2}$. Therefore the new basis is simply

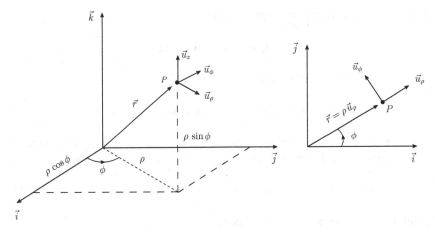

Fig. 2.2 Cartesian and cylindrical (left) or polar (right) coordinates of a given point P. The basis $\{\mathbf{i}, \mathbf{j}, \mathbf{k}\}$ is not drawn to scale

$$
\begin{aligned}
\mathbf{u}_\rho &= \cos\phi\,\mathbf{i} + \sin\phi\,\mathbf{j}, \\
\mathbf{u}_\phi &= -\sin\phi\,\mathbf{i} + \cos\phi\,\mathbf{j}, \\
\mathbf{u}_z &= \mathbf{k}.
\end{aligned}
\tag{2.28}
$$

Written in matrix form

$$
\begin{pmatrix} \mathbf{u}_\rho \\ \mathbf{u}_\phi \\ \mathbf{u}_z \end{pmatrix} = R \begin{pmatrix} \mathbf{i} \\ \mathbf{j} \\ \mathbf{k} \end{pmatrix}, \qquad
R = \begin{pmatrix} \cos\phi & \sin\phi & 0 \\ -\sin\phi & \cos\phi & 0 \\ 0 & 0 & 1 \end{pmatrix}.
\tag{2.29}
$$

The scaling factors in this case are given by $h_\rho = h_z = 1$ and $h_\phi = \rho$. The coordinates of a point P in cylindrical coordinates can thus be expressed as

$$
\mathbf{r} = \rho\,\mathbf{u}_\rho + z\,\mathbf{u}_z.
\tag{2.30}
$$

As we have already mentioned in the previous chapter, it is obvious that the **coordinates do not transform as the components of a vector in cylindrical coordinates** i.e., they do not obey the transformation (1.33) by considering $x' = \rho$, $y' = \phi$ and $z' = z$.

The Cartesian and cylindrical coordinates for a given point P and also the bases are shown graphically in Fig. 2.2 (left). If z is a constant, then we call them *polar* coordinates; this case is also shown in Fig. 2.2 (right).

Let us now find the expression for a differential displacement. Calculating $d\mathbf{u}_\rho$ we obtain

$$
d\mathbf{u}_\rho = (-\sin\phi\,\mathbf{i} + \cos\phi\,\mathbf{j})\,d\phi = \mathbf{u}_\phi\,d\phi.
\tag{2.31}
$$

As $d\mathbf{u_z} = d\mathbf{k} = 0$, $d\mathbf{r}$ can be simply be written as

$$d\mathbf{r} = d\rho\,\mathbf{u}_\rho + \rho\,d\phi\,\mathbf{u}_\phi + dz\,\mathbf{u}_z\,, \tag{2.32}$$

and the velocity vector will thus be

$$\mathbf{v} = \dot{\mathbf{r}} = \dot\rho\,\mathbf{u}_\rho + \rho\,\dot\phi\,\mathbf{u}_\phi + \dot{z}\,\mathbf{u}_z \equiv v_\rho\,\mathbf{u}_\rho + v_\phi\,\mathbf{u}_\phi + v_z\,\mathbf{u}_z\,. \tag{2.33}$$

These last two equations are of course nothing else but (2.10) and (2.11). In order to obtain the expression for the acceleration, we need to additionally calculate $d\mathbf{u}_\phi$

$$d\mathbf{u}_\phi = (-\cos\phi\,\mathbf{i} - \sin\phi\,\mathbf{j})\,d\phi = -\mathbf{u}_\rho\,d\phi\,. \tag{2.34}$$

Therefore, the acceleration can be written as

$$\begin{aligned}
\mathbf{a} = \dot{\mathbf{v}} &= \ddot\rho\,\mathbf{u}_\rho + \dot\rho\,\dot{\mathbf{u}}_\rho + \dot\rho\,\dot\phi\,\mathbf{u}_\phi + \rho\,\ddot\phi\,\mathbf{u}_\phi + \rho\,\dot\phi\,\dot{\mathbf{u}}_\phi + \ddot{z}\,\mathbf{u}_z \\
&= (\ddot\rho - \rho\,\dot\phi^2)\,\mathbf{u}_\rho + (\rho\,\ddot\phi + 2\,\dot\rho\,\dot\phi)\,\mathbf{u}_\phi + \ddot{z}\,\mathbf{u}_z \\
&\equiv a_\rho\,\mathbf{u}_\rho + a_\phi\,\mathbf{u}_\phi + a_z\,\mathbf{u}_z\,.
\end{aligned} \tag{2.35}$$

Finally, an infinitesimal volume and surface will be given by the following expressions

$$\begin{aligned}
dV &= \rho\,d\rho\,d\phi\,dz\,, \\
d\mathbf{s} &= \rho\,d\rho\,d\phi\,\mathbf{u}_z + d\rho\,dz\,\mathbf{u}_\phi + \rho\,d\phi\,dz\,\mathbf{u}_\rho\,.
\end{aligned} \tag{2.36}$$

2.4   Spherical Coordinates

For the spherical coordinates we have $(\alpha,\,\beta,\,\gamma) = (r,\,\theta,\,\phi)$. The relation between Cartesian an cylindrical coordinates is given by

$$\begin{aligned}
x &= r\,\sin\theta\,\cos\phi\,, \\
y &= r\,\sin\theta\,\sin\phi\,, \\
z &= r\,\cos\theta\,,
\end{aligned} \tag{2.37}$$

with $0 \leqslant \phi < 2\pi$ and $0 \leqslant \theta \leqslant \pi$ and with $r = \sqrt{x^2 + y^2 + z^2}$. The components of the basis read

$$\begin{aligned}
\mathbf{u}_r &= \sin\theta\,\cos\phi\,\mathbf{i} + \sin\theta\,\sin\phi\,\mathbf{j} + \cos\theta\,\mathbf{k}\,, \\
\mathbf{u}_\theta &= \cos\theta\,\cos\phi\,\mathbf{i} + \cos\theta\,\sin\phi\,\mathbf{j} - \sin\theta\,\mathbf{k}\,, \\
\mathbf{u}_\phi &= -\sin\phi\,\mathbf{i} + \cos\phi\,\mathbf{j}\,,
\end{aligned} \tag{2.38}$$

Fig. 2.3 Spherical
coordinates of a given point
P. The basis $\{\mathbf{i}, \mathbf{j}, \mathbf{k}\}$ is not
drawn to scale

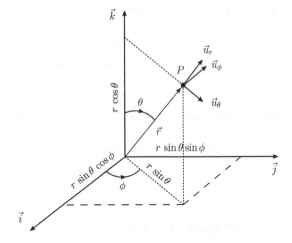

or, written in matrix form

$$
\begin{pmatrix} \mathbf{u}_r \\ \mathbf{u}_\theta \\ \mathbf{u}_\phi \end{pmatrix} = R \begin{pmatrix} \mathbf{i} \\ \mathbf{j} \\ \mathbf{k} \end{pmatrix}, \quad R = \begin{pmatrix} \sin\theta\,\cos\phi & \sin\theta\,\sin\phi & \cos\theta \\ \cos\theta\,\cos\phi & \cos\theta\,\sin\phi & -\sin\theta \\ -\sin\phi & \cos\phi & 0 \end{pmatrix}. \quad (2.39)
$$

The scaling factors are given by $h_r = 1$, $h_\theta = r$ and $h_\phi = r \sin\theta$. The coordinates of a point P in spherical coordinates will be expressed as

$$
\mathbf{r} = r\,\mathbf{u}_r. \quad (2.40)
$$

Again, the transformation (1.33) does not hold for $\mathbf{r}$. The Cartesian and spherical coordinates for a given point P are shown schematically in Fig. 2.3. The velocity vector (2.11), will have the following expression in our particular case

$$
\mathbf{v} = \dot{r}\,\mathbf{u}_r + r\,\dot{\theta}\,\mathbf{u}_\theta + r\,\sin\theta\,\dot{\phi}\,\mathbf{u}_\phi \equiv v_r\,\mathbf{u}_r + v_\theta\,\mathbf{u}_\theta + v_\phi\,\mathbf{u}_\phi. \quad (2.41)
$$

In order to obtain the expression for the acceleration we need to calculate $\dot{\mathbf{u}}_r$, $\dot{\mathbf{u}}_\theta$ and $\dot{\mathbf{u}}_\phi$. The calculation is straightforward (the reader is of course invited to perform the explicit calculations)

$$
\begin{aligned}
\dot{\mathbf{u}}_r &= \dot{\theta}\,\mathbf{u}_\theta + \dot{\phi}\,\sin\theta\,\mathbf{u}_\phi, \\
\dot{\mathbf{u}}_\theta &= -\dot{\theta}\,\mathbf{u}_r + \dot{\phi}\,\cos\theta\,\mathbf{u}_\phi, \\
\dot{\mathbf{u}}_\phi &= -\dot{\phi}\,\sin\theta\,\mathbf{u}_r - \dot{\phi}\,\cos\theta\,\mathbf{u}_\theta.
\end{aligned} \quad (2.42)
$$

Thus, the acceleration takes the form

$$
\begin{aligned}
\mathbf{a} &= \left(\ddot{r} - r\dot{\theta}^2 - r\dot{\phi}^2 \sin^2\theta\right)\mathbf{u}_r + \left(r\ddot{\theta} + 2\dot{r}\dot{\theta} - r\dot{\phi}^2 \cos\theta \sin\theta\right)\mathbf{u}_\theta \\
&\quad + \left(r\ddot{\phi}\sin\theta + 2\dot{r}\dot{\phi}\sin\theta + 2r\dot{\theta}\dot{\phi}\cos\theta\right)\mathbf{u}_\phi \\
&\equiv a_r\,\mathbf{u}_r + a_\theta\,\mathbf{u}_\theta + a_\phi\,\mathbf{u}_\phi \,.
\end{aligned}
\tag{2.43}
$$

Finally, a differential volume and surface will be given by the following expressions

$$
\begin{aligned}
dV &= r^2\,\sin\theta\,dr\,d\theta\,d\phi\,, \\
d\mathbf{s} &= r\,\sin\theta\,dr\,d\phi\,\mathbf{u}_\theta + r\,dr\,d\theta\,\mathbf{u}_\phi + r^2\,\sin\theta\,d\theta\,d\phi\,\mathbf{u}_r\,.
\end{aligned}
\tag{2.44}
$$

2.5   Proposed Exercises

2.1 Check the orthogonality and right handedness properties (2.5) of the bases in cylindrical and spherical coordinates.

Solution: Expressing the components of the cylindrical basis in Cartesian coordinates we have

$$
\mathbf{u}_\rho \times \mathbf{u}_\phi = \begin{vmatrix} \mathbf{i} & \mathbf{j} & \mathbf{k} \\ \cos\phi & \sin\phi & 0 \\ -\sin\phi & \cos\phi & 0 \end{vmatrix} = (\cos^2\phi + \sin^2\phi)\,\mathbf{k} = \mathbf{u}_z\,,
\tag{2.45}
$$

$$
\mathbf{u}_\phi \times \mathbf{u}_z = \begin{vmatrix} \mathbf{i} & \mathbf{j} & \mathbf{k} \\ -\sin\phi & \cos\phi & 0 \\ 0 & 0 & 1 \end{vmatrix} = \cos\phi\,\mathbf{i} + \sin\phi\,\mathbf{j} = \mathbf{u}_\rho\,,
\tag{2.46}
$$

$$
\mathbf{u}_z \times \mathbf{u}_\rho = \begin{vmatrix} \mathbf{i} & \mathbf{j} & \mathbf{k} \\ 0 & 0 & 1 \\ \cos\phi & \sin\phi & 0 \end{vmatrix} = -\sin\phi\,\mathbf{i} + \cos\phi\,\mathbf{j} = \mathbf{u}_\phi\,.
\tag{2.47}
$$

Proceeding the same in the same manner for the spherical basis we obtain

$$
\begin{aligned}
\mathbf{u}_r \times \mathbf{u}_\theta &= \begin{vmatrix} \mathbf{i} & \mathbf{j} & \mathbf{k} \\ \sin\theta\cos\phi & \sin\theta\sin\phi & \cos\theta \\ \cos\theta\cos\phi & \cos\theta\sin\phi & -\sin\theta \end{vmatrix} \\
&= \mathbf{i}\left(-\sin^2\theta\sin\phi - \cos^2\theta\sin\phi\right) + \mathbf{j}\left(\sin^2\theta\cos\phi + \cos^2\theta\cos\phi\right) \\
&= \mathbf{u}_\phi\,.
\end{aligned}
\tag{2.48}
$$

The remaining products are equally straightforward and are left for the reader.

2.2 Given a point P with Cartesian coordinates (x, y, z), obtain its spherical and cylindrical coordinates. Apply the obtained results to the case $(x, y, z) = (3, 3, 3)$.

Solution: In the case of spherical coordinates, from the last line of (2.37) we obtain

$$\theta = \arccos\left(\frac{z}{r}\right). \tag{2.49}$$

Taking the quotient of y and x from the second and first line of (2.37) we obtain

$$\phi = \arctan\left(\frac{y}{x}\right). \tag{2.50}$$

Finally we know that $r = \sqrt{x^2 + y^2 + z^2}$, and together with the conditions $0 \leqslant \theta \leqslant \pi$, $0 \leqslant \phi < 2\pi$, the coordinates (r, θ, ϕ) are uniquely defined.

In the case of cylindrical coordinates we simply have

$$\phi = \arctan\left(\frac{y}{x}\right), \tag{2.51}$$

with $0 \leqslant \phi < 2\pi$ just as in the previous case, and $\rho = \sqrt{x^2 + y^2}$. Together with $z = z$, the coordinates are uniquely defined.

For the spherical coordinates, applying the obtained results to $(x, y, z) = (3, 3, 3)$ we get

$$r = \sqrt{3 \cdot 3^2} = 3\sqrt{3}, \qquad \theta = \arccos\left(1/\sqrt{3}\right) \approx 0.3041\,\pi,$$

$$\phi = \arctan(1) = \{\pi/4, 5\pi/4\}. \tag{2.52}$$

As x, y and z are all positive and also $\cos\theta$ and $\sin\theta$, the correct solution for ϕ is $\pi/4$. Finally, the cylindrical coordinates are given by

$$z = 3, \qquad \rho = \sqrt{2 \cdot 3^2} = 3\sqrt{2}, \qquad \phi = \{\pi/4, 5\pi/4\}. \tag{2.53}$$

Again, as x and y are positive, the correct solution is $\phi = \pi/4$.

2.3 Obtain the expressions for $\nabla\Phi$, $\nabla \cdot \mathbf{A}$ and $\nabla \times \mathbf{A}$ for a scalar field Φ and for a vector field $\mathbf{A}$, in cylindrical and spherical coordinates.

Solution: Substituting $h_\rho = h_z = 1$ and $h_\phi = \rho$ in the case of cylindrical coordinates we obtain

$$\nabla \Phi = \frac{\partial \Phi}{\partial \rho} \mathbf{u}_\rho + \frac{1}{\rho} \frac{\partial \Phi}{\partial \phi} \mathbf{u}_\phi + \frac{\partial \Phi}{\partial z} \mathbf{u}_z , \tag{2.54}$$

$$\nabla \cdot \mathbf{A} = \frac{1}{\rho} \frac{\partial (\rho A_\rho)}{\partial \rho} + \frac{1}{\rho} \frac{\partial A_\phi}{\partial \phi} + \frac{\partial A_z}{\partial z} . \tag{2.55}$$

Finally, the expression for the curl reads

$$\nabla \times \mathbf{A} = \frac{1}{\rho} \begin{vmatrix} \mathbf{u}_\rho & \rho\,\mathbf{u}_\phi & \mathbf{u}_z \\ \partial/\partial_\rho & \partial/\partial_\phi & \partial/\partial_z \\ A_\rho & \rho\,A_\phi & A_z \end{vmatrix} . \tag{2.56}$$

Expanding

$$\nabla \times \mathbf{A} = \mathbf{u}_\rho \left(\frac{1}{\rho} \frac{\partial A_z}{\partial \phi} - \frac{\partial A_\phi}{\partial z} \right) + \mathbf{u}_\phi \left(\frac{\partial A_\rho}{\partial z} - \frac{\partial A_z}{\partial \rho} \right)$$
$$+ \mathbf{u}_z \left(\frac{\partial (\rho A_\phi)}{\partial \rho} - \frac{\partial A_\rho}{\partial \phi} \right) \frac{1}{\rho} . \tag{2.57}$$

As for the spherical coordinates we have $h_r = 1$, $h_\theta = r$ and $h_\phi = r \sin \theta$. Thus we obtain the following results

$$\nabla \Phi = \frac{\partial \Phi}{\partial r} \mathbf{u}_r + \frac{1}{r} \frac{\partial \Phi}{\partial \theta} \mathbf{u}_\theta + \frac{1}{r \sin \theta} \frac{\partial \Phi}{\partial \phi} \mathbf{u}_\phi , \tag{2.58}$$

$$\nabla \cdot \mathbf{A} = \frac{1}{r^2} \frac{\partial (r^2 A_r)}{\partial r} + \frac{1}{r \sin \theta} \frac{\partial (\sin \theta\, A_\theta)}{\partial \theta} + \frac{1}{r \sin \theta} \frac{\partial A_\phi}{\partial \phi} . \tag{2.59}$$

The curl will be given by

$$\nabla \times \mathbf{A} = \frac{1}{r^2 \sin \theta} \begin{vmatrix} \mathbf{u}_r & r\,\mathbf{u}_\theta & r \sin \theta\,\mathbf{u}_\phi \\ \partial/\partial_r & \partial/\partial_\theta & \partial/\partial_\phi \\ A_r & r\,A_\theta & r \sin \theta\,A_\phi \end{vmatrix} \tag{2.60}$$

$$= \mathbf{u}_r \left(\frac{\partial (\sin \theta\, A_\phi)}{\partial \theta} - \frac{\partial A_\theta}{\partial \phi} \right) \frac{1}{r \sin \theta} + \mathbf{u}_\theta \left(\frac{1}{\sin \theta} \frac{\partial A_r}{\partial \phi} - \frac{\partial (r A_\phi)}{\partial r} \right) \frac{1}{r}$$
$$+ \mathbf{u}_\phi \left(\frac{\partial (r A_\theta)}{\partial r} - \frac{\partial A_r}{\partial \theta} \right) \frac{1}{r} . \tag{2.61}$$

2.4 Consider a scalar and a vector field in Cartesian coordinates given by the following expressions

$$\Phi = 3x^2 y z^3 , \tag{2.62}$$

$$\mathbf{A} = 2xyz\,\mathbf{i} + xz\,\mathbf{j} + 2xyz\,\mathbf{k} . \tag{2.63}$$

Calculate $\nabla\Phi$, $\nabla \cdot \mathbf{A}$ and $\nabla \times \mathbf{A}$ at $(x, y, z) = (3, 3, 3)$ in Cartesian coordinates. Use the matrix R as in (2.9) for cylindrical coordinates and obtain the components of $\mathbf{A}$ in the curvilinear basis. Using this result and the expressions calculated in the previous exercise, re-calculate the gradient, divergence and curl. Check that $\nabla \cdot \mathbf{A}$ is basis invariant. Also check that the components of $\nabla\Phi$ and $\nabla \times \mathbf{A}$ obey the transformation given by the matrix R.

Solution: In Cartesian coordinates we have

$$\nabla\phi = 6xyz^3\mathbf{i} + 3x^2z^3\mathbf{j} + 9x^2yz^2\mathbf{k}$$
$$= 3^6\,(2\mathbf{i} + \mathbf{j} + 3\mathbf{k})\,, \tag{2.64}$$
$$\nabla \cdot \mathbf{A} = 2xy + 2yz = 36\,, \tag{2.65}$$

and,

$$\nabla \times \mathbf{A} = x(2z - 1)\mathbf{i} + 2y(x - z)\mathbf{j} + z(1 - 2x)\mathbf{k} = 15\mathbf{i} - 15\mathbf{k}\,. \tag{2.66}$$

The components of $\mathbf{A}$ in cylindrical coordinates will be given by

$$\begin{pmatrix} A_\rho \\ A_\phi \\ A_z \end{pmatrix} = \begin{pmatrix} \cos\phi & \sin\phi & 0 \\ -\sin\phi & \cos\phi & 0 \\ 0 & 0 & 1 \end{pmatrix} \begin{pmatrix} A_x \\ A_y \\ A_z \end{pmatrix}\,. \tag{2.67}$$

Expressing A_z, A_y and A_z in terms of ρ, ϕ and z we obtain the expression we are looking for

$$A_\rho = \cos\phi A_x + \sin\phi A_y = \rho z \cos\phi\sin\phi + 2z\rho^2\cos^2\phi\sin\phi\,, \tag{2.68}$$
$$A_\phi = -\sin\phi A_x + \cos\phi A_y = \rho z \cos^2\phi - 2z\rho^2\cos\phi\sin^2\phi\,, \tag{2.69}$$
$$A_z = 2z\rho^2\cos\phi\sin\phi\,. \tag{2.70}$$

In cylindrical coordinates Φ simply reads

$$\Phi = 3z^3\rho^3\cos^2\phi\sin\phi\,. \tag{2.71}$$

Its gradient takes the form

$$\nabla\Phi = 9z^3\rho^2\cos^2\phi\sin\phi\,\mathbf{u}_\rho + 3z^3\rho^2\cos\phi\left(\cos^2\phi - 2\sin^2\phi\right)\mathbf{u}_\phi$$
$$+ 9z^2\rho^3\cos^2\phi\sin\phi\,\mathbf{u}_z$$
$$= 3^6\left(\frac{3}{\sqrt{2}}\mathbf{u}_\rho - \frac{1}{\sqrt{2}}\mathbf{u}_\phi + 3\mathbf{u}_z\right)\,. \tag{2.72}$$

The expression for the divergence of $\mathbf{A}$ is given by

$$\nabla \cdot \mathbf{A} = 2\rho(z + \rho \cos \phi) \sin \phi = 36. \tag{2.73}$$

The result is, as expected, basis invariant. Finally the expression for the curl is

$$\nabla \times \mathbf{A} = \rho \left(-2z \sin^2 \phi + (2z - 1) \cos^2 \phi + 2\rho \cos \phi \sin^2 \phi\right) \mathbf{u}_\rho$$
$$+ \rho \cos \phi \sin \phi (1 - 4z + 2\rho \cos \phi) \mathbf{u}_\phi + (z - 2z\rho \cos \phi) \mathbf{u}_z. \tag{2.74}$$

The remaining task is to check that the components of $\nabla \phi$ and $\nabla \times \mathbf{A}$ in the curvilinear basis can be obtained from their components in Cartesian coordinates by using the transformation matrix R. This translates into checking that

$$3^6 \begin{pmatrix} \dfrac{3}{\sqrt{2}} \\ \dfrac{-1}{\sqrt{2}} \\ 3 \end{pmatrix} = 3^6 \begin{pmatrix} \dfrac{1}{\sqrt{2}} & \dfrac{1}{\sqrt{2}} & 0 \\ \dfrac{-1}{\sqrt{2}} & \dfrac{1}{\sqrt{2}} & 0 \\ 0 & 0 & 1 \end{pmatrix} \begin{pmatrix} 2 \\ 1 \\ 3 \end{pmatrix}, \tag{2.75}$$

which is obviously true, and

$$\begin{pmatrix} \dfrac{15}{\sqrt{2}} \\ \dfrac{-15}{\sqrt{2}} \\ -15 \end{pmatrix} = \begin{pmatrix} \dfrac{1}{\sqrt{2}} & \dfrac{1}{\sqrt{2}} & 0 \\ \dfrac{-1}{\sqrt{2}} & \dfrac{1}{\sqrt{2}} & 0 \\ 0 & 0 & 1 \end{pmatrix} \begin{pmatrix} 15 \\ 0 \\ -15 \end{pmatrix}, \tag{2.76}$$

which is also true. One equivalent way of preforming this check is by expressing the $\mathbf{u}_\rho$ and $\mathbf{u}_\phi$ in Cartesian coordinates, in (2.72) and (2.74).

This exercise was supposed to make the reader more familiar with curvilinear coordinates and as a cross check for the transformation of the basis components of vectors when working with these bases. One should also realize that it was awfully hard to work in cylindrical coordinates. Thus one should make use of the adequate coordinates in each case depending on the symmetries of the problem.

2.5 Repeat the previous exercise (only for $\nabla \Phi$ and $\nabla \times \mathbf{A}$) for spherical coordinates.

Solution: The components of $\mathbf{A}$ in the spherical basis is given by the following

$$\begin{pmatrix} A_r \\ A_\theta \\ A_\phi \end{pmatrix} = \begin{pmatrix} \sin \theta \cos \phi & \sin \theta \sin \phi & \cos \theta \\ \cos \theta \cos \phi & \cos \theta \sin \phi & -\sin \theta \\ -\sin \phi & \cos \phi & 0 \end{pmatrix} \begin{pmatrix} A_x \\ A_y \\ A_z \end{pmatrix}. \tag{2.77}$$

Expressing A_x, A_y and A_z in terms of r, θ and ϕ we obtain

$$
\begin{aligned}
A_r &= A_x \sin\theta\cos\phi + A_y \sin\theta\sin\phi + A_z \cos\theta \\
&= r^2 \cos\theta \sin^2\theta \cos\phi \sin\phi\, (1 + 2r(\cos\theta + \cos\phi\sin\theta))\,, \quad (2.78) \\
A_\theta &= A_x \cos\theta\cos\phi + A_y \cos\theta\sin\phi - A_z \sin\theta \\
&= r^2 \cos\theta \sin\theta \cos\phi \sin\phi \left(\cos\theta - 2r\sin^2\theta + r\cos\phi\sin 2\theta\right)\,, \quad (2.79)
\end{aligned}
$$

and

$$
\begin{aligned}
A_\phi &= -A_x \sin\phi + A_y \cos\phi \\
&= r^2 \cos\theta \sin\theta \cos\phi \left(\cos\phi - 2r\sin\theta\sin^2\phi\right)\,. \quad (2.80)
\end{aligned}
$$

The expression for Φ reads

$$
\Phi = 3r^6 \cos^3\theta \cos^2\phi \sin^3\theta \sin\phi\,. \quad (2.81)
$$

Thus $\nabla\Phi$ and $\nabla\cdot\mathbf{A}$ take the form

$$
\begin{aligned}
\nabla\Phi &= 18\,r^5 \cos^3\theta \sin^3\theta \cos^2\phi \sin\phi\,\mathbf{u}_r \\
&\quad + 9\,r^5 \cos^2\theta \sin^2\theta \cos^2\phi \sin\phi(\cos^2\theta - \sin^2\theta)\,\mathbf{u}_\theta \\
&\quad + 3\,r^5 \cos^3\theta \sin^2\theta \cos\phi(\cos^2\phi - 2\sin^2\phi)\,\mathbf{u}_\phi \\
&= 3^6 \left(2\sqrt{3}\,\mathbf{u}_r - \sqrt{\frac{3}{2}}\,\mathbf{u}_\theta - \frac{1}{\sqrt{2}}\,\mathbf{u}_\phi\right), \quad (2.82)
\end{aligned}
$$

$$
\begin{aligned}
\nabla\cdot\mathbf{A} &= 2\,r^2 \sin\theta \sin\phi(\cos\theta + \sin\theta\cos\phi) \\
&= 36\,. \quad (2.83)
\end{aligned}
$$

The only thing left is to check that the components of $\nabla\phi$ behave properly

$$
3^6 \begin{pmatrix} 2\sqrt{3} \\ -\sqrt{\dfrac{3}{2}} \\ -\dfrac{1}{\sqrt{2}} \end{pmatrix} = 3^6 \begin{pmatrix} \dfrac{1}{\sqrt{3}} & \dfrac{1}{\sqrt{3}} & \dfrac{1}{\sqrt{3}} \\ \dfrac{1}{\sqrt{6}} & \dfrac{1}{\sqrt{6}} & -\sqrt{\dfrac{2}{3}} \\ \dfrac{-1}{\sqrt{2}} & \dfrac{1}{\sqrt{2}} & 0 \end{pmatrix} \begin{pmatrix} 2 \\ 1 \\ 3 \end{pmatrix}, \quad (2.84)
$$

which is correct.

2.6 Calculate $\nabla\Phi$ with $\Phi = \ln r$, in Cartesian, cylindrical and spherical coordinates for the plane defined by $z = 0$, with r given by $r = \sqrt{x^2 + y^2 + z^2}$ as usual.

Solution: In spherical coordinates the result is rather trivial

$$\nabla \Phi = \frac{\partial \Phi}{\partial r} \mathbf{u}_r = \frac{1}{r} \mathbf{u}_r. \tag{2.85}$$

Particularizing to our case ($z = 0$) and writing $\mathbf{u}_r = \mathbf{r}/r$ in Cartesian coordinates

$$\nabla \Phi(z = 0) = \frac{1}{x^2 + y^2}(x\,\mathbf{i} + y\,\mathbf{j}). \tag{2.86}$$

In cylindrical coordinates $\Phi = \ln\left(\sqrt{\rho^2 + z^2}\right)$ and its gradient reads

$$\nabla \Phi = \frac{\partial \Phi}{\partial \rho} \mathbf{u}_\rho + \frac{\partial \Phi}{\partial z} \mathbf{u}_z = \frac{\rho}{\rho^2 + z^2} \mathbf{u}_\rho + \frac{z}{\rho^2 + z^2} \mathbf{u}_z. \tag{2.87}$$

For $z = 0$ we have

$$\nabla \Phi(z = 0) = \frac{\rho}{\rho^2} \mathbf{u}_\rho = \frac{\rho}{\rho^2}(\cos\phi\,\mathbf{i} + \sin\phi\,\mathbf{j}) = \frac{1}{x^2 + y^2}(x\,\mathbf{i} + y\,\mathbf{j}), \tag{2.88}$$

which is obviously the same result as previously. In Cartesian coordinates it gets somewhat harder. The gradient is given by

$$\begin{aligned}
\nabla \Phi &= \frac{1}{r}(\partial_x r\,\mathbf{i} + \partial_y r\,\mathbf{j} + \partial_z r\,\mathbf{k}) \\
&= \frac{1}{r}(\partial_x\,\mathbf{i} + \partial_y\,\mathbf{j} + \partial_z\,\mathbf{k})\sqrt{x^2 + y^2 + z^2} \\
&= \frac{1}{2r^2}(2x\,\mathbf{i} + 2y\,\mathbf{j} + 2z\,\mathbf{k}),
\end{aligned} \tag{2.89}$$

thus,

$$\nabla \Phi(z = 0) = \frac{1}{x^2 + y^2}(x\,\mathbf{i} + y\,\mathbf{j}). \tag{2.90}$$

It is worth mentioning that this exercise hides a typical mistake that many tend to commit in order to simplify the calculations before hand. Many (including me) are tempted to consider $z = 0$ before calculating the gradient. This is obviously wrong. In this case it would have made no difference, however this does not happen in general.

2.7 Check that components of the velocity and the acceleration of a moving particle, in cylindrical coordinates, obey the vector transformation rule:

$$\begin{pmatrix} v_\rho \\ v_\phi \\ v_z \end{pmatrix} = R \begin{pmatrix} v_x \\ v_y \\ v_z \end{pmatrix}, \qquad \begin{pmatrix} a_\rho \\ a_\phi \\ a_z \end{pmatrix} = R \begin{pmatrix} a_x \\ a_y \\ a_z \end{pmatrix}, \qquad (2.91)$$

with the matrix R given by the expression (2.29).

Solution: In cylindrical coordinates the coordinate z is the same as in Cartesian coordinates, thus the problem gets reduced to checking that

$$\begin{pmatrix} \dot\rho \\ \rho\dot\phi \end{pmatrix} = \begin{pmatrix} \cos\phi & \sin\phi \\ -\sin\phi & \cos\phi \end{pmatrix} \begin{pmatrix} \dot x \\ \dot y \end{pmatrix}, \qquad (2.92)$$

$$\begin{pmatrix} \ddot\rho - \rho\dot\phi^2 \\ \rho\ddot\phi + 2\dot\rho\dot\phi \end{pmatrix} = \begin{pmatrix} \cos\phi & \sin\phi \\ -\sin\phi & \cos\phi \end{pmatrix} \begin{pmatrix} \ddot x \\ \ddot y \end{pmatrix}. \qquad (2.93)$$

Let's consider the first equality. Introducing the explicit expressions for $\dot x$ and $\dot y$ as functions of ρ and ϕ one obtains

$$\begin{aligned} \dot x \cos\phi + \dot y \sin\phi &= \dot\rho \cos^2\phi - \rho\dot\phi \sin\phi \cos\phi + \dot\rho \sin^2\phi + \rho\dot\phi \sin\phi \cos\phi \\ &= \dot\rho \\ &= v_\rho. \end{aligned} \qquad (2.94)$$

The second equality $v_\phi = \rho\dot\phi = -\sin\phi\,\dot x + \cos\phi\,\dot y$, is just as straightforward to check. Moving on to the acceleration, again introducing the explicit expressions for $\ddot x$ and $\ddot y$ we find

$$\begin{aligned} \ddot x \cos\phi + \ddot y \sin\phi &= \cos\phi\,(\ddot\rho \cos\phi - 2\dot\rho\dot\phi \sin\phi - \rho\dot\phi^2 \cos\phi) \\ &\quad + \sin\phi\,(\ddot\rho \sin\phi + 2\dot\rho\dot\phi \cos\phi - \rho\dot\phi^2 \sin\phi) \\ &= (\cos^2\phi + \sin^2\phi)\,(\ddot\rho - \rho\dot\phi^2) \\ &= \ddot\rho - \rho\dot\phi^2 \\ &= a_\rho. \end{aligned} \qquad (2.95)$$

Again, with the same method one can check that the last relation also holds i.e., $a_\phi = \rho\ddot\phi + 2\dot\rho\dot\phi = -\sin\phi\,\ddot x + \cos\phi\,\ddot y$.

Intuitively, we have explained what a vector is and that its intrinsic properties must hold for any reference frame. Mathematically we have checked (in a couple cases) that this holds true i.e., that the components of the velocity and the acceleration indeed obey the vector transformation laws. We have also checked these results applied to vector operators. However, the characteristics of a vector (and this is also true for vector operators) become much more transparent and easy to understand when introducing the **tensor formalism**. One can check the Appendices A and B for more details. Again, we insist that only Appendix A is recommended in order to follow this course (for Special Relativity) and the lecture of Appendix B is only optional for the interested reader. In these two appendices one can find (as simplified as possible) an introduction to tensor analysis in Cartesian and curvilinear coordinates.

2.8 Perform the same calculation as previously for spherical coordinates.

Solution: The resolution of this exercise is slightly more involved. We have to check that

$$
\begin{pmatrix} \dot{r} \\ r\dot{\theta} \\ r\sin\theta\dot{\phi} \end{pmatrix} = \begin{pmatrix} \sin\theta\cos\phi & \sin\theta\sin\phi & \cos\theta \\ \cos\theta\cos\phi & \cos\theta\sin\phi & -\sin\theta \\ -\sin\phi & \cos\phi & 0 \end{pmatrix} \begin{pmatrix} \dot{x} \\ \dot{y} \\ \dot{z} \end{pmatrix}, \tag{2.96}
$$

in the case of the velocity. For the first equality we have

$$
\dot{r} = \sin\theta\cos\phi\,\dot{x} + \sin\theta\sin\phi\,\dot{y} + \cos\theta\,\dot{z}. \tag{2.97}
$$

Expressing $\dot{x}$, $\dot{y}$ and $\dot{z}$ in terms of the new coordinates and grouping terms we obtain

$$
\begin{aligned}
\dot{r} &= \dot{r}(\sin^2\theta\cos^2\phi + \sin^2\theta\sin^2\phi + \cos^2\theta) \\
&\quad + r\dot{\theta}(\sin\theta\cos\theta\cos^2\phi + \sin\theta\cos\theta\sin^2\phi - \cos\theta\sin\theta) \\
&\quad + r\dot{\phi}(-\sin^2\theta\sin\phi\cos\phi + \sin^2\theta\sin\phi\cos\phi) \\
&= \dot{r}. \tag{2.98}
\end{aligned}
$$

The remaining equalities have no further complications and are left for the reader. As for the acceleration we have the following expression

$$
\begin{pmatrix} \ddot{r} - r\dot{\theta}^2 - r\dot{\phi}^2\sin^2\theta \\ r\ddot{\theta} + 2\dot{r}\dot{\theta} - r\dot{\phi}^2\cos\theta\sin\theta \\ r\ddot{\phi}\sin\theta + 2\dot{r}\dot{\phi}\sin\theta + 2r\dot{\theta}\dot{\phi}\cos\theta \end{pmatrix}
$$
$$
= \begin{pmatrix} \sin\theta\cos\phi & \sin\theta\sin\phi & \cos\theta \\ \cos\theta\cos\phi & \cos\theta\sin\phi & -\sin\theta \\ -\sin\phi & \cos\phi & 0 \end{pmatrix} \begin{pmatrix} \ddot{x} \\ \ddot{y} \\ \ddot{z} \end{pmatrix}. \tag{2.99}
$$

The first line reads

$$\ddot{r} - r\dot{\theta}^2 - r\dot{\phi}^2 \sin^2\theta = \sin\theta\,\cos\phi\,\ddot{x} + \sin\theta\,\sin\phi\,\ddot{y} + \cos\theta\,\ddot{z}$$
$$= \ddot{r}(\cos^2\theta + \cos^2\phi\,\sin^2\theta + \sin^2\theta\,\sin^2\phi)$$
$$+ r\,\dot{\theta}^2(-\cos^2\theta - \cos^2\phi\,\sin^2\theta - \sin^2\theta\,\sin^2\phi)$$
$$+ r\,\dot{\phi}^2(-\cos^2\phi\,\sin^2\theta - \sin^2\theta\,\sin^2\phi)$$
$$= \ddot{r} - r\dot{\theta}^2 - r\dot{\phi}^2 \sin^2\theta\,. \tag{2.100}$$

Further Reading

1. J.V. José, E.J. Saletan, *Classical Dynamics: A Contemporary Approach*. Cambridge University Press
2. S.T. Thornton, J.B. Marion, *Classical Dynamics of Particles and Systems*
3. H. Goldstein, C. Poole, J. Safko, *Classical Mechanics*, 3rd edn. Addison Wesley
4. J.R. Taylor, *Classical Mechanics*
5. D.T. Greenwood, *Classical Dynamics*. Prentice-Hall Inc.
6. D. Kleppner, R. Kolenkow, *An Introduction to Mechanics*
7. C. Lanczos, *The Variational Principles of Mechanics*. Dover Publications Inc.
8. W. Greiner, *Classical Mechanics: Systems of Particles and Hamiltonian Dynamics*. Springer
9. H.C. Corben, P. Stehle, *Classical Mechanics*, 2nd edn. Dover Publications Inc.
10. T.W.B. Kibble, F.H. Berkshire, *Classical Mechanics*. Imperial College Press
11. M.G. Calkin, *Lagrangian and Hamiltonian Mechanics*
12. A.J. French, M.G. Ebison, *Introduction to Classical Mechanics*

Chapter 3
Kinematics

Abstract This is a short chapter where we introduce the concept of tangent and normal vectors, tangential and normal acceleration, curvature and torsion. Special emphasis is made on the sign convention for the different kinematic magnitudes introduced here in order to avoid possible future confusions (that, in fact, arise very often). We introduce the definition of the Frenet trihedron and deduce step by step the Frenet equations. At the end of this chapter we propose and solve a few relevant exercises that cover most of the aspects presented here.

3.1 Velocity and Acceleration

A parametric curve $\mathbf{r}(t) \subset \mathbb{R}^3$ defined as

$$
\begin{aligned}
\mathbf{r} \;:\; \mathbb{R} \;&\to\; \mathbb{R}^3 \\
t \;&\mapsto\; \mathbf{r}(t),
\end{aligned}
\tag{3.1}
$$

is called regular if $d\mathbf{r}(t)/dt \neq 0, \forall t \in \mathbb{R}$. If it is not regular in all $\mathbb{R}$ but in some open interval $(t_0, t_1) \subset \mathbb{R}$ then we shall say that it is regular in the open subset (t_0, t_1) of $\mathbb{R}$. The following discussion is valid for the subsets in which $\mathbf{r}(t)$ is regular.

Consider that $\mathbf{r}(t)$ is the trajectory of a point-like particle as a function of time in Cartesian or curvilinear coordinates. An infinitesimal displacement will be given by $d\mathbf{r}$. The differential arc length is defined as the magnitude of the previous quantity (which will always be positive by definition) i.e.,

$$
ds \equiv |d\mathbf{r}| > 0.
\tag{3.2}
$$

The vector $d\mathbf{r}$ is tangent to the curve $\mathbf{r}(t)$ for each time instant t, thus we can define a unitary tangent vector simply as

$$
\tau = \frac{d\mathbf{r}}{ds} = \frac{d\mathbf{r}}{dt}\frac{dt}{ds} = \frac{d\mathbf{r}}{dt}\left|\frac{d\mathbf{r}}{dt}\right|^{-1} = \frac{\mathbf{v}}{v},
\tag{3.3}
$$

© Springer Nature Switzerland AG 2020
V. Ilisie, *Lectures in Classical Mechanics*, Undergraduate Lecture Notes in Physics, https://doi.org/10.1007/978-3-030-38585-9_3

where we have considered, as usual $\mathbf{v} = d\mathbf{r}/dt$ and $v \equiv |\mathbf{v}|$. We can therefore write the velocity vector in terms of its magnitude and the tangent vector as

$$\mathbf{v} = v\boldsymbol{\tau} = \frac{ds}{dt}\boldsymbol{\tau}. \tag{3.4}$$

Note that, as $ds > 0$ and $dt > 0$ always, then $ds/dt > 0$, which is of course, consistent with the fact that v is the magnitude of a vector (positive by definition).

The expression for the (total) acceleration $\mathbf{a}$ can thus be separated in two terms, the tangential and the normal acceleration, as follows

$$\mathbf{a} = \frac{d\mathbf{v}}{dt} = \frac{dv}{dt}\boldsymbol{\tau} + v\frac{d\boldsymbol{\tau}}{dt} = \mathbf{a}_\tau + \mathbf{a}_n = a_\tau\boldsymbol{\tau} + a_n\mathbf{n}. \tag{3.5}$$

It should be obvious from the previous equation that, the tangential acceleration accounts for the variation of the magnitude of the velocity vector and, the normal acceleration accounts for the change of the direction of the moving body. The unitary normal vector $\mathbf{n}$, which is orthogonal to $\boldsymbol{\tau}$ is defined as $d\boldsymbol{\tau}/|d\boldsymbol{\tau}|$. Introducing ds

$$\mathbf{n} = \frac{d\boldsymbol{\tau}}{|d\boldsymbol{\tau}|} = \frac{d\boldsymbol{\tau}}{ds}\frac{ds}{|d\boldsymbol{\tau}|} \equiv \frac{d\boldsymbol{\tau}}{ds}\frac{1}{\kappa} \equiv \frac{d\boldsymbol{\tau}}{ds}R_c, \tag{3.6}$$

where we have defined κ and R_c as

$$\kappa \equiv \frac{1}{R_c} \equiv \frac{|d\boldsymbol{\tau}|}{ds}. \tag{3.7}$$

They are called *instantaneous curvature* and *instantaneous radius of curvature* of the trajectory. The instantaneous radius of curvature R_c is the radius of the circle whose tangent vector at the point given by $\mathbf{r}(t)$ is $\boldsymbol{\tau}$, and whose diameter lies along the line determined by $\mathbf{n}$, as shown in Fig. 3.1. This circle is obviously unique. Also, by definition, the curvature of a circle of radius r is $1/r$ which justifies the term instantaneous curvature that we have introduced.

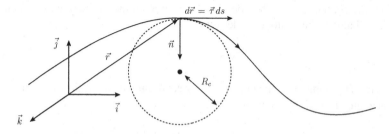

Fig. 3.1 Representation of the trajectory of a point-like particle $\mathbf{r}(t)$ together with the tangent and normal vectors, and the osculating circle of radius $R_c = 1/\kappa$

The normal acceleration can be thus written as

$$\mathbf{a}_n = v \frac{d\tau}{dt} = v \frac{d\tau}{ds}\frac{ds}{dt} = \frac{v^2}{R_c}\mathbf{n}. \tag{3.8}$$

From the previous expression we deduce that $a_n > 0$ thus, $\mathbf{a}_n$ and $\mathbf{n}$ will always have the same sense. This is not the case of a_τ. From its definition $a_\tau = dv/dt$ we conclude that it can be both positive (if v increases with t) or negative (if v decreases with t). Therefore $\mathbf{a}_\tau$ and τ will not necessarily have the same sense, and thus a_τ cannot be (strictly speaking) interpreted as the magnitude of the tangent acceleration.

At this stage the reader might also have serious doubts about why the vector $\mathbf{n}$ defined as previously is orthogonal to τ and why $ds/|d\tau| = R_c$. We shall answer these two questions in the following. Given any unitary vector, in this case τ, the quantity $d\tau$ is orthogonal to τ. The demonstration is quite straightforward. If τ unitary, then $\tau \cdot \tau = 1$ and so

$$d(\tau \cdot \tau) = 0 = 2(d\tau \cdot \tau). \tag{3.9}$$

As for the second question, as we previously mentioned, the curvature κ of a circle is equal to $1/r$, where r is the radius of the circle. Thus, given κ we can always find a circle with radius $1/\kappa$, whose tangent vector is τ and whose diameter lies along the direction given by $\mathbf{n}$. This circle is called the *osculating circle* and we have called its radius R_c.[1]

As τ and $\mathbf{n}$ are orthogonal, the expression for the magnitude of the total acceleration a can be written as

$$a = \left(a_\tau^2 + a_n^2\right)^{1/2} = \left(\dot{v}^2 + \frac{v^4}{R_c^2}\right)^{1/2}. \tag{3.10}$$

Having defined τ and $\mathbf{n}$ we can define a third unitary (pseudo) vector $\mathbf{b}$ (called bi-normal), orthogonal to both, as the vector product

$$\mathbf{b} = \tau \times \mathbf{n}. \tag{3.11}$$

The frame $\{\tau, \mathbf{n}, \mathbf{b}\}$ is called Frenet trihedron. Note that for the formalism we have just introduced we haven't used any specific coordinates, therefore we are free to use the most convenient ones in each case. Same is valid for the following section.

---

[1]Consider the differential arc length for a circle of radius R_c. It will be given by $ds = R_c\, d\theta$ (if $d\theta > 0$, i.e., counter-clockwise motion). Thus the radius can be expressed as $R_c = ds/d\theta$. Comparing this expression to the definition of the instantaneous radius of curvature, we obtain that $|d\tau| = d\theta$. More generically $|d\tau| = |d\theta|$.

3.2 Frenet Equations

We will now deduce the three Frenet equations that relate the three vectors of the Frenet triedron with curvature and torsion. The first equation has already been introduced, namely

$$\boxed{\frac{d\boldsymbol{\tau}}{ds} = \frac{1}{R_c}\mathbf{n}}. \tag{3.12}$$

The second one can be obtained by differentiating $\mathbf{b}$. Remember that, as $\mathbf{b}$ is a unitary vector, $d\mathbf{b}$ is orthogonal to $\mathbf{b}$. Thus we can express $d\mathbf{b}$ as a linear combination of $\boldsymbol{\tau}$ and $\mathbf{n}$

$$\frac{d\mathbf{b}}{ds} = \alpha\boldsymbol{\tau} + \beta\mathbf{n}. \tag{3.13}$$

By taking the dot product with $\boldsymbol{\tau}$ on both sides of the previous equation, we obtain

$$\boldsymbol{\tau} \cdot \frac{d\mathbf{b}}{ds} = \alpha = \frac{d}{ds}(\mathbf{b}\cdot\boldsymbol{\tau}) - \mathbf{b}\cdot\frac{d\boldsymbol{\tau}}{ds} = 0 - \mathbf{b}\cdot\mathbf{n}\frac{1}{R_c} = 0. \tag{3.14}$$

By defining $\sigma \equiv -1/\beta$ (where the minus sign is just a matter of convention) we obtain the second Frenet equation which reads

$$\boxed{\frac{d\mathbf{b}}{ds} = -\frac{1}{\sigma}\mathbf{n}}. \tag{3.15}$$

The quantity σ is called radius of torsion and $1/\sigma$ is called simply torsion, by analogy with the curvature. If the curvature can be regarded as a measurement of the deviation of a curve from a straight line, the torsion measures the rate at which the plane defined by $\boldsymbol{\tau}$ and $\mathbf{n}$ rotates as we move along the curve $\mathbf{r}(t)$. If a curve has a zero curvature $(R_c \to \infty)$ then it is a straight line. In a similar way, if the curve has zero torsion $(\sigma \to \infty)$ then it belongs to a fixed plane.

Note that, while $\kappa \geq 0$ (and therefore $R_c \geq 0$) by definition, the torsion can take any value. However, its sign is of no special interest to us.

We shall now obtain the third and last formula. Similar to the previous case, we shall differentiate $\mathbf{n}$ and write down the result as a linear combination of $\boldsymbol{\tau}$ and $\mathbf{b}$ as follows

$$\frac{d\mathbf{n}}{ds} = \alpha\mathbf{b} + \beta\boldsymbol{\tau}. \tag{3.16}$$

By taking the dot product with $\mathbf{b}$ on both sides of the previous equation we get

$$\mathbf{b}\cdot\frac{d\mathbf{n}}{ds} = \alpha = \frac{d}{ds}(\mathbf{b}\cdot\mathbf{n}) - \mathbf{n}\cdot\frac{d\mathbf{b}}{ds} = 0 + \mathbf{n}\cdot\mathbf{n}\frac{1}{\sigma} = \frac{1}{\sigma}. \tag{3.17}$$

By taking the dot product this time with $\boldsymbol{\tau}$ we obtain

$$\boldsymbol{\tau} \cdot \frac{d\mathbf{n}}{ds} = \beta = \frac{d}{ds}(\boldsymbol{\tau} \cdot \mathbf{n}) - \mathbf{n} \cdot \frac{d\boldsymbol{\tau}}{ds} = 0 - \mathbf{n} \cdot \mathbf{n}\frac{1}{R_c} = -\frac{1}{R_c}. \qquad (3.18)$$

Thus, the third Frenet equation then reads

$$\boxed{\frac{d\mathbf{n}}{ds} = \frac{1}{\sigma}\mathbf{b} - \frac{1}{R_c}\boldsymbol{\tau}}. \qquad (3.19)$$

3.3 Proposed Exercises

3.1 Given a point-like particle that describes a circular trajectory

$$x(t) = r\cos\theta(t), \qquad y(t) = r\sin\theta(t), \qquad (3.20)$$

with r a constant and $\dot{\theta}(t) > 0$ (counter-clockwise motion), calculate $\mathbf{v}$, v, $\boldsymbol{\tau}$, κ, $\mathbf{a}$, a_n, a_τ and $\mathbf{n}$.

Solution: The trajectory $\mathbf{r}(t)$ is given by

$$\mathbf{r}(t) = r\cos\theta(t)\,\mathbf{i} + r\sin\theta(t)\,\mathbf{j}. \qquad (3.21)$$

Thus the expressions for $\mathbf{v}$ and the differential arc length read

$$\mathbf{v} = \frac{d\mathbf{r}}{dt} = r\dot{\theta}(-\sin\theta\,\mathbf{i} + \cos\theta\,\mathbf{j}), \qquad ds = \sqrt{\dot{x}^2 + \dot{y}^2}\,dt = r\,d\theta, \qquad (3.22)$$

where, as usual $\dot{a} \equiv da/dt$. It is obvious from the previous equation that the tangent vector is given by

$$\boldsymbol{\tau} = -\sin\theta\,\mathbf{i} + \cos\theta\,\mathbf{j} = \mathbf{u}_\phi, \qquad (3.23)$$

(where we have identified $\boldsymbol{\tau}$ in polar coordinates with $\mathbf{u}_\phi$) and that $v = r\dot{\theta}$.[2] Using the first Frenet formula, the curvature κ is given by

$$\frac{d\boldsymbol{\tau}}{ds} = \frac{1}{r\,d\theta}(-\cos\theta\,\mathbf{i} - \sin\theta\,\mathbf{j})\,d\theta = \frac{1}{r}(-\cos\theta\,\mathbf{i} - \sin\theta\,\mathbf{j}). \qquad (3.24)$$

---

[2] As we have mentioned earlier $v > 0$ is positive by definition and $d\theta$ can be both positive or negative depending on the motion (counter-clockwise or clockwise), thus strictly speaking $v = r|\dot{\theta}|$.

Fig. 3.2 Schematic
representation of the tangent
and normal vectors
(direction and sense) for the
circular motion

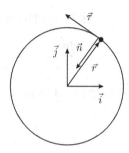

Therefore, as it was already mentioned, for a circle $\kappa = 1/r$. It is also obvious that
the expression for the normal vector is

$$\mathbf{n} = -\cos\theta\,\mathbf{i} - \sin\theta\,\mathbf{j} = -\mathbf{u}_\rho, \tag{3.25}$$

where $\mathbf{u}_\rho$ is the other component of the basis in polar coordinates. Both $\boldsymbol{\tau}$ and $\mathbf{n}$ are
shown for the case $\dot\theta > 0$ in Fig. 3.2. Finally, the acceleration will be given by

$$\begin{aligned}
\mathbf{a} &= r\ddot\theta(-\sin\theta\,\mathbf{i} + \cos\theta\,\mathbf{j}) + r\dot\theta^2(-\cos\theta\,\mathbf{i} - \sin\theta\,\mathbf{j}) \\
&= a_\tau\,\boldsymbol{\tau} + a_n\,\mathbf{n}.
\end{aligned} \tag{3.26}$$

One usually defines the angular velocity as $\omega \equiv \dot\theta$ and the angular acceleration as
$\alpha \equiv \ddot\theta$. Thus we can express v, a_τ and a_n as

$$v = r\omega, \qquad a_\tau = r\alpha, \qquad a_n = r\omega^2 = \frac{v^2}{r}. \tag{3.27}$$

We can observe that even in the case of constant velocity there is still a non-zero
component of the acceleration, the normal acceleration, which is responsible for the
change in the direction of motion.

3.2 For point-like particle with $\mathbf{r}(t)$ given by

$$r(t) = t\,\mathbf{i} + \frac{t^2}{2}\,\mathbf{j} + \frac{t^3}{3}\,\mathbf{k}, \tag{3.28}$$

calculate R_c using $R_c = ds/|d\tau|$ and check that $a_n = v^2/R_c$.

Solution: The velocity simply reads

$$\mathbf{v} = \frac{d\mathbf{r}}{dt} = \mathbf{i} + t\,\mathbf{j} + t^2\,\mathbf{k}. \tag{3.29}$$

Therefore the tangent vector will be given by

$$\boldsymbol{\tau} = \frac{\mathbf{v}}{v} = \frac{1}{\sqrt{1+t^2+t^4}}(\mathbf{i}+t\mathbf{j}+t^2\mathbf{k}), \qquad (3.30)$$

and so, ds is equal to

$$ds = v\,dt = \sqrt{1+t^2+t^4}\,dt. \qquad (3.31)$$

For the modulus $|d\boldsymbol{\tau}|$ one obtains

$$|d\boldsymbol{\tau}| = \frac{\sqrt{1+4t^2+t^4}}{1+t^2+t^4}\,dt. \qquad (3.32)$$

The instantaneous radius of curvature then given by the following expression

$$R_c = \frac{ds}{|d\boldsymbol{\tau}|} = \frac{(1+t^2+t^4)^{3/2}}{\sqrt{1+4t^2+t^4}}. \qquad (3.33)$$

Let's now calculate the acceleration

$$\mathbf{a} = \frac{d\mathbf{v}}{dt} = \mathbf{j}+2t\,\mathbf{k} = a_\tau\,\boldsymbol{\tau}+a_n\,\mathbf{n}, \qquad (3.34)$$

with

$$a = \sqrt{1+4t^2}. \qquad (3.35)$$

The tangential accelerations can be obtained as follows

$$a_\tau = \mathbf{a}\cdot\boldsymbol{\tau} = \frac{t+2t^3}{\sqrt{1+t^2+t^4}}. \qquad (3.36)$$

The normal acceleration can be obtained also in a simple way

$$a_n = \sqrt{a^2-a_\tau^2} = \sqrt{\frac{1+4t^2+t^4}{1+t^2+t^4}}. \qquad (3.37)$$

Using the previously obtained results, we obtain the expression for v^2/R_c

$$\frac{v^2}{R_c} = (1+t^2+t^4)\frac{\sqrt{1+4t^2+t^4}}{(1+t^2+t^4)^{3/2}} = a_n. \qquad (3.38)$$

3.3 For a helicoidal trajectory

$$x(t) = a \cos \omega t, \qquad y(t) = a \sin \omega t, \qquad z(t) = h \omega t, \qquad (3.39)$$

with a, h and ω constants, calculate τ. Using the first Frenet equation calculate R_c and $\mathbf{n}$. Obtain $\mathbf{b}$ and, using the second Frenet formula calculate σ. Check that the third Frenet equation holds true.

Solution: The quantities $d\mathbf{r}$ and ds in this case are given by

$$d\mathbf{r} = (-a\omega \sin \omega t \, \mathbf{i} + a\omega \cos \omega t \, \mathbf{j} + h\omega \, \mathbf{k}) \, dt,$$
$$ds = \omega \sqrt{a^2 + h^2} \, dt. \qquad (3.40)$$

Therefore the tangent vector simply reads

$$\tau = \frac{d\mathbf{r}}{ds} = \frac{1}{\sqrt{a^2 + h^2}} (-a \sin \omega t \, \mathbf{i} + a \cos \omega t \, \mathbf{j} + h \, \mathbf{k}). \qquad (3.41)$$

In order to use the first Frenet formula we have to calculate $d\tau$

$$d\tau = \frac{1}{\sqrt{a^2 + h^2}} (-a\omega \cos \omega t \, \mathbf{i} - a\omega \sin \omega t \, \mathbf{j}) \, dt, \qquad (3.42)$$

and therefore

$$\frac{d\tau}{ds} = \frac{a}{a^2 + h^2} (-\cos \omega t \, \mathbf{i} - \sin \omega t \, \mathbf{j}) = \frac{1}{R_c} \mathbf{n}, \qquad (3.43)$$

which is the first Frenet formula. It is obvious from the previous equation that $R_c = (a^2 + h^2)/a$ and $\mathbf{n} = -\cos \omega t \, \mathbf{i} - \sin \omega t \, \mathbf{j}$. The bi-normal vector can be easily obtained

$$\mathbf{b} = \tau \times \mathbf{n} = \frac{1}{\sqrt{a^2 + h^2}} (h \sin \omega t \, \mathbf{i} - h \cos \omega t \, \mathbf{j} + a \, \mathbf{k}). \qquad (3.44)$$

To apply the second Frenet formula we have to calculate $d\mathbf{b}$

$$d\mathbf{b} = \frac{1}{\sqrt{a^2 + h^2}} (h\omega \cos \omega t \, \mathbf{i} + h\omega \sin \omega t \, \mathbf{j}) \, dt, \qquad (3.45)$$

thus

$$\frac{d\mathbf{b}}{ds} = \frac{h}{a^2 + h^2} (\cos \omega t \, \mathbf{i} + \sin \omega t \, \mathbf{j}) \, dt = -\frac{1}{\sigma} \mathbf{n}. \qquad (3.46)$$

Therefore, σ is simply given by $\sigma = (a^2 + h^2)/h$. The only thing left is to check that

$$\frac{d\mathbf{n}}{ds} = \frac{1}{\sigma}\mathbf{b} - \frac{1}{R_c}\boldsymbol{\tau} \,. \tag{3.47}$$

For the left side of the previous equation we have

$$\frac{d\mathbf{n}}{ds} = \frac{1}{\sqrt{a^2 + h^2}}(\sin \omega t \,\mathbf{i} - \cos \omega t \,\mathbf{j}) \,. \tag{3.48}$$

For the right side of the equation we find

$$\begin{aligned}
\frac{1}{\sigma}\mathbf{b} - \frac{1}{R_c}\boldsymbol{\tau} &= \frac{h}{(a^2 + h^2)^{3/2}}(h \sin \omega t \,\mathbf{i} - h \cos \omega t \,\mathbf{j} + a\,\mathbf{k}) \\
&\quad - \frac{a}{(a^2 + h^2)^{3/2}}(-a \sin \omega t \,\mathbf{i} + a \cos \omega t \,\mathbf{j} + h\,\mathbf{k}) \\
&= \frac{1}{\sqrt{h^2 + a^2}}(\sin \omega t \,\mathbf{i} - \cos \omega t \,\mathbf{j}) \,, \tag{3.49}
\end{aligned}$$

which is the result we were looking for.

Further Reading

1. J.V. José, E.J. Saletan, *Classical Dynamics: A Contemporary Approach*. Cambridge University Press
2. S.T. Thornton, J.B. Marion, *Classical Dynamics of Particles and Systems*
3. H. Goldstein, C. Poole, J. Safko, *Classical Mechanics*, 3rd edn. Addison Wesley
4. J.R. Taylor, *Classical Mechanics*
5. D.T. Greenwood, *Classical Dynamics*. Prentice-Hall Inc.
6. D. Kleppner, R. Kolenkow, *An Introduction to Mechanics*
7. C. Lanczos, *The Variational Principles of Mechanics*. Dover Publications Inc.
8. W. Greiner, *Classical Mechanics: Systems of Particles and Hamiltonian Dynamics*. Springer
9. H.C. Corben, P. Stehle, *Classical Mechanics*, 2nd edn. Dover Publications Inc.
10. T.W.B. Kibble, F.H. Berkshire, *Classical Mechanics*. Imperial College Press
11. M.G. Calkin, *Lagrangian and Hamiltonian Mechanics*
12. A.J. French, M.G. Ebison, *Introduction to Classical Mechanics*

Chapter 4
Newton's Laws, Dynamics and Galilean Relativity

Abstract In this chapter we shall talk about the laws that can be considered as the basis of all modern physics. Newton's *Philosophiae Naturalis Principia Mathematica* constitutes the first (known) formal, mathematical formulation of the physical laws. Here we will present Newton's formulation of Classical Mechanics, also the equivalence principle, and introduce the concepts of inertial reference frames, momentum, conservative and non-conservative forces, angular momentum, torque, work, energy and the corresponding conservation laws. We will also introduce Galilean relativity and re-discuss the notion of inertial reference frames.

4.1 Newton's Laws

We shall start directly by stating Newton's laws of mechanics and shortly comment upon them. Afterwards we will deduce their physical implications and the corresponding conservation theorems. The **first law** describes the movement of free particles.

1. Free particles remain at rest (if initially at rest) or move with constant velocity (if initially moving with constant velocity) unless they are compelled to change their state by an external force.

For the second law we need to introduce the physical quantity called momentum usually denoted by **p**. For a particle with mass m and velocity **v** its momentum will given by

$$\mathbf{p} = m\,\mathbf{v}. \tag{4.1}$$

The **second law** then states

*2. The force **F** acting on a body, initially with momentum **p**, is given by*

$$\mathbf{F} = \frac{d\mathbf{p}}{dt} = m\,\mathbf{a} + \frac{dm}{dt}\,\mathbf{v}. \tag{4.2}$$

This law is called the Law of Inertia and the force **F**, *inertial force*. Note that the usual way of writing the force

© Springer Nature Switzerland AG 2020
V. Ilisie, *Lectures in Classical Mechanics*, Undergraduate Lecture
Notes in Physics, https://doi.org/10.1007/978-3-030-38585-9_4

$$\mathbf{F} = m\mathbf{a}\,, \tag{4.3}$$

is only valid for objects with constant mass. Systems with variable mass will be discussed later on in the next chapter. Hence, **unless stated otherwise, we shall always consider objects with constant mass**. As this mass can be regarded as a proportionality constant between the inertial force and the acceleration we shall call it *inertial mass*. Roughly speaking it is a measure of the resistance of an object to change its velocity.

The **third law** is the law of action and reaction and states the following

3. If an object a exerts a force over an object b given by $\mathbf{f}_{ba}$ *then the object b will exert a force* $\mathbf{f}_{ab}$ *over a, of equal magnitude and opposite sense* $\mathbf{f}_{ab} = -\mathbf{f}_{ba}$.

Several comments are required. Note that we have denoted the force that a exerts over b as $\mathbf{f}_{ba}$ and not $\mathbf{f}_{ab}$. This is more or less the standard convention. We normally denote $\mathbf{f}_b$ as the force that is exerted over b, therefore it is natural to interpret $\mathbf{f}_{ba}$ as the force exerted over b by a. Also, when we treat with a system of more than one particle we reserve the capital letters for the total force, angular momentum, etc., of the system, and we use lowercase letters for the individual quantities. The last comment has to do with the term *opposite sense*. Many authors normally use the terms opposite direction. However, in Chap. 1, when we defined the notion of vector we made a clear distinction between direction and sense. This is why the correct use in our case is *opposite sense*.

These three laws we have just introduced here implicitly define a set of reference systems, the *inertial reference systems*. They are defined as the reference system where Newton's laws are valid. However, some further discussion about this concept is clearly needed, and we shall retake it in Sect. 4.5, after introducing Galileo transformations and Galilean relativity.

There is a fourth law which is the Law of Gravitation. We shall introduce it in the following, however we shall dedicate special attention to this topic in the forthcoming chapters. The **fourth law** states

4. The gravitational force exerted by a (point-like) mass m_2 *over a (point-like) mass* m_1 *is given by*

$$\mathbf{f}_{12} = -G\frac{m_1 m_2}{r_{12}^2}\mathbf{u}_{12}\,, \tag{4.4}$$

where $G \simeq 6.673 \times 10^{-11}\,\mathrm{m}^3\,\mathrm{kg}^{-1}\,\mathrm{s}^{-2}$ is Newton's universal gravitational constant, $\mathbf{r}_1$ and $\mathbf{r}_2$ are the coordinates of m_1 and m_2 and,

$$\mathbf{r}_{12} \equiv \mathbf{r}_1 - \mathbf{r}_2\,, \qquad r_{12} = |\mathbf{r}_{12}|\,, \qquad \mathbf{u}_{12} = \mathbf{r}_{12}/r_{12}\,. \tag{4.5}$$

This is schematically shown in Fig. 4.1. This force is obviously attractive. The masses m_1 and m_2 are called *gravitational masses* for obvious reasons. Unlike the inertial mass, the gravitational masses are calculated by comparing gravitational forces. The simplest example is a balance that compares the weight of two masses in a gravitational field.

Fig. 4.1 The coordinates of m_1, m_2, the force $\mathbf{f}_{12}$ and the coordinates $\mathbf{r}_{12}$ are schematically shown

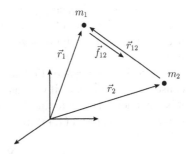

If we call $\overline{m}_1$ the inertial mass of the body whose gravitational mass is m_1, then the acceleration a that it will experiment due to the gravitational attraction is obtained by setting equal the two forces, the gravitational force and the inertial force, as follows

$$-G\frac{m_1 m_2}{r_{12}^2} \mathbf{u}_{12} = -\overline{m}_1 a \mathbf{u}_{12}. \tag{4.6}$$

Newton tested with an experiment with pendulums[1] the so-called *equivalence principle* that states that the gravitational mass is equal to the inertial mass i.e., $m_1 = \overline{m}_1$. Nowadays the equivalence principle is one of the most precise experiments in physics and is one of the building blocks of Einstein's General Theory of Relativity. Thus, by virtue of this principle we can simplify m_1 with $\overline{m}_1$ in the previous expression and, the gravitational acceleration a will be simply given by

$$a = g, \tag{4.7}$$

where we have defined

$$g \equiv G\frac{m_2}{r_{12}^2}. \tag{4.8}$$

Unless stated otherwise we shall always assume that the inertial mass is equal to the gravitational mass.

4.2 Conservative and Central Forces

We say that $\mathbf{F}$ is a **conservative force** if there exists a scalar field U, also called **potential energy**, so that

$$\mathbf{F} = -\nabla U. \tag{4.9}$$

---

[1] See Exercise 4.4 for details.

The minus sign is present just for historical reasons. It is obvious from the previous equation that we can only determine U up to a constant. However, this constant is irrelevant as the underlying physics will depend on the difference of U evaluated at different positions in space.

Given $\mathbf{F}$, if we cannot find such a potential energy then we shall call it non-conservative. Of course, if we want to find out if a given force is conservative or not, it is not practical to try to find U or prove that U does not exist. We shall, therefore, make use of the result (1.43) from Chap. 1, which states

$$\nabla \times (\nabla U) = \mathbf{0}. \tag{4.10}$$

The reciprocal result is also true. Given a vector field $\mathbf{F}$, if $\nabla \times \mathbf{F} = \mathbf{0}$ then there exists a scalar filed U so that $\mathbf{F}$ is given by (4.9). Thus, a force $\mathbf{F}$ is conservative if and only if

$$\nabla \times \mathbf{F} = \mathbf{0}. \tag{4.11}$$

We say that $\mathbf{F}$ is a **central force** if it can be expressed as

$$\mathbf{F} = F(r)\,\mathbf{u}_r\,, \tag{4.12}$$

where $F(r)$ is a scalar field, $r = \sqrt{x^2 + y^2 + z^2}$ and $\mathbf{u}_r = \mathbf{r}/r$, as for spherical coordinates. It is trivial to check that a central force is conservative i.e., $\nabla \times \mathbf{r} = \mathbf{0}$. It is also obvious that the gravitational force is a central force.

4.2.1 Gravitational Potential

One can check that the gravitational potential energy U is given by

$$U(r_{12}) = -G\frac{m_1\,m_2}{r_{12}}, \tag{4.13}$$

with $\mathbf{r}_{12} \equiv \mathbf{r}_1 - \mathbf{r}_2$ and $r_{12} = |\mathbf{r}_{12}|$ as previously. The gravitational force $\mathbf{f}_{12}$ exerted over m_1 will be given by

$$\mathbf{f}_{12} = -\nabla_1 U(r_{12}) \tag{4.14}$$

where we define the operator ∇_i as

$$\nabla_i \equiv \frac{\partial}{\partial x_i}\mathbf{i} + \frac{\partial}{\partial y_i}\mathbf{j} + \frac{\partial}{\partial z_i}\mathbf{k}\,, \tag{4.15}$$

with $i = 1, 2$. Given the previous definition one can explicitly check that

$$
\begin{aligned}
-\mathbf{V}_1 U(r_{12}) &= G m_1 m_2 \, \mathbf{V}_1 \left(\frac{1}{r_{12}} \right) \\
&= G m_1 m_2 \, \mathbf{V}_1 \left[(x_1 - x_2)^2 + (y_1 - y_2)^2 + (z_1 - z_2)^2 \right]^{-1/2} \\
&= -G \frac{m_1 m_2}{2 r_{12}^3} \left[2(x_1 - x_2)\mathbf{i} + 2(y_1 - y_2)\mathbf{j} + 2(z_1 - z_2)\mathbf{k} \right] \\
&= -G \frac{m_1 m_2}{r_{12}^2} \mathbf{u}_{12}.
\end{aligned}
\tag{4.16}
$$

Also, the force $\mathbf{f}_{21}$ exerted over m_2 can be expressed as

$$
\mathbf{f}_{21} = -\mathbf{V}_2 U(r_{12}) = -\mathbf{f}_{12} = \mathbf{V}_1 U(r_{12}). \tag{4.17}
$$

Given a mass M one can define the **potential** (function) $\phi(r)$ as the potential energy per unit mass as

$$
\phi(r) = -G \frac{M}{r}. \tag{4.18}
$$

Thus, given a second mass m placed at a distance r from M, the potential energy will simply be given by

$$
U(r) = m \, \phi(r) = -G \frac{M m}{r}. \tag{4.19}
$$

While the potential energy is related to the force, the potential is related to the concept of field. The gravitational field, that we shall denote as $\mathbf{g}$, is expressed as

$$
\mathbf{g} = -\mathbf{V}\phi(r) = -G \frac{M}{r^2} \mathbf{u}_r, \tag{4.20}
$$

where $\mathbf{r}$ is the point with respect to M, where the field is *measured* (or computed), and where $r = |\mathbf{r}|$, $\mathbf{u}_r = \mathbf{r}/r$. This is shown in Fig. 4.2a. Note that the gravitational field $\mathbf{g}$ points towards M meaning that it is an attractive potential.

The gravitational force exerted over a mass m (placed at a point with coordinates $\mathbf{r}$ with respect to M) due to the field $\mathbf{g}$, as shown in Fig. 4.2b, will be given by

$$
\mathbf{f}_{mM} = m \, \mathbf{g} = -G \frac{M m}{r^2} \mathbf{u}_r, \tag{4.21}
$$

which is nothing but the expression (4.4) and where

$$
|\mathbf{g}| = g = G \frac{M}{r^2}. \tag{4.22}
$$

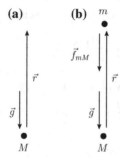

Fig. 4.2 The direction and sense of the gravitational field generated by a mass M, at a point with coordinates $\mathbf{r}$ with respect to M (**a**). Additionally, the direction and sense of the force exerted on a mass m (with coordinates $\mathbf{r}$ with respect to M) by the mass M (**b**)

4.3 Force, Energy, Work, and Energy Conservation

The kinetic energy K of a point-like particle with momentum $\mathbf{p}$ is defined as

$$K = \frac{p^2}{2m} = \frac{1}{2}mv^2 . \tag{4.23}$$

Another closely related, fundamental quantity that we will analyse in the following, and that will naturally give rise to the energy conservation theorem is the *(mechanical) work* W. In its differential form it reads

$$dW = \mathbf{F} \cdot d\mathbf{r} , \tag{4.24}$$

which can roughly be interpreted as the amount of energy transferred by a force.

Let's now consider the physical trajectory of a particle $\mathbf{r} = \mathbf{r}(t)$ and introduce it in the previous expression

$$dW = \mathbf{F} \cdot \frac{d\mathbf{r}}{dt}dt = \mathbf{F} \cdot \mathbf{v}dt = m(a_n\,\mathbf{n} + a_\tau\,\boldsymbol{\tau})v\,\boldsymbol{\tau}dt = m\,a_\tau\,v\,dt . \tag{4.25}$$

One thing that we notice is that the normal part of the force does not contribute to W (as $\mathbf{n} \perp \boldsymbol{\tau}$). Let's now differentiate the kinetic energy (4.23) with respect to time. We obtain

$$dK = mv\frac{dv}{dt} = m\,v\,a_\tau\,dt . \tag{4.26}$$

Thus, we have just deduced that

$$dW = dK , \tag{4.27}$$

Fig. 4.3 Physical path Γ_{AB}, given by the physical trajectory $\mathbf{r}(t)$ of a particle, in between two points A and B

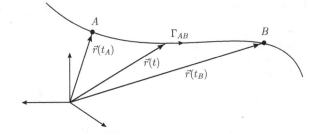

which is a very generic result as we have made no assumptions on the force i.e., $\mathbf{F}$ can have a generic dependence $\mathbf{F} = \mathbf{F}(\mathbf{r}, \mathbf{v}, \mathbf{a}, t)$ with $\mathbf{r} = \mathbf{r}(t)$, $\mathbf{v} = \mathbf{v}(\mathbf{r}, t)$ and $\mathbf{a} = \mathbf{a}(\mathbf{r}, \mathbf{v}, t)$. The finite work in between an initial point A and a final point B will be given by the integral

$$W_{AB} = \int_{\Gamma_{AB}} dK = m \int_{\Gamma_{AB}} v(\mathbf{r}, t)\, a_\tau(\mathbf{r}, \mathbf{v}, t)\, dt\,, \qquad (4.28)$$

along the physical (real) path Γ_{AB}, as schematically shown in Fig. 4.3.

For the previous calculations, for obtaining dW and W_{AB} we have considered, as mentioned previously, the physical trajectory of a particle $\mathbf{r} = \mathbf{r}(t)$. In order to emphasize this aspect, we shall call it **dynamical work**, in contrast with the *virtual work* that we shall define in the next subsection.

4.3.1 Conservative Forces

Let us now turn our attention to conservative forces. Consider a function U so that $\mathbf{F} = -\nabla U$. If we consider that U can be more generic and also depend explicitly on time i.e., of the form $U = U(\mathbf{r}, t)$, then dU can be written as

$$dU = \nabla U \cdot d\mathbf{r} + \frac{\partial U}{\partial t} dt\,. \qquad (4.29)$$

In this case dW takes the form

$$dW = \mathbf{F} \cdot d\mathbf{r} = -\nabla U \cdot d\mathbf{r} = -dU + \frac{\partial U}{\partial t} dt\,. \qquad (4.30)$$

If $\partial U/\partial t = 0$ then dW can be simply expressed as

$$dW = -dU(\mathbf{r})\,. \qquad (4.31)$$

Fig. 4.4 W_{AB} evaluated for a real (physical) and an arbitrary path. For a conservative force (with no explicit time-dependence) $W_{AB}^{\text{virtual}} = W_{AB}^{\text{dynamical}}$

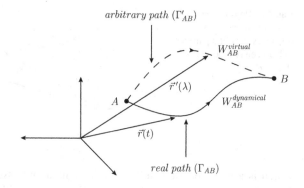

Note that there is a big difference between this expression and (4.27) because here U only depends on **r** and therefore

$$W_{AB} = -\int_{\Gamma'_{AB}} dU = U(\mathbf{r}_A) - U(\mathbf{r}_B) = U\big(\mathbf{r}(t_A)\big) - U\big(\mathbf{r}(t_B)\big), \qquad (4.32)$$

is path independent (it only depends on the end points, and not on the path Γ'_{AB}). Thus, in this case Γ'_{AB} can be totally arbitrary.

We conclude that for a conservative force, if the potential does not depend explicitly on time i.e., $\partial U / \partial t = 0$, the work W_{AB} is path independent (W_{AB} will give the same result by taking the physical path Γ_{AB} given by the physical trajectory $\mathbf{r}(t)$, or any other path in between the same initial and final points A and B). The previous result can be stated equivalently as: the work of a conservative force for a closed trajectory Γ_C is zero

$$W_C = -\oint_{\Gamma_C} dU = 0. \qquad (4.33)$$

Given a generic force **F** (conservative or not) and an arbitrary path Γ'_{AB} given by the curve $\mathbf{r}'(\lambda)$ (with λ the parameter of the curve), we shall call the work in between two points A and B along Γ'_{AB}, **virtual work**, in order to differentiate it from the dynamical work, which is a term we have reserved for the physical path only. Schematically this is shown in Fig. 4.4. Thus, for a conservative force with $\partial U / \partial t = 0$,

$$W_{AB}^{\text{virtual}} = W_{AB}^{\text{dynamical}}. \qquad (4.34)$$

As (4.27) is a generic result, we can conclude that for a conservative force, with $\partial U / \partial t$,

$$-dU = dK \qquad \Rightarrow \qquad dK + dU = 0. \qquad (4.35)$$

By defining the total energy as

$$E = K + U, \tag{4.36}$$

we obtain the energy conservation theorem, which states that for a conservative force with $\partial U / \partial t = 0$ the total energy is conserved i.e.,

$$dE = dU + dK = 0. \tag{4.37}$$

4.3.2 Conservative and Non-conservative Forces

Let us now analyse the case in which we have both a conservative force $\mathbf{F}$ (originated from a potential U) and a non-conservative force $\mathbf{F}_{nc}$. We shall denote the total force as $\mathbf{F}_t = \mathbf{F} + \mathbf{F}_{nc}$. In this case, the total work (in differential form) will be given by

$$dW = dK = \mathbf{F}_t \cdot d\mathbf{r} = \mathbf{F} \cdot d\mathbf{r} + \mathbf{F}_{nc} \cdot d\mathbf{r} = -dU + dW_{nc}. \tag{4.38}$$

Again, if $dE = dK + dU$ we obtain that the conservation law simply reads

$$dE = dK + dU = dW_{nc} = \mathbf{F}_{nc} \cdot d\mathbf{r}. \tag{4.39}$$

Therefore, the variation of the total energy will be given by the work corresponding to the non-conservative force. This result is the generalization of the previous conservation theorem and it will turn out to be useful, especially for the analysis of inelastic collisions in Chap. 7.

The simplest example of a non-conservative force is a constant friction force (drag) of the form $|\mathbf{F}_f| = \mu |\mathbf{N}|$, where $\mathbf{N}$ is the normal force and μ a constant coefficient $(0 \leq \mu \leq 1)$.[2] The normal force (exerted on an object that lays on a surface, in the presence of a gravitational field) is the force exerted by the surface in the opposite sense of the gravitational force, as shown in Fig. 4.5 with $N = mg$.

The previously defined friction force can usually be separated into two categories, as static and dynamical. The static friction force satisfies

$$|\mathbf{F}_s| = \mu_s |\mathbf{N}|, \tag{4.40}$$

where μ_s is called the static friction coefficient. By applying an external force $\mathbf{F}_{app}$ (orthogonal to $\mathbf{N}$) over an object, for $\mathbf{F}_{app} \leq \mathbf{F}_s$, the friction force will have the same magnitude and direction as $\mathbf{F}_{app}$, but opposite sense. Once surpassed this limit, the object will start moving, however the friction force will be smaller in magnitude than the corresponding static friction force, and will be given by

---

[2]There have also bee reported cases where μ was slightly greater than one.

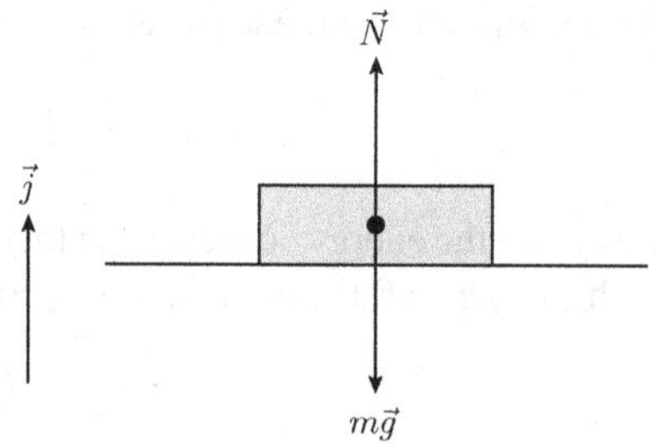

Fig. 4.5 The normal and the gravitational force exerted over on object at rest on a horizontal surface

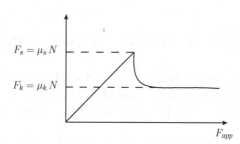

Fig. 4.6 The friction force as a function of the (external) applied force $\mathbf{F}_{app}$ over an object

$$|\mathbf{F}_k| = \mu_k |\mathbf{N}|, \qquad (4.41)$$

(kinetic friction force) with $\mu_k < \mu_s$. This is shown in Fig. 4.6. This is a well studied phenomenon and it is better understood at microscopic level. Friction forces are not fundamental forces, however at a macroscopic level we can effectively describe them in many cases with these simple approximations. The μ_s and μ_k coefficients at macroscopic level cannot be calculated from basic principles. They are empirically determined for each system. More examples will be analysed in the Proposed Exercises section.

4.4 Angular Momentum, Torque and Conservation

Another highly useful quantity we need to define in physics is angular momentum

$$\mathbf{L} = \mathbf{r} \times \mathbf{p} = m\,\mathbf{r} \times \mathbf{v}. \qquad (4.42)$$

It is obvious that, following the definitions given here, $\mathbf{L}$ is not a vector (nor a pseudo-vector) strictly speaking, as it depends on $\mathbf{r}$ (it is however, invariant under rotations). Even so, it is a quantity that turns out to be very relevant, and whose conservation is crucial in physics (just as linear momentum conservation). Given the angular momentum, we define the *torque* τ as

$$\tau = \frac{d\mathbf{L}}{dt} = m\,\mathbf{v} \times \mathbf{p} + m\,\mathbf{r} \times \mathbf{F} = m\,\mathbf{r} \times \mathbf{F}. \qquad (4.43)$$

Thus we can state the angular momentum conservation theorem in a simple manner as follows: if the net force **F** (the sum of all the forces) that acts over an object is zero or, it is a central force, then the angular momentum of the body is conserved over time.[3]

It is worth pointing out that if the angular momentum is conserved then the movement takes place in a fixed plane. This result will be of fundamental importance when studying planetary motion and Kepler's laws.

4.5 Galilean Relativity and Inertial Reference Frames

Consider the translation of a reference frame (with no rotation of the axes, for simplicity) as shown in Fig. 4.7. If **r** are the coordinates of a point P in the reference system $\{\mathcal{O}, \mathbf{i}, \mathbf{j}, \mathbf{k}\}$, **r**′ in the $\{\mathcal{O}', \mathbf{i}, \mathbf{j}, \mathbf{k}\}$ system and, **R** is given by $\overline{\mathcal{O}\mathcal{O}'}$, then

$$\mathbf{r} = \mathbf{R} + \mathbf{r}'. \tag{4.44}$$

Let's now consider that **r**, **r**′ and **R** depend on time i.e., the point P moves with time and the two reference frames have relative motion. Taking the time derivatives on both sides of the previous equation we obtain

$$\mathbf{v} = \mathbf{V} + \mathbf{v}'. \tag{4.45}$$

These are the Galilean transformations of the velocities for a reference system $\mathcal{O}'$ that moves with velocity **V** with respect $\mathcal{O}$, where **v** and **v**′ are the velocities of a moving body with respect to $\mathcal{O}$ and $\mathcal{O}'$. As the axes are the same, the vector components will transform as

$$\begin{pmatrix} v_x \\ v_y \\ v_z \end{pmatrix} = \begin{pmatrix} V_x \\ V_y \\ V_z \end{pmatrix} + \begin{pmatrix} v'_x \\ v'_y \\ v'_z \end{pmatrix}, \tag{4.46}$$

in Cartesian coordinates.

If, on the other hand, P is a fixed point with coordinates x, y and z in the $\mathcal{O}$ frame, then the Galilean transformations of the coordinates will be simply given by

$$\begin{pmatrix} x \\ y \\ z \end{pmatrix} = \begin{pmatrix} R_{0,x} \\ R_{0,y} \\ R_{0,z} \end{pmatrix} + \begin{pmatrix} V_x t \\ V_y t \\ V_z t \end{pmatrix} + \begin{pmatrix} x' \\ y' \\ z' \end{pmatrix}, \tag{4.47}$$

where $\mathbf{R}_0$ is the initial *distance* between the two origins.

---

[3] This last case should be obvious, if $\mathbf{F} = F(r)\mathbf{u}_r$ then $\mathbf{r} \times \mathbf{u}_r = \mathbf{0}$.

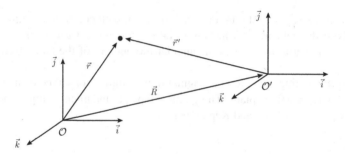

Fig. 4.7 Translation (with no rotation) of a reference system in Cartesian coordinates

From the point of view of an observer at rest in $\mathcal{O}'$, the reference system $\mathcal{O}$ moves with velocity $-\mathbf{V}$ with respect to $\mathcal{O}'$. Thus, the previous expressions in the primed reference system are simply obtained by re-writing (4.46) and (4.47) as follows

$$\begin{pmatrix} v'_x \\ v'_y \\ v'_z \end{pmatrix} = \begin{pmatrix} v_x - V_x \\ v_y - V_y \\ v_z - V_z \end{pmatrix}, \quad \begin{pmatrix} x' \\ y' \\ z' \end{pmatrix} = \begin{pmatrix} x - R_{0,x} - V_x\, t \\ y - R_{0,y} - V_y\, t \\ z - R_{0,z} - V_z\, t \end{pmatrix}. \tag{4.48}$$

As we have seen, one can simply deduce Galileo's transformations of the coordinates and velocities from a simple translation.

Why are these transformations so important one might ask. Consider that $\mathbf{V}$ is constant. In this case, by taking again the time derivatives of (4.45) we obtain

$$\mathbf{a} = \mathbf{a}' \quad \Rightarrow \quad m\,\mathbf{a} = m\,\mathbf{a}' \quad \Rightarrow \quad \mathbf{F} = \mathbf{F}'. \tag{4.49}$$

Therefore, if Newton's laws are valid in $\mathcal{O}$ they will also be valid in $\mathcal{O}'$. This means that a reference frame that moves with constant velocity with respect to an inertial reference frame, is also an inertial reference frame.

Let's now consider that $\mathbf{V}$ is not constant. Then $\mathbf{a} = \mathbf{A} + \mathbf{a}'$ and so $\mathbf{F} \neq \mathbf{F}'$, thus and observer from $\mathcal{O}$ would say that $\mathcal{O}'$ is not an inertial reference system. However, from the point of view of an observer from $\mathcal{O}'$, it is $\mathcal{O}$ that moves with an acceleration $-\mathbf{A}$ with respect to $\mathcal{O}'$, and therefore such an observer would affirm that $\mathcal{O}$ is non-inertial. Apparently, in order to solve this ambiguity one needs to consider the existence of an absolute inertial reference system that stands still and with respect to which, all other inertial reference frames are stationary or they move with constant velocity. This was in fact Newton's proposal. Many centuries after, it was Einstein's General Relativity that revealed that one does not have to make such strong assumptions and showed that in fact there is no need for absolute space or time. He, in fact, managed to re-invent gravitation and mechanics as his equations are valid in any arbitrary reference system.

However, we know from day-to-day experience and from many experiments that Newton's laws have a wide range of validity, and one does not have to make use of

General Relativity except in special cases (extreme velocities, strong gravitational fields, description of the Universe at large scales, Cosmology, etc.). There is an ad-hoc solution which is quite simple. The far away stars and galaxies can be considered as fixed as they change their position approximately 1 arc-s/century. Therefore any reference frame that is stationary or moving with constant velocity with respect to this reference system, can be considered an inertial reference frame. A more *modern* description of non-inertial reference frames comes from Einstein and, they are referred to, as *the reference systems in which the laws of motion hold in their simplest form*. For example, Newton's equations of motion in their simplest form state that $\mathbf{F} = d\mathbf{p}/dt$. When moving in non-inertial reference frames (as we shall see in Chap. 9) additional fictitious forces will be present and the previous (simple) equation will no longer be true.

The Earth can be approximately considered as an inertial reference frame. However, for large time periods or for long distances this approximation starts to fail, as the Earth rotates around its axis, orbits around the Sun, etc. We shall analyse some of these interesting effects in Chap. 9 where we shall study non-inertial reference systems in great detail.

4.6 Proposed Exercises

4.1 Given a central force $\mathbf{F} = F(r)\mathbf{u}_r$, obtain the expression (in integral form) of the corresponding central potential energy (the potential energy whose gradient gives rise to the force $\mathbf{F}$).

Solution: The solution is straightforward as follows

$$-dU = -\nabla U \cdot d\mathbf{r} = \mathbf{F} \cdot d\mathbf{r} = F(r)\mathbf{u}_r \cdot d\mathbf{r} = F(r)\,dr\,, \qquad (4.50)$$

where we have used

$$dr = d\sqrt{x^2 + y^2 + z^2} = \frac{xdx + ydy + zdz}{r} = \frac{\mathbf{r}}{r} \cdot d\mathbf{r} = \mathbf{u}_r \cdot d\mathbf{r}\,. \qquad (4.51)$$

Integrating, (up to a constant) we obtain

$$U = -\int F(r)\,dr\,. \qquad (4.52)$$

The integration constant is not relevant as the meaningful physical quantities will never depend on it. They will always depend only on the difference of the potential energy evaluated in different points in space.

4.2 For the uniformly accelerated circular motion from Exercise 3.1 (from the previous chapter), for a body with mass m calculate $\mathbf{F}$, $\mathbf{L}$ and the torque.

Solution: In Exercise 3.1 we have obtained the expressions for the two components of the acceleration (normal and tangential)

$$\mathbf{a}_\tau = r\alpha\,\boldsymbol{\tau}\,,\qquad \mathbf{a}_n = r\omega^2\,\mathbf{n} = \frac{v^2}{r}\,\mathbf{n}\,,\tag{4.53}$$

with $\mathbf{r} = r\cos\theta\,\mathbf{i} + r\sin\theta\,\mathbf{j}$, $\omega \equiv \dot{\theta}$, $\alpha \equiv \dot{\omega}$, $\mathbf{v} = r\omega\,\boldsymbol{\tau}$ ($\omega > 0$) and r constant. Thus the total force can also be separated in two components

$$\mathbf{F}_c = m\,\mathbf{a}_n = m\frac{v^2}{r}\,\mathbf{n}\,,\tag{4.54}$$

which is called the **centripetal** force and which is responsible for the change in the direction of motion, and

$$\mathbf{F}_\tau = mr\alpha\,\boldsymbol{\tau}\,,\tag{4.55}$$

which is the tangential force responsible for the change of the magnitude v. They are schematically shown in Fig. 4.8. The angular momentum will simply be given by

$$\mathbf{L} = m\,\mathbf{r} \times \mathbf{v} = mrv\,(\mathbf{u}_\rho \times \mathbf{u}_\tau) = mrv\,\mathbf{k} = mr^2\omega\,\mathbf{k}\,.\tag{4.56}$$

The quantity

$$I \equiv mr^2\,,\tag{4.57}$$

is called the **momentum of inertia** of the mass m with respect to the z axis. For a generic movement I is a 3×3 matrix and it is called *inertia tensor*. We shall

Fig. 4.8 Direction and sense of the tangential and centripetal force for the uniformly accelerated circular motion

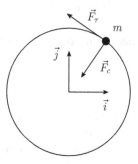

explicitly analyse this topic later on, in Chap. 10. Thus $L = |\mathbf{L}| = I\omega$. As for the torque, we simply have

$$\mathbf{N} = m\,\mathbf{r} \times (a_n\,\mathbf{n} + a_\tau\,\boldsymbol{\tau}) = ma_\tau\,\mathbf{r} \times \boldsymbol{\tau} = mr^2\alpha\,\mathbf{k}. \qquad (4.58)$$

Here we have used $\mathbf{N}$ for the torque (and not $\boldsymbol{\tau}$) in order to distinguish it from the tangent vector. The normal acceleration does not contribute in this case as $\mathbf{n} \parallel \mathbf{r}$. Note that the magnitude of the torque can also be written in terms of the momentum of inertia $N = |\mathbf{N}| = I|\alpha|$.

4.3 As we shall see in Chap. 8, the gravitational potential $\phi(r)$ for a uniform spherical mass distribution (with total mass M), for $r \geq R$, where R is the radius of the sphere, is given by

$$\phi(r) = -G\frac{M}{r}. \qquad (4.59)$$

This means that on the surface of the sphere and outside, the gravitational potential behaves as if the mass distribution were a point-like particle with all its mass concentrated in the center of the sphere. Assuming that the Earth is approximately spherical with $R = 6.371 \times 10^6$ (m) and that its mass density is approximately uniform (with $M = 5.972 \times 10^{24}$ kg), calculate the gravitational acceleration g on the surface of the Earth and argue why it is approximately constant close to the surface. For a second (point-like) mass $m \ll M$, show that the acceleration A of M (due to the gravitational pull of m) can be neglected. Calculate also the potential energy as a function of the height h (measured with respect to the surface of the Earth).

Solution: As we have seen previously in (4.8), the gravitational acceleration of a body due to the presence of a gravitational field generated by a mass M is given by

$$g = G\frac{M}{R^2}, \qquad (4.60)$$

where R is the distance between the two bodies. In our particular case M and R are the mass and the radius of the Earth. Thus $g \simeq 9.82$ m/s^2. This is the acceleration that any mass m will suffer due to the gravitational field of M (and it is independent of m). Close to the surface (at a distance r from the surface) the gravitational acceleration will be given by

$$g' = G\frac{M}{(R+r)^2} \simeq G\frac{M}{R^2} = g, \qquad (4.61)$$

where we have considered $r \ll R$. Obviously this is why close to the surface of the Earth, g can be considered as constant. However, it depends on the precision of the experiments we are dealing with, for example $g' \simeq 9.79 \, \text{m/s}^2$ for $r = 10$ km.

A second mass m also generates a gravitational field and the corresponding acceleration A of M will be given by

$$A = G\frac{m}{R^2}. \qquad (4.62)$$

Note that these two accelerations, A and g, are given with respect to an external inertial reference system. The relative acceleration will be greater in magnitude. It will be given by the sum $g + A$. However, in our case

$$\frac{A}{g} = \frac{m}{M} \ll 1, \qquad (4.63)$$

(because $m \ll M$), and therefore A can be neglected.

We can thus consider that the spherical body with mass M generates a central force field close to its surface, of the form

$$\mathbf{F} = -mg\,\mathbf{u}_r, \qquad (4.64)$$

with $\mathbf{u}_r$ pointing outwards the surface. Using the result from Exercise 4.1, we obtain the potential energy as a function of the height

$$U = mgh + C, \qquad (4.65)$$

with C an unphysical constant that we can set to 0 (as we have already mentioned the physical quantities will depend on the difference on the potential energy, for different heights in this case).

4.4 Given a simple pendulum (a mass hanging from a string of length l with neglectable mass and a fixed end) whose gravitational mass is m and inertial mass $\overline{m}$, find the equations of motion for the angle $\theta(t)$ for small oscillations ($\sin \theta \simeq \theta$) and argue what kind of experiment could be suitable to find a possible difference between m and $\overline{m}$. Hint: instead of writing the components of the forces in terms of the basis $\{\mathbf{i}, \mathbf{j}\}$, write them in terms of the normal and the tangent vectors, as shown in Fig. 4.9, in order to find a simple version of the equations of motion.

Solution: Just as previously in Exercise 3.1, for $\dot{\theta} > 0$ (counter-clockwise), $\mathbf{n}$ and $\boldsymbol{\tau}$ have the direction and sense shown in Fig. 4.9 (before θ reaches $\theta_{\max}$). The tangential force has the opposite sense (with respect to $\boldsymbol{\tau}$) because the mass of the pendulum

Fig. 4.9 Representation of
the normal and tangent
vectors together with the
gravitational force (left) and
representation of all the
forces with their direction
and sense (right) (**T** stands
for the tension of the string)

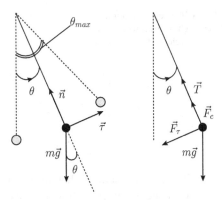

reaches its maximum velocity for $\theta = 0$ and slows down as $\theta \to \theta_{\max}$. Similar considerations can be made when moving in the opposite sense. However, in all cases we will always write down the following equation

$$-mg\sin\theta\,\tau = \overline{m}\,a_\tau\,\tau, \tag{4.66}$$

with $a_\tau < 0$ and where g is the strength of the gravitational field $g = GM/R^2 \approx$ 9.81 m/s^2 (as calculated previously). Simplifying and re-writing a_τ as a function of $\ddot{\theta}$ we obtain

$$\ddot{\theta} + \frac{m}{\overline{m}}\frac{g}{l}\sin\theta = 0. \tag{4.67}$$

For small oscillations $\sin\theta \simeq \theta$, the previous expression reads

$$\ddot{\theta} + \Omega^2\theta = 0, \qquad \Omega^2 \equiv \frac{m}{\overline{m}}\frac{g}{l}\ (\text{constant}). \tag{4.68}$$

The solution to this equation is straightforward

$$\theta(t) = \theta_{\max}\cos(\Omega t + \phi), \tag{4.69}$$

with ϕ, an integration constant. Here Ω is the angular frequency and the period of motion will be thus given by

$$T = \frac{2\pi}{\Omega} = 2\pi\sqrt{\left(\frac{\overline{m}}{m}\right)\frac{l}{g}}. \tag{4.70}$$

If one assumes the equivalence principle then the previous expression of the period is mass independent. Newton's idea was exactly to exploit this feature. He thought that for different mass configurations and for different materials, if m and $\overline{m}$ were not equivalent, then a difference in T would be obtained. Newton performed such

Fig. 4.10 Representation of
the height h from which the
mass is released

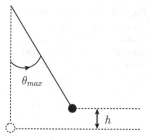

experiments and concluded that $(\overline{m}/m)$ does not differ from one substance to another more than one part in 10^3. More precise experiments were performed afterwards.[4] It is also worth mentioning that Galileo was really the first to observe that pendulums of equal length have the same oscillating period independently of their mass.

4.5 For the previous exercise, assuming $m = \overline{m}$, calculate the tension T exerted on the string. What can we say about T? Calculate the position of the mass for which the tension is maximum $T = T_{\max}$. Express $T_{\max}$ as a function of h, where h is the height from which the mass is released, as shown in Fig. 4.10.

Solution: Looking at Fig. 4.9 we can write down the forces along the normal axis

$$m\, a_n\, \mathbf{n} = F_c\, \mathbf{n} = T\, \mathbf{n} - mg\cos\theta\, \mathbf{n}. \tag{4.71}$$

Thus, the modulus of the tension is given by

$$T = \frac{mv^2}{l} + mg\cos\theta. \tag{4.72}$$

The peculiar thing we notice about T is that its value varies with v. As both terms of the previous sum are positive, T reaches its maximum value for $v = v_{\max}$ and $\cos\theta = 1$, which occurs for $\theta = 0$. We can now use energy conservation in order to express $v_{\max}$ as a function of h.

When the mass reaches $\theta_{\max}$, its total energy (the sum of the kinetic and the potential energy) will be given by

$$E = K + U = 0 + mgh, \tag{4.73}$$

---

[4]For a brief historical introduction read Chap. I, Sect. 2 from Steven Weinberg, *Gravitation end Cosmology: Principles and Applications of the General Theory of Relativity.*

where we have set C (the constant of integration of the potential energy) to 0 (without loss of generality). When the mass reaches $\theta = 0$, its total energy reads

$$E = K + U = \frac{1}{2}mv_{max}^2 + 0. \tag{4.74}$$

As the energy is conserved we obtain

$$v_{max} = \sqrt{2gh}. \tag{4.75}$$

The maximum value of the tension T_{max} can thus be expressed as a function of h as follows

$$T_{max} = \frac{mv_{max}^2}{l} + mg = mg\left(1 + \frac{2h}{l}\right). \tag{4.76}$$

If $h = 0$, then the mass is initially at rest and it remains at rest with $T = mg$ as expected.

4.6 Given the potential energy $V(\mathbf{r}) = 2x + yz$ and the points A and B with (Cartesian) coordinates $\mathbf{r}_A = (1, 1, 0)$, $\mathbf{r}_B = (3, 2, 1)$ calculate the work $W_{AB} = V(\mathbf{r}_A) - V(\mathbf{r}_B)$. Calculate the corresponding force $\mathbf{F} = -\nabla \cdot V$. Check that the virtual work along the path Γ_{AB} given by the sum of two segments $\Gamma_{AB} = \Gamma_{AC} + \Gamma_{CB}$ where $\Gamma_{AC} \equiv \overline{AC}, \Gamma_{CB} \equiv \overline{CB}$ and $\mathbf{r}_C = (0, 0, 0)$, is equal to the previously calculated W_{AB}.

Solution: For the first part, the answer is trivial

$$W_{AB} = V_A - V_B = 2 - 8 = -6. \tag{4.77}$$

The force is equally straightforward to calculate

$$\mathbf{F} = -\nabla \cdot V = -2\mathbf{i} - z\mathbf{j} - y\mathbf{k}. \tag{4.78}$$

Let us now write down the path Γ_{AB} in parametric form. For the first segment we define

$$\mathbf{r}_{AC} \equiv \mathbf{r}_C - \mathbf{r}_A = -\mathbf{i} - \mathbf{j}, \tag{4.79}$$

and therefore $\mathbf{r}(\lambda)$ will simply be given by

$$\mathbf{r}(\lambda) = \mathbf{r}_A + \lambda\mathbf{r}_{AC} = (1 - \lambda)\mathbf{i} + (1 - \lambda)\mathbf{j} = x(\lambda)\mathbf{i} + y(\lambda)\mathbf{j} + z(\lambda)\mathbf{k}, \tag{4.80}$$

with $\lambda \in [0, 1]$ and $z(\lambda) = 0$. Thus $d\mathbf{r}(\lambda)$ reads

$$d\mathbf{r}(\lambda) = -d\lambda\mathbf{i} - d\lambda\mathbf{j}. \tag{4.81}$$

Parametrizing also the components of $\mathbf{F}$ as functions of λ we obtain

$$\mathbf{F} = -2\mathbf{i} - (1 - \lambda)\mathbf{k}. \tag{4.82}$$

Thus, along this path, the only contribution to W_{AC} is

$$W_{AC} = \int_{\Gamma_{AC}} \mathbf{F}(\lambda) \cdot d\mathbf{r}(\lambda) = 2\int_0^1 d\lambda = 2. \tag{4.83}$$

As for the second segment it is obvious that $\mathbf{r}(\lambda)$ can be parametrized as

$$\mathbf{r}(\lambda) = 3\lambda\mathbf{i} + 2\lambda\mathbf{j} + \lambda\mathbf{k}, \tag{4.84}$$

again with $\lambda \in [0, 1]$. We obtain $d\mathbf{r} = (3\mathbf{i} + 2\mathbf{j} + \mathbf{k})d\lambda$ and $\mathbf{F} = -2\mathbf{i} - \lambda\mathbf{j} - 2\lambda\mathbf{k}$ and so

$$W_{CB} = \int_{\Gamma_{CB}} \mathbf{F} \cdot d\mathbf{r} = \int_0^1 (-6 - 2\lambda - 2\lambda) \, d\lambda = -8. \tag{4.85}$$

As expected the result for the total contribution of the virtual work $W_{AC} + W_{CB}$ is the same as the one calculated previously.

4.7 Consider the force field $\mathbf{F} = 3xyz\,\mathbf{i} + 2\mathbf{j}$ and the points A, B and B' with (Cartesian) coordinates $\mathbf{r}_A = (2, 1, 2)$, $\mathbf{r}_B = (1, 0, 3)$ and $\mathbf{r}_{B'} = (3, 2, 0)$. Calculate the virtual work W_{AB} along the paths Γ_{AB} and Γ'_{AB} given by $\Gamma_{AB} \equiv \overline{AB}$ and $\Gamma'_{AB} = \Gamma_{AB'} + \Gamma_{B'B}$ with $\Gamma_{AB'} \equiv \overline{AB'}$ and $\Gamma_{B'B} \equiv \overline{B'B}$. What can we conclude? Check that the answer is correct by calculating $\nabla \times \mathbf{F}$.

Solution: As in the previous exercise we start by calculating the parametric curve corresponding to the first path

$$\mathbf{r}_{AB} = \mathbf{r}_B - \mathbf{r}_A = -\mathbf{i} - \mathbf{j} + \mathbf{k}, \tag{4.86}$$

and so, the path Γ_{AB} can be parametrized by

$$\mathbf{r}(\lambda) = \mathbf{r}_A + \lambda\mathbf{r}_{AB} = (2 - \lambda)\mathbf{i} + (1 - \lambda)\mathbf{j} + (2 + \lambda)\mathbf{k}, \tag{4.87}$$

with $\lambda \in [0, 1]$. Therefore $d\mathbf{r}$ simply reads

$$d\mathbf{r} = (-\mathbf{i} - \mathbf{j} + \mathbf{k})d\lambda. \tag{4.88}$$

The force field along the given path can be expressed as a function of λ as

$$\mathbf{F} = 3(2 - \lambda)(1 - \lambda)(2 + \lambda)\mathbf{i} + 2\mathbf{j}. \tag{4.89}$$

The virtual work along the previously given path finally gives

$$W_{AB} = \int_0^1 \left[-3(2 - \lambda)(1 - \lambda)(2 + \lambda) - 2 \right] d\lambda = -\frac{31}{4}. \tag{4.90}$$

For the first part $\Gamma_{AB'}$ of the second path, we have

$$\mathbf{r}(\lambda) = (2 + \lambda)\mathbf{i} + (1 + \lambda)\mathbf{j} + (2 - 2\lambda)\mathbf{k}, \quad d\mathbf{r} = (\mathbf{i} + \mathbf{j} - 2\mathbf{k})d\lambda,$$
$$\mathbf{F} = 3(2 + \lambda)(1 + \lambda)(2 - 2\lambda)\mathbf{i} + 2\mathbf{j}, \tag{4.91}$$

and so $W_{AB'}$ is given by the following integral

$$W_{AB'} = \int_0^1 \left[3(2 + \lambda)(1 + \lambda)(2 - 2\lambda) + 2 \right] d\lambda = \frac{23}{2}. \tag{4.92}$$

Finally, for the second part $\Gamma_{B'B}$ of the second path, we have

$$\mathbf{r}(\lambda) = (3 - 2\lambda)\mathbf{i} + (2 - 2\lambda)\mathbf{j} + 3\lambda\mathbf{k}, \quad d\mathbf{r} = (-2\mathbf{i} - 2\mathbf{j} + 3\mathbf{k})d\lambda,$$
$$\mathbf{F} = 9(3 - 2\lambda)(2 - 2\lambda)\lambda\mathbf{i} + 2\mathbf{j}, \tag{4.93}$$

and so $W_{B'B}$ is given by the following integral

$$W_{AB'} = \int_0^1 \left[-18(3 - 2\lambda)(2 - 2\lambda)\lambda - 4 \right] d\lambda = -16, \tag{4.94}$$

thus, the total work along the second path gives

$$W_{AB} = -\frac{9}{2}, \tag{4.95}$$

which is not the previously obtained result. We conclude thus, that the force $\mathbf{F}$ is not conservative. We can straightforwardly check this result by calculating the curl

$$\nabla \times \mathbf{F} = 3xy\,\mathbf{j} - 3xz\,\mathbf{k} \tag{4.96}$$

which is different from $\mathbf{0}$ in general.

Fig. 4.11 A body of mass m_1 laying over the surface of a triangular wedge, over which a force $\mathbf{F}_{app}$ is applied along the positive **i** axis

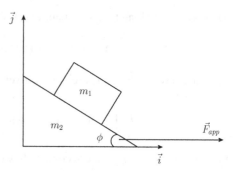

4.8 Consider a body of mass m_1 laying over an inclined surface of a triangular wedge with slope ϕ ($\phi < 90°$) and mass m_2. Over this wedge we apply an external constant force $\mathbf{F}_{app} = F_{app}\,\mathbf{i}$ along the positive **i** axis ($F_{app} > 0$) as shown in Fig. 4.11. If the static friction force of the object (of mass m_1) with the surface of the wedge is given by $|\mathbf{F}_s| = \mu_s\,|\mathbf{N}|$, obtain the maximum and minimum value of F_{app} in order to maintain the object m_1 at rest with respect to the wedge. How does F_{app}^{min} depend on μ_s and ϕ?

Solution: We will first calculate the minimum force F_{app} necessary to prevent m_1 to slide downward (along the positive **i** axis). The key to the problem consists in realizing that the applied force over the wedge will accelerate both masses and the value of this acceleration will be given by

$$(m_1 + m_2)\,a = F_{app}\,. \tag{4.97}$$

In our particular case we are looking for $F_{app}^{min} = (m_1 + m_2)\,a_{min}$. The complete diagram of all the forces is shown in Fig. 4.12. Let us consider the balance of the forces on m_1 along the **i** axis

$$N\,\sin\phi - F_s\,\cos\phi - m_1\,a_{min} = 0\,. \tag{4.98}$$

Manipulating the expression we obtain

$$m_1\,\frac{F_{app}^{min}}{m_1 + m_2} = N\,\sin\phi - \mu_s\,N\,\cos\phi\,. \tag{4.99}$$

The forces that act on m_1 along the **j** axis are given by

$$N\,\cos\phi + F_s\,\sin\phi - m_1\,g = 0\,, \tag{4.100}$$

Fig. 4.12 Diagram of all
forces acting on m_1 and m_2

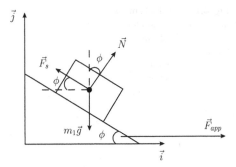

where we have considered that m_1 must not move along the **j** axis. Manipulating the
previous expression we obtain

$$N = \frac{m_1\, g}{\cos\phi + \mu_s\, \sin\phi}. \tag{4.101}$$

Substituting this result in (4.99) we obtain

$$F_{app}^{min} = (m_1 + m_2)\, g\, \frac{\sin\phi - \mu_s\, \cos\phi}{\cos\phi + \mu_s\, \sin\phi}. \tag{4.102}$$

Let us now consider the limit $F_{app}^{min} = 0$. This means that if we place m_1 over the
surface of the wedge, m_1 will remain at rest (it will not slide forward) only due to
the action of the friction force. No minimum applied force would be necessary to
maintain m_1 at rest with respect to m_2. This limiting case is given by

$$\mu_s = \tan\phi \equiv \mu_s^{lim}. \tag{4.103}$$

If $\mu_s < \tan\phi$, then F_{app}^{min} is positive. This means that the friction force is not enough to
maintain m_1 at rest with respect to the wedge and thus we need to apply a minimum
force along the positive **i** axis in order to achieve it.

As for the maximum force that we can apply (along the positive **i** axis) preventing
m_1 to slide upward, it can be obtained by simply changing the sense of $\mathbf{F}_s$ in Fig. 4.12,
which effectively translates in changing the sign of μ_s in (4.102):

$$F_{app}^{max} = (m_1 + m_2)\, g\, \frac{\sin\phi + \mu_s\, \cos\phi}{\cos\phi - \mu_s\, \sin\phi}. \tag{4.104}$$

Fig. 4.13 A vehicle moving
in uniform circular motion is
schematically shown
together with the radial
friction force with its
direction and sense

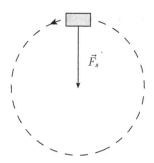

4.9 Consider a vehicle of mass m moving in uniform circular motion (with constant v). The static friction $|\mathbf{F}_s| = \mu_s\,|\mathbf{N}|$ of the tires with the pavement, has a radial direction and sense pointing towards the center of the circle (as shown in Fig. 4.13). This is the force that prevents the vehicle from sliding, helping it maintain the circular motion (do not confuse it with the tangential friction force that is responsible for the spinning of the wheels). Obtain the maximum velocity that the vehicle can reach without sliding. How does this velocity get modified if we triple the vehicle's mass? What angle should the pavement be inclined, in order to have a maximum velocity that triples the previously calculated one?

Solution: The maximum velocity that the vehicle can reach is given by the condition $\mathbf{F}_c = \mathbf{F}_s$, where $\mathbf{F}_c$ is the centripetal force. If $\mathbf{F}_c > \mathbf{F}_s$ then the vehicle will start sliding. On the contrary if $\mathbf{F}_c \leq \mathbf{F}_s$ the vehicle maintains its circular motion without sliding. Therefore

$$m\frac{v_{\text{max}}^2}{r} = \mu_s\,m\,g\,, \tag{4.105}$$

where r is the radius of the circle. We thus obtain

$$v_{\text{max}} = \sqrt{\mu_s\,g\,r}\,, \tag{4.106}$$

which is independent of m. We can conclude that, as the maximum velocity does not depend on m, if we modify the vehicle's mass, it will remain the same.

In order to answer the last question consider the configuration shown in Fig. 4.14. We now have two contributions to the maximum velocity, one proportional to the projection of the normal force along the **i** axis and another one proportional to the projection of the friction force along the same axis

$$F_s\,\cos\phi + N\,\sin\phi = \frac{m v_{\text{max}}^2}{r}\,. \tag{4.107}$$

Fig. 4.14 Diagram of the forces acting on m for an inclined pavement of slope ϕ

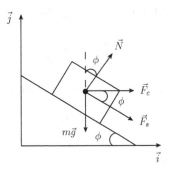

A second equation is obtained by imposing that the vehicle does not move along the **j** axis

$$N \cos \phi - m g - F_s \sin \phi = 0. \qquad (4.108)$$

From these two equations we obtain

$$v_{\text{max}} = \sqrt{\frac{r g (\mu_s \cos \phi + \sin \phi)}{\cos \phi - \mu_s \sin \phi}}. \qquad (4.109)$$

If we wish this velocity to be three times larger than the one calculated previously in (4.106), the condition we need to impose is

$$3\sqrt{\mu_s} = \sqrt{\frac{\mu_s \cos \phi + \sin \phi}{\cos \phi - \mu_s \sin \phi}}, \qquad (4.110)$$

which can be solved numerically.

4.10 Determine the expression for $x(t)$ for a particle moving in a straight line along the **i** axis of some inertial reference frame, under the influence of a constant force $\mathbf{F} = m\,\mathbf{a} = m\,a\,\mathbf{i}$.

Solution: This is nothing but the rectilinear uniformly accelerated motion that one normally studies in high school. We shall use this exercise as a reminder and move on to more complex cases in the following exercises. The differential equation we have to solve is

$$m\,\ddot{x} = m\,\dot{v} = m\frac{dv}{dt} = m\,a. \qquad (4.111)$$

Therefore, the expression for $v(t)$ will be given by

$$\int_{v_0}^{v} dv' = \int_{t_0}^{t} a \, dt' \quad \Rightarrow \quad v(t) = v_0 + a(t - t_0). \tag{4.112}$$

The expression for $x(t)$ finally reads

$$\int_{x_0}^{x} dx' = \int_{t_0}^{t} v(t') \, dt' = \int_{t_0}^{t} \left(v_0 + a(t' - t_0) \right) dt', \tag{4.113}$$

thus

$$x(t) = x_0 + v_0(t - t_0) + \frac{1}{2} a(t - t_0)^2. \tag{4.114}$$

4.11 Consider a free falling body (due to the gravitational pull $\mathbf{F}_g = -m \, g \, \mathbf{j}$) and a frictional force (with the air) given by $\mathbf{F}_f = b \, v^2 \, \mathbf{j}$ with b a positive constant, as shown in Fig. 4.15. Check that there is a maximum velocity, called terminal (or limit) velocity v_l, that the body reaches under these conditions. Obtain its expression. Also obtain the expressions for $v(t)$ and $h(t)$ considering that the initial height is h_0 at $t_0 = 0$ (s) and that the object was initially at rest.

Solution: Let's start by writing down the total force exerted over the falling body

$$\mathbf{F} = \mathbf{F}_g + \mathbf{F}_f = (-mg + bv^2) \, \mathbf{j}. \tag{4.115}$$

The body starts falling at some height h_0 with no initial velocity. Due to the gravitational pull it accelerates and therefore it starts gaining velocity. As the velocity grows, the frictional force also grows as $\sim v^2$. The absolute value of the maximum velocity v_l that the object can reach, will be thus given by the condition $|\mathbf{F}_g| = |\mathbf{F}_f|$:

$$mg = bv_l^2 \quad \Rightarrow \quad v_l = \sqrt{mg/b}. \tag{4.116}$$

Let us now obtain $v(t)$. The equations of motion in this case read

$$F = m \frac{dv}{dt} = -mg + bv^2. \tag{4.117}$$

We can re-write the previous differential equation in terms of the terminal velocity as follows

$$\frac{dv}{v^2 - v_l^2} = \frac{g}{v_l^2} \, dt \quad \Rightarrow \quad \frac{dv}{1 - v^2/v_l^2} = -g \, dt. \tag{4.118}$$

Fig. 4.15 Free falling body
under a gravitational field
with a frictional force bv^2

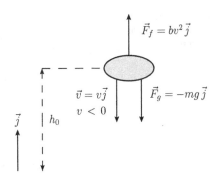

The integration is straightforward and the result is the inverse hyperbolic function

$$\int_0^v \frac{dv'}{1-(v'^2/v_l^2)} = -\int_0^t g\,dt' \quad \Rightarrow \quad v_l \operatorname{arctanh}(v/v_l) = -gt. \quad (4.119)$$

Thus $v(t)$ can simply be written as

$$v(t) = v_l \tanh(-gt/v_l) = -v_l \tanh(gt/v_l) = -v_l \frac{1-e^{-2gt/v_l}}{1+e^{-2gt/v_l}}, \quad (4.120)$$

where the minus sign is simply due to the fact that the object moves along the negative
j axis i.e., we have considered $\mathbf{v} = v(t)\,\mathbf{j}$ with $v(t) < 0$ and we have taken v_l as the
absolute value of the terminal velocity. Finally, the height will be given by

$$\int_{h_0}^h dh' = \int_0^t v(t')dt' \quad \Rightarrow \quad h(t) = h_0 + v_l t - \frac{v_l^2}{g}\ln\left(\frac{1+e^{2gt/v_l}}{2}\right). \quad (4.121)$$

In the limit $b \to 0$ or equivalently $v_l \to \infty$ one recovers the usual expression for
a free falling body with no friction

$$\lim_{v_l \to \infty} v(t) = -gt, \qquad \lim_{v_l \to \infty} h(t) = h_0 - \frac{1}{2}gt^2. \quad (4.122)$$

4.12 For the previous exercise, for $b = 0.5$ $(\mathrm{N\,s^2/m^2})$ and $m = 5$ (kg) obtain
the time instants t_l and $t_{0.99}$ in which the object reaches its terminal velocity
and 99% of its terminal velocity. Calculate also the correspondent heights $h(t_l)$
and $h(t_{0.99})$ given the initial height $h_0 = 100$ (m). What height will it reach for
$t' = t_{0.99}$ in the limit $b \to 0$? Take $g = 9.81$ $(\mathrm{m/s^2})$.

Fig. 4.16 The hyperbolic function tanh(x) as a function of x

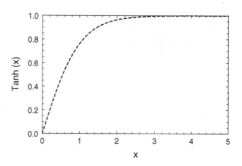

Solution: We have seen that the absolute value of the velocity behaves like $|v(t)| \sim$ tanh(kt), with k some positive constant. The behaviour of this function is plotted in Fig. 4.16.

Thus $\lim_{x\to\infty}$ tanh$(x) = 1$ and therefore the object will reach its terminal velocity v_l for $t_l \to \infty$. However, we can observe that it is a function that grows very quickly thus we expect it to reach 0.99 v_l in a few seconds. We are looking for the solution to the equation

$$\frac{1 - e^{-2gt_{0.99}/v_l}}{1 + e^{-2gt_{0.99}/v_l}} = 0.99 \quad \Rightarrow \quad t_{0.99} = -\frac{v_l}{2g} \ln\frac{0.01}{1.99}, \qquad (4.123)$$

with v_l simply given by

$$v_l = \sqrt{5 \cdot 9.81/0.5} \simeq 9.9\,\text{m/s}. \qquad (4.124)$$

Therefore $t_{0.99} = 2.67$ s. The height at the time instant t_l will be obviously zero i.e., $h(t_l) = 0$ (it reaches the ground within a few seconds and $t_l = \infty$). As for $h(t_{0.99})$ it can be easily calculated $h(t_{0.99}) = 80.44$ m. As for the limit $b \to 0$ for $t' = 2.67$ s we obtain $h(t') = 65$ m. This is obviously an intuitive result, as in the limit $b \to 0$ the air offers no motion resistance and therefore the object reaches the ground faster.

4.13 A body of mass m moves along the **i** axis with initial velocity $\mathbf{v_0} = v_0\mathbf{i}$ (with $v_0 > 0$) with a friction force $\mathbf{F}_f = -b\mathbf{v}$ acting upon it (with b a positive constant) as shown in Fig. 4.17. Obtain the time interval $(t - t_0)$ that the object takes to come to a stop (to reach $v = 0$). What distance does the object travel in this time interval?

Solution: The equations of motion are given by the following differential equation

$$m\frac{dv}{dt} = -bv \quad \Rightarrow \quad -\frac{m}{b}\frac{dv}{v} = dt. \qquad (4.125)$$

Fig. 4.17 An object moving with initial velocity v_0 along the positive **i** axis with a friction force proportional to v

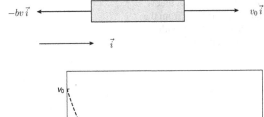

Fig. 4.18 Plot of v as a function of t

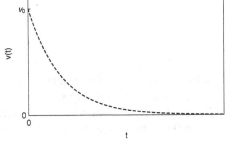

Taking $t_0 = 0$ and integrating we obtain $t = t(v)$

$$\int_0^t dt' = -\frac{m}{b} \int_{v_0}^v \frac{dv'}{v'} \quad \Rightarrow \quad t = -\frac{m}{b} \ln\left(\frac{v}{v_0}\right). \tag{4.126}$$

From the previous expression, it is straightforward to obtain $v(t)$

$$v(t) = v_0 \, e^{-\frac{b}{m}t}. \tag{4.127}$$

Thus $v \to 0$ as $t \to \infty$ meaning that it takes an infinite amount of time to come to a stop. The behaviour of $v(t)$ is shown in Fig. 4.18.

One might suspect that as it takes an infinite time to reach $v = 0$, the travelled distance will also be infinite. However, as we can observe in the previous plot, it decays fast enough so that the area below the curve (which is the travelled distance) is finite. Taking the initial point $x_0 = 0$ the travelled distance as a function of t is easily obtained by integrating $v(t)$

$$x(t) = \int_0^t v(t')dt' = \int_0^t v_0 \, e^{-\frac{b}{m}t'} dt' = v_0 \frac{m}{b}\left(1 - e^{-\frac{b}{m}t}\right). \tag{4.128}$$

The total travelled distance is simply given by the limit

$$x(t \to \infty) = v_0 \frac{m}{b}, \tag{4.129}$$

which is obviously finite.

Fig. 4.19 An object that is
launched vertically from the
surface of the Earth with
initial velocity v_0

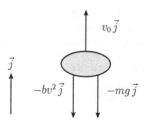

4.14 Consider an object that is vertically launched from the Earth's
surface with initial velocity $\mathbf{v_0} = v_0\,\mathbf{j}$. If the friction with the air is given by
$\mathbf{F}_f = -bv^2\,\mathbf{j}$ (as shown in Fig. 4.19), calculate the amount of time that it takes
for the body to come to a stop. What maximum height will it reach?

Solution: The equation of motion in this case is given by

$$m\frac{dv}{dt} = -bv^2 - mg \quad \Rightarrow \quad \frac{dv}{-g - bv^2/m} = dt\,. \tag{4.130}$$

We can observe that again, this integration will be of the form

$$\int \frac{a}{1 + a^2 x^2}\,dx = \arctan(ax) + \text{constant}\,. \tag{4.131}$$

Thus, manipulating the expression we obtain

$$-\frac{1}{g}\sqrt{\frac{mg}{b}}\int_{v_0}^{v}\frac{\sqrt{b/mg}}{1 + (b/mg)v'^2}\,dv' = \int_0^t dt\,, \tag{4.132}$$

where we have considered $t_0 = 0$. We thus, obtain the expression for $t(v)$

$$t = \frac{1}{g}\sqrt{\frac{mg}{b}}\left[\arctan\left(\sqrt{b/mg}\,v_0\right) - \arctan\left(\sqrt{b/mg}\,v\right)\right]\,. \tag{4.133}$$

The amount of time for the object to reach to a stop will be given by $t_s \equiv t(v = 0)$

$$t_s = \frac{1}{g}\sqrt{\frac{mg}{b}}\arctan\left(\sqrt{b/mg}\,v_0\right)\,. \tag{4.134}$$

In order to calculate the height we need the expression for $v(t)$. It can be easily
obtained from (4.133)

$$v(t) = \sqrt{\frac{mg}{b}} \tan\left(\arctan\left(\sqrt{b/mg}\, v_0\right) - g\sqrt{b/mg}\, t\right). \qquad (4.135)$$

In order to integrate the previous expression we first need to simplify it using the following result

$$\tan(A - C) = \frac{\cos C \sin A - \cos A \sin C}{\cos A \cos C + \sin A \sin C}. \qquad (4.136)$$

Thus, if we define $a \equiv \sqrt{b/mg}$ we obtain

$$\begin{aligned} v(t) &= a^{-1} \tan\left(\arctan(a\, v_0) - g\, a\, t\right) \\ &= a^{-1} \frac{a\, v_0 \cos(g\, a\, t) - \sin(g\, a\, t)}{\cos(g\, a\, t) + a\, v_0 \sin(g\, a\, t)}. \end{aligned} \qquad (4.137)$$

The height will then be given by the integral

$$h(t) = \int_0^t v(t')dt' = \frac{1}{a^2 g} \ln\left(\cos(g\, a\, t) + a\, v_0 \sin(g\, a\, t)\right), \qquad (4.138)$$

and the maximum height will be given by the previous expression evaluated at t_s

$$h(t_s) = \frac{\ln(1 + a^2 v_0^2)}{2a^2 g}, \qquad (4.139)$$

where we have used $\cos(\arctan(x)) = \sin(\arctan(x)) = (1 + x^2)^{-1/2}$

4.15 Repeat the previous exercise for $b = 0$. Compare numerically t_s and $h(t_s)$ for $b = 1/5$ and $b = 0$ ($\mathrm{N\,s^2/m^2}$), where mass and initial velocity are given by $m = 5$ (kg) and $v_0 = 5$ (m/s).

Solution: In this case the equations are terribly simple and the results are obvious

$$v(t) = v_0 - gt, \qquad\qquad t_s = v_0/g,$$
$$h(t) = v_0 t - \frac{1}{2}gt^2, \qquad\qquad h(t_s) = v_0^2/2g. \qquad (4.140)$$

Even if it is not a trivial task to calculate the limit $b \to 0$ for the expressions from the previous exercise, by taking this limit one obtains the results from (4.140). Numerically, for $b = 1/5$ ($\mathrm{N\,s^2/m^2}$) and $g = 9.81$ (m/s^2) we obtain

$$t_s = 0.49\,\mathrm{s}, \qquad h(t_s) = 1.21\,\mathrm{m}. \qquad (4.141)$$

Fig. 4.20 Two free-falling spheres under the gravitational pull, with friction forces proportional to the radii and the velocities

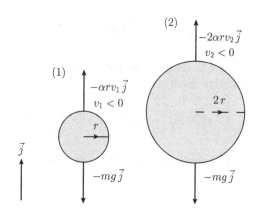

For $b = 0$ we simply obtain

$$t_s = 0.51\,\text{s}, \qquad h(t_s) = 1.27\,\text{m}. \tag{4.142}$$

We can observe that the time interval is very little affected by friction (in this case), and so is the travelled distance.

4.16 Two spherical bodies with the same mass m and radii r and $2r$, are dropped simultaneously from a certain height. If the friction force acting on the bodies is of the form $F_f = \alpha R v$ with α a constant, R the radius of the sphere and v its velocity (as shown in Fig. 4.20), which one of the bodies reaches a greater terminal velocity? Which sphere reaches first, 90% of its terminal velocity, and which reaches the ground first?

Solution: Considering the following sign convention $\mathbf{v}_i = v_i(t)\,\mathbf{j}$ with $v_i(t) < 0$ (with $i = 1, 2$), the equations of motion for the two cases read

$$F_1 = -m\,g - \alpha\,r\,v_1, \qquad F_2 = -m\,g - 2\,\alpha\,r\,v_2. \tag{4.143}$$

The absolute values of the terminal velocities, that we shall call $v_{l,i}$, as explained previously, are given by the condition $F_i = 0$, therefore we obtain

$$v_{l,1} = \frac{m\,g}{\alpha\,r} = 2\,v_{l,2}. \tag{4.144}$$

Thus, the terminal velocity of the smaller sphere is greater than the terminal velocity of the second sphere (in absolute value). This is a quite logical result as the first object offers less resistance as it has a smaller radius. In order to obtain the answer to the second question we need to calculate the velocity of each body as a function of t

$$m\frac{dv_1}{dt} = -mg - \alpha r\, v_1\,, \qquad m\frac{dv_2}{dt} = -mg - 2\alpha r\, v_2\,. \qquad (4.145)$$

Manipulating the expression and expressing it in terms of $v_{l,i}$ we obtain the following simple equation

$$\frac{dv_i}{v_i + v_{l,i}} = -\frac{g}{v_{l,i}}dt\,. \qquad (4.146)$$

Integrating, with the initial conditions $t_0 = 0$ and $v_i(t_0) = 0$ we get to

$$\int_0^{v_i} \frac{dv_i'}{v_i' + v_{l,i}} = -\int_0^t \frac{g}{v_{l,i}}dt' \quad \Rightarrow \quad \ln\left(1 + \frac{v_i}{v_{l,i}}\right) = -\frac{g}{v_{l,i}}t\,, \qquad (4.147)$$

and therefore $v_i(t)$ simply takes the form

$$v_i(t) = -v_{l,i}\left(1 - e^{-\frac{g}{v_{l,i}}t}\right)\,, \qquad (4.148)$$

where again, the minus sign is due to the fact that the objects move along the negative **j** axis and we have taken $v_{l,i}$ as the absolute value. However, in order to find out which object reaches first 90% of its terminal velocity it is easier to use the previous expression (4.147). If we call t_1 the time instant when the first object reaches $0.9\, v_{l,1}$ (in absolute value) and t_2 the time instant when the first object reaches $0.9\, v_{l,2}$ (also in absolute value) we obtain

$$\ln(1 + 0.9) = -\frac{g}{v_{l,1}}t_1\,, \qquad \ln(1 + 0.9) = -\frac{g}{v_{l,2}}t_2\,. \qquad (4.149)$$

Thus, we get to the following relation

$$\frac{t_1}{t_2} = \frac{v_{l,1}}{v_{l,2}} = 2 \quad \Rightarrow \quad t_1 = 2t_2\,, \qquad (4.150)$$

and so, the second object reaches first 90% of its terminal velocity. Again, this is a quite intuitive result, as the second sphere suffers more friction with the air, it slows down faster than the first one.

Given the initial height h_0, the height as a function of t for each sphere is obtained by integrating

$$\int_{h_0}^{h_i} dh = \int_0^t v(t')dt' \quad \Rightarrow \quad h_i = h_0 - v_{l,i}t + \frac{v_{l,i}^2}{g}\left(1 - e^{-\frac{g}{v_{l,i}}t}\right)\,. \qquad (4.151)$$

One can check that the limits for $\alpha \to 0$ ($v_{l,i} \to \infty$) give the correct answer

$$\lim_{v_{l,i}\to\infty} v_i = -gt\,, \qquad \lim_{v_{l,i}\to\infty} h_i = h_0 - \frac{1}{2}gt^2\,. \qquad (4.152)$$

Fig. 4.21 The function $f(t)$ as a function of t

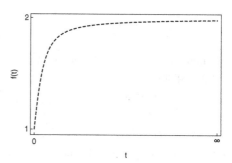

Fig. 4.22 Two reference systems $\mathcal{O}$ and $\mathcal{O}'$ with $\mathcal{O}'$ moving with velocity $\mathbf{u} = u\,\mathbf{i}$ relative to $\mathcal{O}$. The velocity $\mathbf{v}$ of a moving particle with respect to $\mathcal{O}$ is also shown

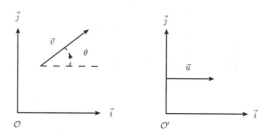

In order to see which object reaches the ground first we shall compare the quantities $h_1 - h_0$ and $h_2 - h_0$

$$\frac{h_1 - h_0}{h_2 - h_0} = \frac{-2v_{l,2}t + \frac{4v_{l,2}^2}{g}\left(1 - e^{-\frac{g}{2v_{l,2}}t}\right)}{-v_{l,2}t + \frac{v_{l,2}^2}{g}\left(1 - e^{-\frac{g}{v_{l,2}}t}\right)} \equiv f(t). \qquad (4.153)$$

The behaviour of the previously defined function is plotted in Fig. 4.21. Therefore $f(t) > 1$ for $t > 0$ and so the value of $h_0 - h_1$ is always greater than $h_0 - h_2$. As a consequence, the first object (as expected) reaches the ground first.

4.17 Consider the $\{\mathcal{O}', \mathbf{i}, \mathbf{j}, \mathbf{k}\}$ reference frame that moves with velocity $\mathbf{u} = u\,\mathbf{i}$ with respect to the $\{\mathcal{O}, \mathbf{i}, \mathbf{j}, \mathbf{k}\}$ frame. Consider a particle that moves with velocity $\mathbf{v} = v_x\,\mathbf{i} + v_y\,\mathbf{j}$ with respect to $\mathcal{O}$ forming an angle θ with the horizontal ($\mathbf{i}$) axis, as shown in Fig. 4.22. Obtain the magnitude of the velocity $\mathbf{v}'$ of the moving body in the $\mathcal{O}'$ reference frame. Obtain θ'.

Solution: Let's start by writing down $\mathbf{v}$ and $\mathbf{v}'$ as functions of their moduli and the angles θ and θ'. In the $\mathcal{O}$ reference frame we have

$$\mathbf{v} = v_x\,\mathbf{i} + v_y\,\mathbf{j}, \qquad v = |\mathbf{v}| = \sqrt{v_x^2 + v_y^2},$$

$$v_x = v\cos\theta, \qquad v_y = v\sin\theta. \qquad (4.154)$$

In the $\mathcal{O}'$ reference frame we have

$$\mathbf{v}' = v'_x \mathbf{i} + v'_y \mathbf{j}, \qquad v' = |\mathbf{v}'| = \sqrt{v'^2_x + v'^2_y},$$
$$v'_x = v' \cos \theta', \qquad v'_y = v' \sin \theta'. \tag{4.155}$$

The Galilean transformations in this case read $\mathbf{v}' = \mathbf{v} - \mathbf{u}$ and therefore

$$v'_x = v_x - u, \qquad v'_y = v_y. \tag{4.156}$$

Introducing the moduli of the vectors and the angles, the previous expressions read

$$v' \cos \theta' = v \cos \theta - u, \qquad v' \sin \theta' = v \sin \theta. \tag{4.157}$$

Taking the quotient of the two previous expressions, we obtain

$$\tan \theta' = \frac{v \sin \theta}{v \cos \theta - u} = \frac{\sin \theta}{\cos \theta - u/v}, \tag{4.158}$$

and therefore θ' will simply be given by

$$\theta' = \arctan \left(\frac{\sin \theta}{\cos \theta - u/v} \right). \tag{4.159}$$

On the other hand, squaring and summing both expressions from (4.157) we obtain

$$v'^2 \sin^2 \theta' + v'^2 \cos^2 \theta' = v^2 \sin^2 \theta + v^2 \cos^2 \theta + u^2 - 2uv \cos \theta, \tag{4.160}$$

which simplifies to

$$v' = \sqrt{u^2 + v^2 - 2uv \cos \theta}, \tag{4.161}$$

which is the remaining expression we were looking for.

4.18 Consider the coordinates x, y and z of a fixed point and, v_x, v_y and v_z the components of the velocity of a moving particle with respect to a reference system $\{\mathcal{O}, \mathbf{i}, \mathbf{j}, \mathbf{k}\}$. Consider a second reference frame $\{\mathcal{O}', \mathbf{i}', \mathbf{j}', \mathbf{k}'\}$ that moves with (constant) velocity $\mathbf{V}$ with respect to $\mathcal{O}$, and where the axes of the primed basis are obtain through a proper orthogonal rotation R (1.28) of the non-primed one. This is schematically shown in Fig. 4.23. Obtain the coordinates x', y' and z' of the fixed point and the components v'_x, v'_y and v'_z of the moving body with respect to $\mathcal{O}'$.

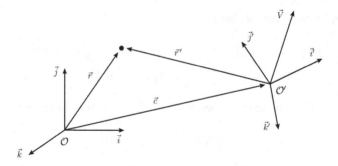

Fig. 4.23 Two reference systems $\{\mathcal{O}, \mathbf{i}, \mathbf{j}, \mathbf{k}\}$, and $\{\mathcal{O}', \mathbf{i}', \mathbf{j}', \mathbf{k}'\}$, where the primed basis $\{\mathbf{i}', \mathbf{j}', \mathbf{k}'\}$ is obtained from the non-primed one through a proper orthogonal rotation, and where $\mathcal{O}'$ moves with velocity $\mathbf{V}$ with respect to $\mathcal{O}$

Solution: Remember the transformation (1.40) for a rotation and a translation:

$$\begin{pmatrix} x' \\ y' \\ z' \end{pmatrix} = R \begin{pmatrix} x - c_x \\ y - c_y \\ z - c_z \end{pmatrix}. \tag{4.162}$$

Considering that R is constant and that $\mathbf{c}$ varies with time i.e.,

$$\frac{d\mathbf{c}}{dt} = \mathbf{V}, \tag{4.163}$$

by taking the time derivatives on both sides of expression (4.162), we obtain

$$\begin{pmatrix} v'_x \\ v'_y \\ v'_z \end{pmatrix} = R \begin{pmatrix} v_x - V_x \\ v_y - V_y \\ v_z - V_z \end{pmatrix}, \tag{4.164}$$

for the transformation of the velocities and

$$\begin{pmatrix} x' \\ y' \\ z' \end{pmatrix} = R \begin{pmatrix} x - c_{0,x} - V_x t \\ y - c_{0,y} - V_y t \\ z - c_{0,z} - V_z t \end{pmatrix}, \tag{4.165}$$

for the coordinates of a fixed point. These transformations are similar to the ones from (4.48) except for the multiplying rotation matrix R.

Further Reading

1. J.V. José, E.J. Saletan, *Classical Dynamics: A Contemporary Approach*. Cambridge University Press

2. S.T. Thornton, J.B. Marion, *Classical Dynamics of Particles and Systems*
3. H. Goldstein, C. Poole, J. Safko, *Classical Mechanics*, 3rd edn. Addison Wesley
4. J.R. Taylor, *Classical Mechanics*
5. D.T. Greenwood, *Classical Dynamics*. Prentice-Hall Inc.
6. D. Kleppner, R. Kolenkow, *An Introduction to Mechanics*
7. C. Lanczos, *The Variational Principles of Mechanics*. Dover Publications Inc.
8. W. Greiner, *Classical Mechanics: Systems of Particles and Hamiltonian Dynamics*. Springer
9. H.C. Corben, P. Stehle, *Classical Mechanics*, 2nd edn. Dover Publications Inc.
10. T.W.B. Kibble, F.H. Berkshire, *Classical Mechanics*. Imperial College Press
11. M.G. Calkin, *Lagrangian and Hamiltonian Mechanics*
12. A.J. French, M.G. Ebison, *Introduction to Classical Mechanics*

Chapter 5
Systems of Particles and Variable Mass

Abstract After introducing the needed defining properties of a N-particle system (such as the center of mass, motion of the center of mass and relative motion with respect to the center of mass, total kinetic energy, force, angular momentum etc.), together with the notions introduced in the previous chapter, and with a little guidance, along this chapter, one will be able to prove many related theorems and deduce interesting results for such systems. Among others the reader will prove König's theorems as an exercise, energy, momentum and angular momentum conservation, and also analyse simple systems of two particles with their peculiarities. Systems with variable mass will also be treated in this chapter together with related exercises.

5.1 Point-Like Particle Systems

Consider a system of N point-like particles ($N \geq 2$) with coordinates $\{\mathbf{r}_i\}_{i=1}^N$ with respect to a Cartesian reference system $\{\mathcal{O}, \mathbf{i}, \mathbf{j}, \mathbf{k}\}$. With respect to this reference system, the coordinates of the center of mass (CM) are given by $\mathcal{R}$:

$$\mathcal{R} = \frac{1}{M} \sum_i m_i \, \mathbf{r}_i \,, \tag{5.1}$$

where M is the total mass of the system $M = \sum_i m_i$. Another reference frame that we can naturally define is the CM, given by $\{\mathcal{O}_{CM}, \mathbf{i}, \mathbf{j}, \mathbf{k}\}$, as shown in Fig. 5.1. In this reference system[1] the coordinates of the particles will be given by $\{\mathbf{r}_i'\}_{i=1}^N$. They are called the relative coordinates to the CM, and they obey the transformation law

$$\mathbf{r}_i = \mathcal{R} + \mathbf{r}_i' \,. \tag{5.2}$$

---

[1] We could have defined this reference system with different orientation of the axes, however, we are not interested in adding further complications, as they would be irrelevant in this chapter.

© Springer Nature Switzerland AG 2020
V. Ilisie, *Lectures in Classical Mechanics*, Undergraduate Lecture
Notes in Physics, https://doi.org/10.1007/978-3-030-38585-9_5

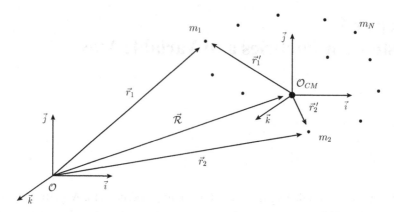

Fig. 5.1 An N-particle system and their corresponding coordinates in an arbitrary reference frame and in the CM frame

Mathematical relations between physical quantities (kinetic energy, angular momentum) described in the $\{\mathcal{O}, \mathbf{i}, \mathbf{j}, \mathbf{k}\}$ reference system and in the CM frame will be derived later on as proposed exercises.

Considering that the system is closed (the total mass is constant i.e., $dM/dt = 0$) and that the individual masses are also constant i.e., $dm_i/dt = 0$, one obtains the velocity of the center of mass by simply taking the time derivative on both sides of (5.1)

$$\mathcal{V} = \frac{1}{M} \sum_i m_i \, \mathbf{v}_i \, , \tag{5.3}$$

where, obviously $\mathbf{v}_i = d\mathbf{r}_i/dt$. Thus, the CM momentum is just the sum of the individual momenta of the particles

$$\mathcal{P} = M\mathcal{V} = \sum_i m_i \, \mathbf{v}_i = \sum_i \mathbf{p}_i \, . \tag{5.4}$$

Taking the time derivative of (5.3) one obtains the CM acceleration

$$\mathcal{A} = \frac{1}{M} \sum_i m_i \, \mathbf{a}_i \, . \tag{5.5}$$

Similarly, one can observe that the total force acting on CM is

$$\mathbf{F} = M\mathcal{A} = \sum_i m_i \, \mathbf{a}_i = \sum_i \mathbf{f}_i \, , \tag{5.6}$$

where $\mathbf{f}_i$ is the total force acting on the particle i. One should observe in the previous expression that the force acting on the CM is equivalent to the total net force acting on the system! Furthermore, the force $\mathbf{f}_i$ can be decomposed into two contributions

Fig. 5.2 A two particle system in an arbitrary reference frame, and the relative coordinates given by the components of **r**

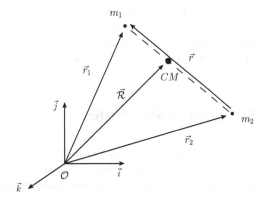

$$\mathbf{f}_i = \mathbf{f}_i^{\text{ext}} + \sum_{j(\neq i)} \mathbf{f}_{ij}^{\text{int}}, \tag{5.7}$$

where $\mathbf{f}_i^{\text{ext}}$ is the total external force, and $\mathbf{f}_{ij}^{\text{int}}$ the internal force exerted on the particle i by the particle j. Thus **F** can be further decomposed into the total external force and the total internal force as

$$\mathbf{F} = \sum_i \mathbf{f}_i^{\text{ext}} + \sum_{i \neq j} \mathbf{f}_{ij}^{\text{int}} \equiv \mathbf{F}^{\text{ext}} + \mathbf{F}^{\text{int}}. \tag{5.8}$$

At this point it is worth explaining the notations we have used (and we will further use) for the sum indices. $\sum_{i \neq j}$ is the sum that runs over both i and j with $i \neq j$. $\sum_{j(\neq i)}$ only runs over j with the condition $j \neq i$. Later on we shall also use $\sum_{i < j}$ which runs on both i and j with the condition $i < j$.

Given a system of N particles, one can define its total kinetic energy K as the sum of all individual kinetic energies and, the total angular momentum **L** as the sum of all the angular momenta i.e.,

$$K = \sum_i \frac{1}{2} m_i v_i^2, \qquad \mathbf{L} = \sum_i \mathbf{r}_i \times \mathbf{p}_i. \tag{5.9}$$

In the particular case of a two-particle system, instead of defining the coordinates relative to the CM frame, one can alternatively define the relative coordinates (to each other) given by

$$\mathbf{r} \equiv \mathbf{r}_1 - \mathbf{r}_2, \tag{5.10}$$

as shown in Fig. 5.2. Thus, taking the time derivatives, one can also define a relative velocity and a relative acceleration

$$\mathbf{v} = d\mathbf{r}/dt = \mathbf{v}_1 - \mathbf{v}_2, \qquad \mathbf{a} = d\mathbf{v}/dt = \mathbf{a}_1 - \mathbf{a}_2. \tag{5.11}$$

These relative quantities will turn out to be very useful later on, for the analysis of simple two-particle systems moving in a central potential.

5.2 Variable Mass Systems

Imagine that we are interested in the momentum variation of a body with variable mass m_1. In order to correctly solve this problem, we need to consider a slightly more general configuration. We will consider a closed system (with constant total mass) formed by two bodies with initial masses m_1 and m_2, thus with $M = m_1 + m_2$ constant. The total momentum will be given by

$$\mathcal{P} = M\mathcal{V} = \mathbf{p}_1 + \mathbf{p}_2. \tag{5.12}$$

Considering that the internal forces can be neglected, its time derivative will be equal to the total external force exerted over the system

$$\mathbf{F}^{\text{ext}} = \frac{d\mathcal{P}}{dt} = M\mathcal{A}. \tag{5.13}$$

In terms of the momenta $\mathbf{p}_1$ and $\mathbf{p}_2$ we have

$$\frac{d\mathbf{p}_1}{dt} + \frac{d\mathbf{p}_2}{dt} = m_1\mathbf{a}_1 + m_2\mathbf{a}_2 + \mathbf{v}_1\frac{dm_1}{dt} + \mathbf{v}_2\frac{dm_2}{dt} = \mathbf{F}^{\text{ext}}. \tag{5.14}$$

As M constant $dM = 0$, therefore $dm_2/dt = -dm_1/dt$ and our previous expression becomes

$$\mathbf{F}^{\text{ext}} = m_1\mathbf{a}_1 + m_2\mathbf{a}_2 + \mathbf{v}\frac{dm_1}{dt}, \tag{5.15}$$

where $\mathbf{v} = \mathbf{v}_1 - \mathbf{v}_2$ is the relative velocity. The previous expression is all we need to solve a few interesting exercises that will help us more intuitively understand the underlying concepts. We can however, still deduce some general results. Let us separate $\mathbf{F}^{\text{ext}}$ as

$$\mathbf{F}^{\text{ext}} = \mathbf{F}_1^{\text{ext}} + \mathbf{F}_2^{\text{ext}}, \tag{5.16}$$

with

$$\mathbf{F}_2^{\text{ext}} = m_2\mathbf{a}_2, \qquad \mathbf{F}_1^{\text{ext}} = m_1\mathbf{a}_1 + \mathbf{v}\frac{dm_1}{dt}. \tag{5.17}$$

Hence, if we are interested in the total force acting upon the system 1 we have to employ the previous expression corresponding to $\mathbf{F}_1^{\text{ext}}$

$$\mathbf{F}_1^{\text{ext}} = m_1 \mathbf{a}_1 + \mathbf{v} \frac{dm_1}{dt}, \qquad (5.18)$$

which is the Galilean invariant generalization of (4.2), whose validity is limited to the reference frame where the relative velocity $\mathbf{v}$ is equal to $\mathbf{v}_1$. In order to better understand the previous formula, one can check an alternative derivation in Appendix E. Also a set of relevant exercises will be solved at the end of the following section.

It might seem to the reader, that the theoretical introduction of this chapter is rather short and even poor. This is not at all true. After going through all the previous chapters, with the minimum amount of information provided in the current one, the reader should be more than prepared to solve all the proposed exercises, without any further complication, as most of them can be solved by direct application of the definitions provided in the previous sections.

5.3 Proposed Exercises

5.1 Given a two-particle system, and given $\mathbf{r}$ as defined in (5.10), prove that the following relations hold true:

$$\mathbf{r}_1 = \mathcal{R} + \frac{m_2}{M} \mathbf{r}, \qquad \mathbf{r}_2 = \mathcal{R} - \frac{m_1}{M} \mathbf{r}. \qquad (5.19)$$

Solution: From the definition of $\mathcal{R}$ and $\mathbf{r}$, for the case of a two-particle system, one obtains

$$\mathcal{R} = \frac{m_1 \mathbf{r}_1 + m_2 \mathbf{r}_2}{M} = \frac{m_1 \mathbf{r}_1 + m_2 (\mathbf{r}_1 - \mathbf{r})}{M} = \mathbf{r}_1 - \frac{m_2}{M} \mathbf{r}. \qquad (5.20)$$

thus

$$\mathbf{r}_1 = \mathcal{R} + \frac{m_2}{M} \mathbf{r}. \qquad (5.21)$$

The solution for $\mathbf{r}_2$ is just as straightforward. One must also note, that by taking the time derivatives in the previous expression, one obtains similar relations for the velocity and the acceleration i.e.,

$$\mathbf{v}_1 = \mathcal{V} + \frac{m_2}{M} \mathbf{v}, \qquad \mathbf{a}_1 = \mathcal{A} + \frac{m_2}{M} \mathbf{a}, \qquad (5.22)$$

and similar for $\mathbf{v}_2$ and $\mathbf{a}_2$.

5.2 Prove that the following relation holds true for a two-particle system (König's theorem for the kinetic energy)

$$K = K_{CM} + K_{rel},\qquad(5.23)$$

where K is total the kinetic energy of the system, K_{CM} the CM kinetic energy and K_{rel} the relative kinetic energy:

$$K = \frac{1}{2}m_1 v_1^2 + \frac{1}{2}m_2 v_2^2, \qquad K_{CM} = \frac{1}{2}M\mathcal{V}^2,$$

$$K_{rel} = \frac{1}{2}\mu v^2,\qquad(5.24)$$

and where μ is the reduced mass of the system defined as $\mu \equiv m_1 m_2/M$.

Solution: Using the results from the previous exercise, one can re-express v_1 and v_2 in terms of the relative velocity v and the CM velocity $\mathcal{V}$, thus the total kinetic energy will have the following expression

$$
\begin{aligned}
K &= \frac{1}{2}m_1\left(\mathcal{V} + \frac{m_2}{M}v\right)^2 + \frac{1}{2}m_2\left(\mathcal{V} - \frac{m_1}{M}v\right)^2 \\
&= \frac{1}{2}(m_1 + m_2)\mathcal{V}^2 + \frac{1}{2}m_1\frac{m_2^2}{M^2}v^2 + \frac{1}{2}m_2\frac{m_1^2}{M^2}v^2.
\end{aligned}\qquad(5.25)
$$

The previous equation simplifies to

$$K = \frac{1}{2}M\mathcal{V}^2 + \frac{1}{2}\mu v^2.\qquad(5.26)$$

Note that the crossed terms ($\sim v\cdot\mathcal{V}$) cancel each-other.

5.3 Prove that the following relation holds true for a two-particle system (König's theorem for the angular momentum)

$$\mathbf{L} = \mathbf{L}_{CM} + \mathbf{L}_{rel},\qquad(5.27)$$

where $\mathbf{L}$ is total angular momentum of the system, $\mathbf{L}_{CM}$ the CM angular momentum and $\mathbf{L}_{rel}$ the relative angular momentum:

$$\mathbf{L} = m_1\mathbf{r}_1 \times \mathbf{v}_1 + m_2\mathbf{r}_2 \times \mathbf{v}_2, \qquad \mathbf{L}_{CM} = M\mathcal{R} \times \mathcal{V},$$

$$\mathbf{L}_{rel} = \mu\mathbf{r} \times \mathbf{v}.\qquad(5.28)$$

Solution: The solution is just as straightforward as for the previous exercise. Re-expressing the needed terms, one obtains

$$\mathbf{L} = m_1 \left(\mathcal{R} + \frac{m_2}{M} \mathbf{r} \right) \times \left(\mathcal{V} + \frac{m_2}{M} \mathbf{v} \right) + m_2 \left(\mathcal{R} - \frac{m_1}{M} \mathbf{r} \right) \times \left(\mathcal{V} - \frac{m_1}{M} \mathbf{v} \right)$$

$$= (m_1 + m_2) \mathcal{R} \times \mathcal{V} + \frac{m_1 m_2^2}{M^2} \mathbf{r} \times \mathbf{v} + \frac{m_1^2 m_2}{M^2} \mathbf{r} \times \mathbf{v}$$

$$= \mu \mathbf{r} \times \mathbf{v} + \mathbf{L}_{CM} .\tag{5.29}$$

5.4 Prove that $\mathcal{R}' = 0$, $\mathcal{V}' = 0$ and $\mathcal{A}' = 0$, where

$$\mathcal{R}' = \frac{1}{M} \sum_i m_i \mathbf{r}'_i , \qquad \mathcal{V}' = \frac{1}{M} \sum_i m_i \mathbf{v}'_i ,$$

$$\mathcal{A}' = \frac{1}{M} \sum_i m_i \mathbf{a}'_i ,\tag{5.30}$$

are the position, velocity and acceleration of the CM expressed in the CM reference frame $\{\mathcal{O}_{CM}, \mathbf{i}, \mathbf{j}, \mathbf{k}\}$.

Solution: These statements are quite intuitive and can be proved straightforwardly by multiplying (5.2) by m_i on both sides and summing over i

$$\sum_i m_i \mathbf{r}_i = \sum_i m_i \mathcal{R} + \sum_i m_i \mathbf{r}'_i \tag{5.31}$$

and therefore

$$M \mathcal{R} = M \mathcal{R} + M \mathcal{R}' ,\tag{5.32}$$

thus, $\mathcal{R}' = 0$. The proof for the other two statements follows immediately by taking time derivatives on both sides of (5.31). These results will turn out to be very useful next, when proving König's theorem for the kinetic energy and the angular momentum for a generic N-particle system.

5.5 Prove that the following relation holds true for a generic N-particle system

$$K = K_{CM} + K' ,\tag{5.33}$$

where K is total the kinetic energy of the system, K_{CM} the CM kinetic energy and K' is the kinetic energy relative to the CM reference system:

$$K = \sum_i \frac{1}{2} m_i \mathbf{v}_i^2, \qquad K_{CM} = \frac{1}{2} M \mathcal{V}^2,$$

$$K' = \sum_i \frac{1}{2} m_i \mathbf{v}_i'^2. \tag{5.34}$$

Solution: Taking the time derivative of (5.2) and inserting the resulting expression into the corresponding expression of the kinetic energy, one obtains

$$\sum_i \frac{1}{2} m_i \mathbf{v}_i^2 = \sum_i \frac{1}{2} m_i \left(\mathcal{V} + \mathbf{v}_i' \right)^2$$

$$= \frac{1}{2} \mathcal{V}^2 \sum_i m_i + \sum_i \frac{1}{2} m_i \mathbf{v}_i'^2 + \mathcal{V} \sum_i m_i \mathbf{v}_i'$$

$$= \frac{1}{2} M \mathcal{V}^2 + \sum_i \frac{1}{2} m_i \mathbf{v}_i'^2. \tag{5.35}$$

5.6 Prove that the following relation holds true for a generic N-particle system

$$\mathbf{L} = \mathbf{L}_{CM} + \mathbf{L}', \tag{5.36}$$

where $\mathbf{L}$ is total angular momentum of the system, $\mathbf{L}_{CM}$ the CM angular momentum and $\mathbf{L}'$ the angular momentum relative to the CM reference frame:

$$\mathbf{L} = \sum_i m_i \, \mathbf{r}_i \times \mathbf{v}_i, \qquad \mathbf{L}_{CM} = M \mathcal{R} \times \mathcal{V},$$

$$\mathbf{L}' = \sum_i m_i \mathbf{r}_i' \times \mathbf{v}_i'. \tag{5.37}$$

Solution: Similar to the previous case, by substituting $\mathbf{r}_i$ and $\mathbf{v}_i$ in the expression for the angular momentum we get

$$\sum_i m_i \, \mathbf{r}_i \times \mathbf{v}_i = \sum_i m_i \left(\mathcal{R} + \mathbf{r}_i' \right) \times \left(\mathcal{V} + \mathbf{v}_i' \right)$$

$$= \mathcal{R} \times \mathcal{P} + \sum_i m_i \, \mathbf{r}_i' \times \mathbf{v}_i' + \mathcal{R} \times \mathcal{P}' + \mathcal{R}' \times \mathcal{P}$$

$$= \mathbf{L}_{CM} + \mathbf{L}'. \tag{5.38}$$

5.7 Show that for a system under the action of an internal central potential i.e., the potential between two particles i an j is given by $U_{ij} = U_{ij}\left(\left|\mathbf{r}_i - \mathbf{r}_j\right|\right)$, the total force acting on the system is given by the total external force i.e., $\mathbf{F} = \mathbf{F}^{\text{ext}}$.

Solution: First one should not get confused by the previous expression of the potential. It should be interpreted the following way. U_{ij} is the potential between a pair of particles i and j, and it may be different for each pair. We explicitly write down the dependence $|\mathbf{r}_i - \mathbf{r}_j|$ in order to emphasize that it only depends on the absolute value of the relative distance between the two particles (this is the reason why it is called central). Also remember that the gradient of a central potential gives rise to a central force, as seen in the previous chapter.

Thus, the internal force exerted over the particle i by the particle j (due to the potential U_{ij}) is given by

$$\mathbf{f}_{ij}^{\text{int}} = -\nabla_i U_{ij}\left(\left|\mathbf{r}_i - \mathbf{r}_j\right|\right) , \qquad (5.39)$$

where $\mathbf{r}_i = (x_i, y_i, z_i)$ and with ∇_i given by

$$\nabla_i \equiv \frac{\partial}{\partial x_i}\mathbf{i} + \frac{\partial}{\partial y_i}\mathbf{j} + \frac{\partial}{\partial z_i}\mathbf{k} . \qquad (5.40)$$

Due to the symmetry of the potential one can check that[2]

$$\begin{aligned} \mathbf{f}_{ji}^{\text{int}} &= -\nabla_j U_{ij}\left(\left|\mathbf{r}_i - \mathbf{r}_j\right|\right) \\ &= \nabla_i U_{ij}\left(\left|\mathbf{r}_i - \mathbf{r}_j\right|\right) = -\mathbf{f}_{ij}^{\text{int}} . \end{aligned} \qquad (5.41)$$

Therefore, when summing over all internal forces, they will all cancel by pairs given a net result equal to zero

$$\mathbf{F}^{\text{int}} = \sum_{i \neq j} \mathbf{f}_{ij}^{\text{int}} = \sum_{i<j} (\mathbf{f}_{ij}^{\text{int}} + \mathbf{f}_{ji}^{\text{int}}) = \mathbf{0} . \qquad (5.42)$$

In conclusion, the motion of the center of mass for a system under the action of a central internal potential is only perturbed by external forces. This is quite a strong statement. It means that if the internal forces of a system are not central, the CM can suffer a net acceleration without any external intervention! However, this is (at least for now) not a realistic case, as all the known fundamental forces in Nature are central.

5.8 Calculate the torque $\boldsymbol{\tau} = d\mathbf{L}/dt$ for a system moving under the action of an internal central potential.

---

[2]If one doubts about the following result, a simple way of intuitively understand it is to consider a simple potential such as $U_{ij} = \sqrt{(\mathbf{r}_i - \mathbf{r}_j)^2}$, and check that (5.41) holds true.

Solution: The explicit expression for the torque, in this case, is simply given by

$$\tau = d\mathbf{L}/dt = \sum_{i=1} \mathbf{r}_i \times \mathbf{f}_i . \tag{5.43}$$

with $\mathbf{f}_i = \mathbf{f}_i^{\text{ext}} + \sum_{j(\neq i)} \mathbf{f}_{ij}^{\text{int}}$. For the case of central internal forces one can simply write $\mathbf{f}_{ij}^{\text{int}}$ as

$$\mathbf{f}_{ij}^{\text{int}} \equiv f_{ij}^{\text{int}}(r_{ij}) \, \mathbf{r}_{ij} , \tag{5.44}$$

with $\mathbf{r}_{ij} \equiv \mathbf{r}_i - \mathbf{r}_j$ and $r_{ij} = |\mathbf{r}_{ij}|$, and where $f_{ij}^{\text{int}}(r_{ij}) < 0$ if the force is attractive and $f_{ij}^{\text{int}}(r_{ij}) > 0$ if it is repulsive. Therefore, separating external and internal forces, τ takes the form

$$\begin{aligned}
\tau &= \sum_i \mathbf{r}_i \times \mathbf{f}_i^{\text{ext}} + \sum_{i \neq j} \mathbf{r}_i \times \mathbf{f}_{ij}^{\text{int}} \\
&= \sum_i \mathbf{r}_i \times \mathbf{f}_i^{\text{ext}} + \sum_{i<j} \left(\mathbf{r}_i \times \mathbf{f}_{ij}^{\text{int}} + \mathbf{r}_j \times \mathbf{f}_{ji}^{\text{int}} \right) \\
&= \sum_i \mathbf{r}_i \times \mathbf{f}_i^{\text{ext}} + \sum_{i<j} \mathbf{r}_{ij} \times \mathbf{f}_{ij}^{\text{int}} \\
&= \sum_i \mathbf{r}_i \times \mathbf{f}_i^{\text{ext}} .
\end{aligned} \tag{5.45}$$

Again, as a conclusion, if the internal forces of a system are central, the angular momentum is perturbed only by external forces.

5.9 In (5.8) we have expressed the total force acting on a system as

$$\mathbf{F} = \sum_i \mathbf{f}_i^{\text{ext}} + \sum_{i \neq j} \mathbf{f}_{ij}^{\text{int}} \equiv \mathbf{F}^{\text{ext}} + \mathbf{F}^{\text{int}} . \tag{5.46}$$

Consider that each $\mathbf{f}_i^{\text{ext}}$ and $\mathbf{f}_{ij}^{\text{int}}$ have a conservative and a non-conservative component i.e.,

$$\mathbf{f}_i^{\text{ext}} = \mathbf{f}_{i,c}^{\text{ext}} + \mathbf{f}_{i,nc}^{\text{ext}} , \tag{5.47}$$

$$\mathbf{f}_{ij}^{\text{int}} = \mathbf{f}_{ij,c}^{\text{int}} + \mathbf{f}_{ij,nc}^{\text{int}} , \tag{5.48}$$

and that $\mathbf{f}_{ij,c}^{\text{int}}$ are central. Deduce the energy conservation theorem for such a case.

Solution: On one hand, in the previous chapter in (4.27) we have found that

$$dW = dK. \tag{5.49}$$

Let us now consider that the kinetic energy is given by the sum

$$K = \sum_i K_i, \tag{5.50}$$

where K_i are the individual kinetic energies. Therefore dW (the total infinitesimal work of the system) will be given by

$$dW = dK = \sum_i dK_i. \tag{5.51}$$

On the other hand

$$dW = \sum_i \mathbf{f}_{i,c}^{\text{ext}} \cdot d\mathbf{r}_i + \sum_i \mathbf{f}_{i,nc}^{\text{ext}} \cdot d\mathbf{r}_i$$
$$+ \sum_{i \neq j} \mathbf{f}_{ij,c}^{\text{int}} \cdot d\mathbf{r}_i + \sum_{i \neq j} \mathbf{f}_{ij,nc}^{\text{int}} \cdot d\mathbf{r}_i. \tag{5.52}$$

Let's now analyse each term of the previous equation. For the first one we have

$$\sum_i \mathbf{f}_{i,c}^{\text{ext}} \cdot d\mathbf{r}_i = -\sum_i dU_i^{\text{ext}} \equiv -dU^{\text{ext}}. \tag{5.53}$$

The second one reads

$$\sum_i \mathbf{f}_{i,nc}^{\text{ext}} \cdot d\mathbf{r}_i = \sum_i dW_{i,nc}^{\text{ext}} \equiv dW_{nc}^{\text{ext}}. \tag{5.54}$$

The calculation for the third term is a little bit more involved, but still rather straightforward[3]

$$\sum_{i \neq j} \mathbf{f}_{ij,c}^{\text{int}} \cdot d\mathbf{r}_i = \sum_{i < j} (\mathbf{f}_{ij,c}^{\text{int}} \cdot d\mathbf{r}_i + \mathbf{f}_{ji,c}^{\text{int}} \cdot d\mathbf{r}_j)$$
$$= \sum_{i < j} \mathbf{f}_{ij,c}^{\text{int}} \cdot d\mathbf{r}_{ij}. \tag{5.55}$$

Introducing the potential

---

[3] Again, if one has any problem in following the calculations presented here for a generic potential $U_{ij}(|\mathbf{r}_i - \mathbf{r}_j|)$ one should use a simple case, such as $U_{ij} = \sqrt{(\mathbf{r}_i - \mathbf{r}_j)^2}$ in order to understand these results more intuitively.

$$\sum_{i \neq j} \mathbf{f}_{ij,c}^{int} \cdot d\mathbf{r}_i = -\sum_{i<j} \boldsymbol{\nabla}_i U_{ij}(|\mathbf{r}_i - \mathbf{r}_j|) \cdot d\mathbf{r}_{ij}$$

$$= -\sum_{i<j} \boldsymbol{\nabla}_i U_{ij}(r_{ij}) \cdot d\mathbf{r}_{ij}$$

$$= -\sum_{i<j} \frac{dU_{ij}(r_{ij})}{dr_{ij}} dr_{ij}$$

$$= -\sum_{i<j} dU_{ij}(r_{ij})$$

$$\equiv -dU^{int}. \tag{5.56}$$

where we have used the definitions from the previous exercise of $\mathbf{r}_{ij}$ and r_{ij}. The last term is simply given by

$$\sum_{i \neq j} \mathbf{f}_{ij,nc}^{int} \cdot d\mathbf{r}_i = \sum_{i \neq j} dW_{ij,nc}^{int} \equiv dW_{nc}^{int}. \tag{5.57}$$

We can therefore express dW of the total system as

$$dW = dK = -dU^{ext} + dW_{nc}^{ext} - dU^{int} + dW_{nc}^{int}, \tag{5.58}$$

and so, the energy conservation theorem for such a system states

$$\boxed{dE = dK + dU^{ext} + dU^{int} = dW_{nc}^{ext} + dW_{nc}^{int}}, \tag{5.59}$$

which is the expected result. This the generic result of the energy conservation theorem that is a fundamental pillar of all physics. We shall use this result, particularized to each case, all along this book.

5.10 Given a central internal force $\mathbf{f}_{ij}^{int}$ find the expression for the central potential $U_{ij}^{int}(r_{ij})$.

Solution: The solution is straightforward

$$-dU_{ij}^{int}(r_{ij}) = \mathbf{f}_{ij}^{int} \cdot \mathbf{r}_{ij}$$

$$= f_{ij}^{int}(r_{ij})\mathbf{r}_{ij} \cdot \mathbf{r}_{ij}$$

$$= \frac{f_{ij}^{int}(r_{ij})}{r_{ij}} \mathbf{u}_{ij} \cdot \mathbf{r}_{ij}$$

$$= \frac{f_{ij}^{int}(r_{ij})}{r_{ij}} dr_{ij}, \tag{5.60}$$

where we used (5.44) and we have defined the unitary vector $\mathbf{u}_{ij} \equiv \mathbf{r}_{ij}/r_{ij}$. Integrating, (up to a constant) we obtain

$$U_{ij}^{\text{int}}(r_{ij}) = -\int \frac{f_{ij}^{\text{int}}(r_{ij})}{r_{ij}} dr_{ij} . \qquad (5.61)$$

This expression might seem different from (4.52), but this is only because we have used different definitions for $\mathbf{F}$ and $\mathbf{f}_{ij}^{\text{int}}$ i.e., $\mathbf{F} = F(r)\mathbf{u}_r$ with $\mathbf{u}_r$ unitary, and $\mathbf{f}_{ij}^{\text{int}} = f_{ij}^{\text{int}}(r_{ij})\mathbf{r}_{ij}$ with $\mathbf{r}_{ij}$ not unitary. Again, the integration constant is irrelevant as the physical quantities will never depend on it.

5.11 Consider a two-particle system with masses $m_1 = 5$ kg and $m_2 = 10$ kg, with initial positions given by $\mathbf{r}_1^0 = 0.4\,\mathbf{i}$ m and $\mathbf{r}_2^0 = -0.2\,\mathbf{i}$ m, and with initial velocities given by $\mathbf{v}_1^0 = (1/45) \cdot 10^{-3}\,\mathbf{j}$ m/s and $\mathbf{v}_2^0 = (-1/90) \cdot 10^{-3}\,\mathbf{j}$ m/s. If the only force acting on the system is the mutual gravitational attraction, calculate the initial position of the center of mass $\mathcal{R}^0$ and the velocity of the center of mass $\mathcal{V}(t)$ as a function of time. What can we say about $\mathcal{R}(t)$? Calculate the total angular momentum of the system $\mathbf{L}(t)$ and check König's theorem for the angular momentum. Calculate the initial kinetic energy K^0 of the system and check König's theorem. Calculate the total initial energy.

Solution: The initial position of the center of mass will be given by

$$\mathcal{R}^0 = \frac{1}{m_1 + m_2}(m_1 \mathbf{r}_1^0 + m_2 \mathbf{r}_2^0) = \mathbf{0} . \qquad (5.62)$$

As there are no external forces acting on the system the velocity of the CM is conserved thus

$$\mathcal{V}(t) = \mathcal{V}^0 = \frac{1}{m_1 + m_2}(m_1 \mathbf{v}_1^0 + m_2 \mathbf{v}_2^0) = \mathbf{0} , \qquad (5.63)$$

and therefore $\mathcal{R}(t) = \mathcal{R}^0 = \mathbf{0}$. As for the angular momentum, the only forces acting on the system are central internal forces, thus it is obviously conserved and,

$$\mathbf{L}(t) = \mathbf{L}^0 = \mathbf{r}_1^0 \times \mathbf{p}_1^0 + \mathbf{r}_2^0 \times \mathbf{p}_2^0$$

$$= m_1 \begin{vmatrix} \mathbf{i} & \mathbf{j} & \mathbf{k} \\ 0.4 & 0 & 0 \\ 0 & (1/45) \cdot 10^{-3} & 0 \end{vmatrix} + m_2 \begin{vmatrix} \mathbf{i} & \mathbf{j} & \mathbf{k} \\ -0.2 & 0 & 0 \\ 0 & (-1/90) \cdot 10^{-3} & 0 \end{vmatrix}$$

$$= \frac{1}{15} \cdot 10^{-3}\,\mathbf{k} \; (\text{m}^2\,\text{kg/s}) . \qquad (5.64)$$

As the CM angular momentum is zero ($\mathcal{V} = \mathbf{0}$) the previous result should be equal to $\mu \mathbf{r}^0 \times \mathbf{v}^0$, where μ is the reduced mass

$$\mu = m_1 m_2/(m_1 + m_2) = 10/3 \text{ (kg)}, \tag{5.65}$$

$\mathbf{r}^0 = \mathbf{r}_1^0 - \mathbf{r}_2^0 = 0.6\mathbf{i}$ (m) are the initial relative coordinates and $\mathbf{v}^0 = \mathbf{v}_1^0 - \mathbf{v}_2^0 = (1/30) \cdot 10^{-3}\mathbf{j}$ (m/s) is the initial relative velocity. We obtain

$$\mathbf{L}^0 = \mu \begin{vmatrix} \mathbf{i} & \mathbf{j} & \mathbf{k} \\ 0.6 & 0 & 0 \\ 0 & (1/30) \cdot 10^{-3} & 0 \end{vmatrix} = \frac{1}{15} \cdot 10^{-3}\mathbf{k} \text{ (m}^2\text{ kg/s)},$$

which is the expected result. The initial kinetic energy is given by

$$\begin{aligned} K^0 &= \frac{1}{2}m_1(\mathbf{v}_1^0)^2 + \frac{1}{2}m_2(\mathbf{v}_2^0)^2 \\ &= (1/54) \cdot 10^{-7} \text{ (kg m}^2/\text{s}^2) \\ &= \frac{1}{2}\mu(\mathbf{v}^0)^2. \end{aligned} \tag{5.66}$$

Finally the initial gravitational potential energy will be given by the following expression

$$U^0 = -G\frac{m_1 m_2}{|\mathbf{r}_1^0|} = -5.56 \cdot 10^{-9} \text{ (kg m}^2/\text{s}^2), \tag{5.67}$$

and thus the total energy is

$$E = K^0 + U^0 = -3.71 \cdot 10^{-9} \text{ (kg m}^2/\text{s}^2). \tag{5.68}$$

5.12 Consider the Atwood machine shown in Fig. 5.3, where the masses m_1 and m_2 (with $m_2 > m_1$) are initially at rest (and the mass of the string is negligible). Calculate $\mathbf{r}_{1,2}(t)$, the coordinates of the CM given by $\mathcal{R}(t)$, its velocity $\mathcal{V}(t)$ and acceleration $\mathcal{A}(t)$. Calculate the total force exerted over the system. Calculate the total kinetic energy and check König's theorem for the kinetic energy. Do the same for the angular momentum.

Solution: The equations of motion for each mass will be given by the following system

$$\begin{aligned} -m_1 g + T &= m_1 a, \\ -m_2 g + T &= -m_2 a, \end{aligned} \tag{5.69}$$

with $g, a > 0$. Therefore the expression for the acceleration is given by

Fig. 5.3 The Atwood machine with $m_2 > m_1$ (left) and diagrams of the forces acting on each mass (right) where T stands for the tension of the string

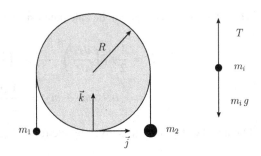

$$a = \frac{m_2 - m_1}{M} g, \tag{5.70}$$

where $M = m_1 + m_2$, the total mass of the system. The expressions for $\mathbf{r}_i(t)$ (with $i = 1, 2$) can thus be trivially calculated

$$\mathbf{r}_1(t) = -R\mathbf{j} + \frac{1}{2}at^2\mathbf{k},$$
$$\mathbf{r}_2(t) = R\mathbf{j} - \frac{1}{2}at^2\mathbf{k} = -\mathbf{r}_1(t), \tag{5.71}$$

where we have set $t_0 = 0$. Thus $\mathcal{R}$ simply reads

$$\mathcal{R} = \frac{m_1\mathbf{r}_1 + m_2\mathbf{r}_2}{M} = \frac{m_2 - m_1}{M}\left(R\mathbf{j} - \frac{1}{2}\frac{m_2 - m_1}{M}gt^2\mathbf{k}\right). \tag{5.72}$$

By taking the time derivative on both sides we obtain the expression for $\mathcal{V}$,

$$\mathcal{V} = -\frac{(m_2 - m_1)^2}{M^2}gt\,\mathbf{k}. \tag{5.73}$$

Finally the acceleration simply reads

$$\mathcal{A} = -\frac{(m_2 - m_1)^2}{M^2}g\,\mathbf{k}. \tag{5.74}$$

The total external force exerted over the system is given by the following expression

$$\begin{aligned}
\mathbf{F} &= m_1\frac{d^2\mathbf{r}_1}{dt^2} + m_2\frac{d^2\mathbf{r}_2}{dt^2} \\
&= (m_1 - m_2)a\,\mathbf{k} \\
&= -\frac{(m_2 - m_1)^2}{M}g\,\mathbf{k} \\
&= M\mathcal{A}.
\end{aligned} \tag{5.75}$$

The total kinetic energy can be calculated straightforwardly

$$K = \frac{1}{2}m_1 \left(\frac{d\mathbf{r}_1}{dt}\right)^2 + \frac{1}{2}m_2 \left(\frac{d\mathbf{r}_2}{dt}\right)^2$$
$$= \frac{1}{2}a^2t^2(m_1 + m_2) = \frac{1}{2}\frac{(m_2 - m_1)^2}{M}g^2t^2 . \tag{5.76}$$

On the other hand K_{CM} is given by the following expression

$$K_{CM} = \frac{1}{2}M\mathcal{V}^2 = \frac{1}{2}\frac{(m_2 - m_1)^4}{M^3}g^2t^2 . \tag{5.77}$$

The expression for relative velocity can be also calculated straightforwardly

$$\mathbf{v}(t) = \frac{d}{dt}(\mathbf{r}_1(t) - \mathbf{r}_2(t)) = 2at\,\mathbf{k} . \tag{5.78}$$

Thus, the relative kinetic energy is simply given by

$$K_{rel} = \frac{1}{2}\mu\mathbf{v}^2$$
$$= \frac{1}{2}\frac{m_1 m_2}{m_1 + m_2}4a^2t^2$$
$$= 2m_1 m_2 \frac{(m_2 - m_1)^2}{M^3}g^2t^2 . \tag{5.79}$$

Even if it may not be obvious at first sight, by summing K_{rel} and K_{CM} and simplifying the resulting expression, one obtains K. The corresponding calculation for the angular momentum is similar and left for the reader.

5.13 A projectile of mass M is launched from the ground with initial velocity $\mathbf{v}_0$. At some instant during its parabolic trajectory it explodes into two pieces, with masses m_1 and m_2, that fly away horizontally. This is schematically shown in Fig. 5.4. Knowing that the mass m_1 reaches the ground at a distance d_0 from the origin, find out the coordinates of the point where m_2 reaches the ground. Neglect the air resistance.

Solution: Let us first calculate the point where the mass M would have reached the ground (if it hadn't explode). The equations of motion are straightforward (this is nothing but the typical exercise for the parabolic trajectory that is solved in high school):

$$M\frac{d^2y}{dt^2} = -Mg , \qquad M\frac{d^2x}{dt^2} = 0 . \tag{5.80}$$

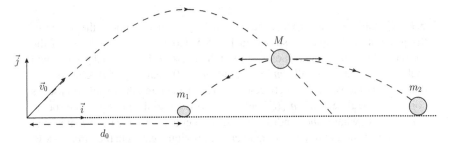

Fig. 5.4 Parabolic trajectory of a projectile of mass M that explodes along its trajectory into two masses m_1 and m_2 that fly away horizontally

The solution is

$$v_y(t) = v_{0,y} - gt, \qquad v_x(t) = v_{0,x}, \qquad (5.81)$$

for the velocity and

$$y(t) = v_{0,y}t - \frac{1}{2}gt^2, \qquad x(t) = v_{0,x}t, \qquad (5.82)$$

for the upward and horizontal distance, where we have considered $t_0 = 0$. The time interval $t_f - t_0 = t_f$ for the particle to reach the maximum height is given by the condition $v_y(t_f) = 0$ thus $t_f = v_{0,y}/g$. Therefore it takes $2t_f$ for the mass M (without explosion) to reach the ground. The horizontal distance will be given by

$$d = v_x(2t_f) = 2v_{0,x}t_f = \frac{v_{0,x}v_{0,y}}{g}. \qquad (5.83)$$

As the explosion is due to internal forces, and there is no other external force acting on the projectile but the gravitational force, the impact point where the projectile would have reached the ground (without exploding) is the coordinate of the CM i.e., $X_{CM} = d$. Thus

$$MX_{CM} = m_1x_1 + m_2x_2 = m_1d_0 + m_2x_2, \qquad (5.84)$$

and so, the point where m_2 hits the ground is

$$x_2 = \frac{M}{m_2}X_{CM} - \frac{m_1}{m_2}x_1 = \frac{M}{m_2}\frac{v_{0,x}v_{0,y}}{g} - \frac{m_1}{m_2}d_0. \qquad (5.85)$$

There are other ways of solving this problem but this is the simplest one, because we have made use of the properties of the motion of the center of mass. Note this was possible because, after the explosion, the two masses fly away horizontally. If this weren't the case, the solution would have been slightly more involved.

5.14 Consider a space rocket that is launched vertically under the action of the gravitational force, as shown in Fig. 5.5. Consider that the mass of the rocket and the (unburned) fuel is m_1 and the mass of the expelled gas (burned fuel) is m_2. Thus, initially $m_1 = M$ and $m_2 = 0$. Using (5.15), considering that the relative velocity $\mathbf{v}$ between the space rocket and the expelled gases is constant, calculate $v_1(t)$, $m_1(t)$ and $m_2(t)$. If the initial velocity of the rocket is $v_1(0) = 0$, what is the condition for the rocket to overcome the gravitational pull and gain momentum? Comment all the previous results in the limit $g \to 0$.

Solution: By using expression (5.15), we obtain the following equation of motion

$$-M g = -(m_1 + m_2) g = m_1 \frac{dv_1}{dt} + m_2 \frac{dv_2}{dt} + v \frac{dm_1}{dt} . \tag{5.86}$$

This expression can be separated in two independent parts

$$-m_2 g = m_2 \frac{dv_2}{dt} , \tag{5.87}$$

which describes the evolution of the sub-system 2 (of the expelled gases) and

$$-m_1 g = m_1 \frac{dv_1}{dt} + v \frac{dm_1}{dt} , \tag{5.88}$$

which describes the sub-system 1 given by the rocket and the fuel. If one does not clearly see this separation, one can check an alternative derivation of the previous equation in Appendix E.

Here we are interested in the evolution of the sub-system 1. Its equation of motion can be re-written as

$$dv_1 = -g \, dt - v \frac{dm_1}{m_1} . \tag{5.89}$$

Integrating, taking $t_0 = 0$ we obtain $v_1(t)$

$$v_1(t) = v_1(0) - gt - v \ln \left(\frac{m_1(t)}{m_1(0)} \right) . \tag{5.90}$$

Manipulating the previous expression we obtain $m_1(t)$

$$m_1(t) = m_1(0) \, e^{-\frac{v_1(t) - v_1(0) + gt}{v}} . \tag{5.91}$$

We can now calculate the mass of the expelled gas $m_2(t)$. We know that $M = m_1 + m_2$ is a constant magnitude and that initially $M = m_1(0)$, thus we can straightforwardly deduce that

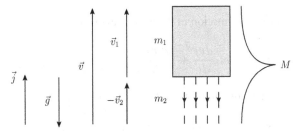

Fig. 5.5 A space rocket (containing unburned fuel) of mass m_1. The mass of the propellant (burned gases) is given by m_2. The total mass $M = m_1 + m_2$ is constant i.e., $dm_1 = -dm_2$. The corresponding vector quantities are also shown

$$M = m_1(0) = m_1(t) + m_2(t),$$ (5.92)

and so

$$m_2(t) = m_1(0) - m_1(t) = m_1(0)\left(1 - e^{-\frac{v_1(t) - v_1(0) + gt}{v}}\right).$$ (5.93)

As a cross check we can observe that $m_2(t \to 0) \to 0$. If the initial velocity $v_1(0) = 0$, then the necessary condition for the rocket to take off is simply given by $v_1(t) > 0$ which translates into

$$-gt - v\ln\left(\frac{m_1(t)}{m_1(0)}\right) > 0 \quad \Rightarrow \quad m_1(t) < m_1(0)\,e^{-\frac{gt}{v}}.$$ (5.94)

Re-expressing the previous condition in terms of the mass of the propellant $m_2(t)$ we obtain

$$m_2(t) > M\left(1 - e^{-\frac{gt}{v}}\right),$$ (5.95)

The limit $g \to 0$ is straightforward and it doesn't introduce any bad behaviour in the equations. For the velocity we have

$$v_1(t) = -v\ln\left(\frac{m_1(t)}{m_1(0)}\right),$$ (5.96)

which is, of course, positive because $m_1(0) > m_1(t)$. We can observe from the previous expression that the rocket starts moving instantaneously as there is no gravitational pull.

5.15 Repeat the previous exercise considering a constant mass variation $dm_1/dt = -C$ (with C a positive constant).

Solution: The equations of motion in this case are given by

$$-m_1 g = m_1 \frac{dv_1}{dt} + v \frac{dm_1}{dt}, \qquad \frac{dm_1}{dt} = -C. \tag{5.97}$$

We can easily solve the equation for $m_1(t)$. The solution is

$$m_1(t) = M - Ct, \tag{5.98}$$

therefore the mass of the propellant $m_2(t)$ is simply given by

$$m_2(t) = Ct. \tag{5.99}$$

The remaining equation then reads

$$m_1(t) \frac{dv_1}{dt} = -m_1(t)g + vC, \tag{5.100}$$

which can be re-arranged as

$$dv_1 = -g\,dt + \frac{v}{M/C - t}\,dt. \tag{5.101}$$

Integrating we obtain

$$v_1(t) = v_1(0) - gt + v \ln\left(\frac{M}{M - Ct}\right). \tag{5.102}$$

We can observe that in order to take off, for $v_1(0) = 0$, the fuel burning must fulfil

$$v \ln\left(\frac{M}{M - Ct}\right) > gt. \tag{5.103}$$

Manipulating the expression we obtain

$$Ct > M(1 - e^{-\frac{gt}{v}}). \tag{5.104}$$

In the absence of the gravitational force the previous condition becomes $Ct > 0$ which obviously holds true, as both terms are positive.

5.16 Repeat the Exercise (5.14) considering a friction force with the atmosphere $\mathbf{F} = -b\, m_1\, v_1^2\, \mathbf{j}$ with b a positive constant and, with a mass variation given by $m_1(t) = M e^{-ct}$ (with c a positive constant). Can it be solved analytically? Why?

Solution: Let us first calculate the mass variation dm_1/dt. We obtain

$$\frac{dm_1}{dt} = -cMe^{-ct} = -cm_1(t).$$ (5.105)

The equations of motion in this case will be given by

$$-b\,m_1 v_1^2 - m_1 g = m_1 \frac{dv_1}{dt} + v\frac{dm_1}{dt},$$ (5.106)

which translates into

$$\frac{dv_1}{dt} = -bv_1^2 - g + cv.$$ (5.107)

We can observe that the object reaches a terminal velocity (as in Exercise 4.11 from the previous chapter). Its expression can be easily obtained by setting $dv_1/dt = 0$. Thus

$$v_{l,1} = \sqrt{\frac{-g + cv}{b}}.$$ (5.108)

Note that necessarily $cv > g$. We can now write (5.107) in terms of $v_{l,1}$

$$\frac{dv_1}{dt} = b\left(v_{l,1}^2 - v_1^2\right) \quad \Rightarrow \quad \frac{dv_1}{1 - v_1^2/v_{l,1}^2} = b\,v_{l,1}^2.$$ (5.109)

The previous expression can be solved just as in Exercise 4.11

$$v_{l,1} \operatorname{arctanh}\left(v_1/v_{l,1}\right) = b\,v_{l,1}^2\,t.$$ (5.110)

Thus

$$v_1(t) = v_{l,1} \tanh(b\,v_{l,1}\,t).$$ (5.111)

Note that we have been able to solve this exercise analytically because we have explicitly chosen a friction force proportional to m_1 and also because dm_1 was proportional to m_1. This way we were able to eliminate the $m_1(t)$ dependence of the equation of motion and the remaining equation was trivial to solve. This is not in general the case and therefore, more *realistic* problems necessarily have to be solved numerically. Further complications are also introduced due to the fact that g varies with the height, and the height varies with time. Thus, we end up having to solve a system of coupled differential equations which is not trivial in most cases. However the previous examples can give us a generic idea of the physics behind rocket propulsion which is not far away from the truth.

Fig. 5.6 A tank that expels liquid horizontally at constant velocity v through a pipe placed at the bottom

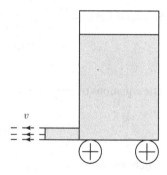

5.17 A tank filled with a liquid of constant density ρ is provided with a pump that expels the liquid horizontally at constant relative velocity v, through a cylindrical pipe (of radius r) placed at the bottom, as shown in Fig. 5.6. If the total initial mass (of the liquid and the tank) is $M = m_1(0)$ (and the mass of the expelled liquid is given by m_2), calculate dm_1/dt and integrate the obtained expression. Obtain the expression for the velocity $v_1(t)$ and for the horizontal displacement $x_1(t)$ of the tank. Neglect all friction forces.

Solution: The differential mass for a fluid with constant density (which in this case is dm_2) is given by

$$dm_2 = \rho dV \,, \tag{5.112}$$

where dV is the infinitesimal volume occupied by dm_2. We can further express dV for the liquid going through the pipe as

$$dV = \pi r^2 v dt \,. \tag{5.113}$$

Therefore the expression we are looking for is

$$\frac{dm_1}{dt} = -\frac{dm_2}{dt} = -\pi r^2 \rho v \quad \Rightarrow \quad m_1(t) = M - \pi r^2 \rho v t \,. \tag{5.114}$$

The equations of motion are simpler in this case. They are given by

$$m_1 \frac{dv_1}{dt} = -v \frac{dm_1}{dt} \quad \Rightarrow \quad dv_1 = \frac{\pi r^2 \rho v^2}{M - \pi r^2 \rho v t} dt \,. \tag{5.115}$$

The solution for $v_1(t)$ with $t_0 = 0$ is simply

$$v_1(t) = v_1(0) + v \ln\left(\frac{M}{m_1(t)}\right) \,. \tag{5.116}$$

Integrating once again we obtain the travelled distance $x_1(t)$

$$x_1(t) = x_1(0) + v_1(0)t + vt - \frac{M - \pi r^2 \rho v t}{\pi r^2 \rho} \ln\left(\frac{M}{m_1(t)}\right). \qquad (5.117)$$

The forces that appear in problems with objects with variable mass, should not be regarded as some kind of mysterious phenomena that pops out of the equations and that one cannot fully understand.[4] Microscopically millions of particles are being expelled with their corresponding momenta $\mathbf{p}_i$ in some direction. Thus, by momentum conservation, the rocket must gain momentum in the same direction but opposite sense. As it is physically impossible to keep track of all the involved particles, we use an effective (macroscopic) approach, which is mathematically equivalent, in order to solve the problem. As mentioned earlier, a possibly more intuitive deduction of the *rocket equation* based on momentum conservation is given in Appendix E.

Further Reading

1. J.V. José, E.J. Saletan, *Classical Dynamics: A Contemporary Approach*. Cambridge University Press
2. S.T. Thornton, J.B. Marion, *Classical Dynamics of Particles and Systems*
3. H. Goldstein, C. Poole, J. Safko, *Classical Mechanics*, 3rd edn. Addison Wesley
4. J.R. Taylor, *Classical Mechanics*
5. D.T. Greenwood, *Classical Dynamics*. Prentice-Hall Inc.
6. D. Kleppner, R. Kolenkow, *An Introduction to Mechanics*
7. C. Lanczos, *The Variational Principles of Mechanics*. Dover Publications Inc.
8. W. Greiner, *Classical Mechanics: Systems of Particles and Hamiltonian Dynamics*. Springer
9. H.C. Corben, P. Stehle, *Classical Mechanics*, 2nd edn. Dover Publications Inc.
10. T.W.B. Kibble, F.H. Berkshire, *Classical Mechanics*. Imperial College Press
11. M.G. Calkin, *Lagrangian and Hamiltonian Mechanics*
12. A.J. French, M.G. Ebison, *Introduction to Classical Mechanics*

---

[4]This is not Quantum Mechanics!

Chapter 6
One-Dimensional Potentials and Two-Dimensional Central Potentials

Abstract In this chapter we are going to start analysing the simplest model of one-dimensional particle motion, which corresponds to the motion under the influence of a one-dimensional potential and no friction forces. This model is of special interest as it can be analytically solved in many cases. Motion in more than one dimension, most of the times, has no analytical solution and must be solved numerically. There are some exceptions however, and here we are going to analyse the two-dimensional motion under a central potential, and in particular we are going to analyse Kepler's potential. The analytic solution for the motion under the influence of Kepler's potential is derived explicitly by solving the corresponding differential equations of motion. Within the broad class of central potentials we shall also analyse potentials of the type kr^n for integer powers of n. As we shall observe, due to their particular functional dependence, we shall be able to treat central potentials in most aspects as if they were one-dimensional. We are also going to justify mathematically that the two-body problem with an internal central potential can be studied as a one dimensional problem.

6.1 One-Dimensional Potentials

Consider the energy conservation theorem (5.59) for our particular case: one-dimensional motion under the influence of an external one-dimensional potential energy U^{ext}. Suppressing the ext upper label, the expression reads

$$dE = dK + dU(x) = 0, \tag{6.1}$$

as there are no non-conservative forces that act on the system. This means that the total energy

$$E = \frac{1}{2}mv^2 + U(x), \tag{6.2}$$

© Springer Nature Switzerland AG 2020
V. Ilisie, *Lectures in Classical Mechanics*, Undergraduate Lecture Notes in Physics, https://doi.org/10.1007/978-3-030-38585-9_6

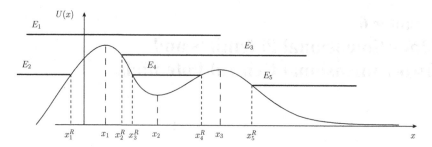

Fig. 6.1 Representation of: the potential energy curve, different total energy levels, maxima and minima and turning points

is conserved,[1] where $v = dx/dt$. By knowing the total energy we can calculate v as a function of x

$$v(x) = \pm\sqrt{\frac{2(E - U(x))}{m}}, \tag{6.3}$$

where the sign corresponds to the sense of motion.

The behaviour of the system can be characterized by simply analysing the potential energy $U(x)$ as follows. Consider the curve $(x, U(x))$ shown in Fig. 6.1, where x_1 is the global maximum, x_3 a local maximum and x_2, a local minimum. The behaviour of $U(x)$ at $\pm\infty$ is given by

$$\lim_{x \to +\infty} U(x) = 0, \qquad \lim_{x \to -\infty} U(x) = -\infty. \tag{6.4}$$

The straight lines labelled as E_i ($i = 1, \dots, 5$) are different values for the total energy, that correspond to different regions of interest. The points labelled as x_i^R ($i = 1, \dots, 5$) are called turning points, and correspond to points where the motion reverses its sense, as we shall see next. Let us thus analyse each case in more detail.

- If the total energy of the system is $E_1 \in (U(x_1), \infty)$ then we say the motion is unbounded. As E_1 is always greater than the global maximum of $U(x)$, the magnitude $|v| > 0$ and so, the particle will continue its motion towards $x \to +\infty$ if it is initially moving towards $+\infty$, or towards $x \to -\infty$ if initially moving towards $-\infty$.
- For the second case we have $E_2 \in (U(-\infty), U(x_1)]$ meaning that $x \in (-\infty, x_1]$. Here the motion is right-bounded. If the particle is initially moving towards $x \to -\infty$, the particle will not change its sense. If, on the other hand, it is initially moving towards $x \to +\infty$, it will reach the turning point x_1^R for which $E_2 = U(x_1^R)$ (and so $v = 0$ at x_1^R). At this point the particle reverses the sense of its motion. As we can observe, $v = 0$ at a turning point. We can thus obtain the turning points from the condition

---

[1] The previous result is valid up to an integration constant which we normally re-absorb into $U(x)$.

$$U(x_i^R) = E . \tag{6.5}$$

Note that as E is constant, in the region where $U \to -\infty$, we have $v \to \infty$. Thus, if we consider the limits imposed by Special Relativity (the speed of light, which is finite, is the maximum allowed speed in Nature), this type of regions of the potential are unphysical, as there would be a region where $v > c$.

- For the third case we have $E_3 \in [U(x_1), U(x_3))$ which means that $x \in [x_1, +\infty)$. This motion is left-bounded and it has a turning point which is given by x_2^R.
- For the fourth case we have $E_4 \in [U(x_2), U(x_3)]$ with $x \in [x_3^R, x_4^R]$. This motion is both left and right bounded. In this case we simply call it bounded. As it should result obvious, in this region, the particle is trapped in between the two turning points.
- As for the fifth and last case, the motion is left-bounded, $E_5 \in [U(x_3), 0)$ and $x \in [x_3, \infty)$.

It should be mentioned that there are also a few special cases. They occur when the particle starts its motion at some extrema i.e. at x_i with initial energy $E = U(x_i)$. This means that $K = 0$ and the particle will remain trapped at that position for an infinite amount of time, unless some small perturbation is introduced. In that case, if the particle was initially trapped at some local minimum, its motion will consist in small oscillations about the corresponding minimum. If on the contrary, the particle was trapped at some maximum i.e., at x_1 with initial energy $E = U(x_1)$, a small perturbation (in x or in v) will make the particle's motion fall into one of the previously described cases i.e., E_1, E_2 or E_3. This is why we refer to local maxima as unstable points.

Last, if not already clear, the points corresponding to the region $E < U$ are forbidden as they give rise to non-physical complex velocities.

Roughly speaking, we have presented all the information that one needs in order to completely solve the one dimensional potential problem. There can be potentials whose expressions are more complex and that need to be solved numerically, however this only represents a minor additional complication, and we shall not focus our attention on such cases. Summarizing, the information we need in order to fully characterize the system is

- local minima/maxima (that we have denoted by x_i):

$$\left. \frac{dU}{dx} \right|_{x=x_i} = 0 \, (\text{maximum or minimum}) ,$$

$$\left. \frac{d^2U}{dx^2} \right|_{x=x_i} > 0 \, (\text{minimum}) , \qquad \left. \frac{d^2U}{dx^2} \right|_{x=x_i} < 0 \, (\text{maximum}) . \tag{6.6}$$

- behaviour in the limit $\pm\infty$:

$$\lim_{x \to \pm\infty} U(x) . \tag{6.7}$$

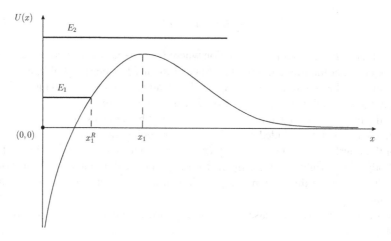

Fig. 6.2 Bounded potential with one turning point

- turning points x_i^R i.e., points where the particle reverses the sense of its motion:

$$U(x_i^R) = E. \qquad (6.8)$$

Our discussion can be trivially extended to potentials with saddle points, but we shall not be concerned with those cases here.

In the following we are going to briefly discuss a potential that presents some exception to the previous rules. It presents regions where the motion is bounded (both right and left) however, it only presents one turning point.[2] This kind of potentials present regions that are not physically allowed (by Special Relativity), but even so it is interesting to analyse them mathematically. An example of such potential is shown in Fig. 6.2.

Consider the region $E_1 \in (U(x \to 0), U(x_1)]$. Here $x \in (0, x_1]$ and we have a bounded motion with only one turning point x_1^R. If the particle is initially moving towards $x \to 0$, it will keep moving towards this limit for an infinite amount of time. Its kinetic energy will continue to increase towards $+\infty$ and it will never reach $x = 0$. Thus this motion is bounded and the bounds are 0 and x_1^R. If the particle is initially moving towards the right, it will reach the turning point and it will start moving towards $x \to 0$. Similar considerations can be made for $E_2 \in (U(x_1), +\infty)$ where $x \in (0, +\infty)$. The motion in this region is thus left-bounded.

In the following we are going to analyse a special class of three-dimensional potentials, that due to their symmetry can be reduced to a two-dimensional motion. Furthermore, if one is only interested in the relative motion, this problem can be reduced to a one-dimensional equivalent potential. This turns out to be specially useful for solving many problems regarding planetary motion.

---

[2]Similarly other potentials can present regions where the motion is right or left bounded with no turning points, or even both right and left bounded regions, with no turning points.

6.2 Central Potentials

Consider a point-like particle of mass m moving under the influence of a central potential energy of the form $U^{ext}(r)$. If we consider that no other force acts upon the system, the energy conservation theorem, just as in the previous case, states

$$dE = dK + dU^{ext} = 0. \tag{6.9}$$

Suppressing the *ext* upper index and integrating, we obtain

$$E = \frac{1}{2}mv^2 + U(r), \tag{6.10}$$

where E stands for the total energy which is a constant of motion. As the potential (energy) is central, the angular momentum is also conserved (see Chap. 5). Thus, the motion takes place in a plane. Without loss of generality we can always choose this plane to be (x, y).[3] We can thus express all the physical quantities we are interested in analysing, in polar coordinates. The coordinates and the velocity read

$$\mathbf{r} = r\,\mathbf{u}_r, \qquad \mathbf{v} = \dot{r}\,\mathbf{u}_r + r\dot{\phi}\,\mathbf{u}_\phi. \tag{6.11}$$

The angular momentum will be thus given by

$$\mathbf{l} = m\mathbf{r} \times \mathbf{v} = mr^2\dot{\phi}\,\mathbf{k}, \tag{6.12}$$

with $l = |\mathbf{l}| = mr^2\dot{\phi}$, a constant of motion. Let us now express the kinetic energy in polar coordinates. We have

$$K = \frac{1}{2}mv^2 = \frac{1}{2}m(\dot{r}\,\mathbf{u}_r + r\dot{\phi}\,\mathbf{u}_\phi)^2 = \frac{1}{2}m\dot{r}^2 + \frac{1}{2}mr^2\dot{\phi}^2. \tag{6.13}$$

We observe that K can be split into two pieces, the radial part, and the *angular* part

$$K = K_r + K_\phi, \tag{6.14}$$

with

$$K_r = \frac{1}{2}m\dot{r}^2, \qquad K_\phi = \frac{1}{2}mr^2\dot{\phi}^2 = \frac{l^2}{2mr^2}. \tag{6.15}$$

The angular term K_ϕ, as in the previous expression, can be conveniently expressed in terms of the magnitude of the angular momentum l and the mass m, which are both constants. Therefore we have $K_\phi = K_\phi(r)$. The total energy of the system finally reads

---

[3] We can always find a reference system where the plane of motion is (x, y).

$$E = \frac{1}{2}m\dot{r}^2 + \frac{l^2}{2mr^2} + U(r).$$ (6.16)

We can thus define and effective potential energy as

$$U_{eff}(r) = \frac{l^2}{2mr^2} + U(r),$$ (6.17)

and so

$$E = \frac{1}{2}m\dot{r}^2 + U_{eff}(r).$$ (6.18)

This expression is identical to the total energy corresponding to a one-dimensional one-body problem with a potential energy given by $U_{eff}(r)$. We can therefore profit from the previously introduced concepts and use them for this analysis. As we shall see in the following subsection, we will be able to further extend this analysis to the two-body problem.

As the motion is two-dimensional, the extrema of U_{eff} will correspond to bounded trajectories. Let us define an effective force as $F_{eff} = -dU_{eff}/dr$. This means that $m\ddot{r} = -dU_{eff}/dr$. Consider now the extrema r_i of U_{eff} i.e. the set of real values r_i that fulfil the condition

$$\left.\frac{dU_{eff}(r)}{dr}\right|_{r_i} = 0.$$ (6.19)

For these values the effective force is zero, therefore $\dot{r} = 0$, which means that there is no radial motion. However as $l \neq 0$ in general, the particle has an angular motion. We therefore conclude that for the extrema r_i of the effective potential energy, if $l \neq 0$ the motion is circular. Without getting into further details, the circular motion is stable if r_i is a minimum and unstable if r_i is a maximum.[4]

Depending on the shape of U_{eff} and the total energy of the system, there can be regions where the motion is bounded without being circular i.e., close to a minimum, the motion can be elliptic (as will shall see in the following subsection).

The turning points r_i^R can be calculated similarly to the one dimensional case

$$E = U_{eff}(r_i^R),$$ (6.20)

where E is the total energy of the system.

To conclude this part of the analysis, there is another typical constant of motion associated to central potentials, and that is the *areolar velocity*. Consider the trajectory of a point-like particle of mass m. The quantities $\mathbf{r}$, $d\mathbf{r}$ and $\mathbf{r} + d\mathbf{r}$ form a triangle, as shown in Fig. 6.3. The area of the triangle is given by

---

[4]With similar arguments as for the one dimensional case, if r_i is a minimum, by introducing a small perturbation the motion will still be bounded, and it will present small oscillations about the circular trajectory. This is not, in general, true for the maxima, and therefore they are unstable points.

Fig. 6.3 The trajectory of a
point-like mass together with
the triangle formed by $\mathbf{r}$, $d\mathbf{r}$
and $\mathbf{r} + d\mathbf{r}$

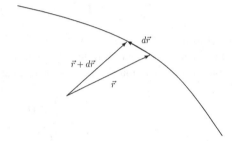

$$dA = \frac{1}{2}|\mathbf{r} \times d\mathbf{r}| = \frac{1}{2}|\mathbf{r} \times \mathbf{v}\,dt| = \frac{l}{2m}dt \,. \tag{6.21}$$

As the angular momentum is conserved (for a central potential), the areolar velocity
defined as

$$\frac{dA}{dt} = \frac{l}{2m}\,, \tag{6.22}$$

which is the area covered by $\mathbf{r}$ per unit time, is a constant of motion.

6.2.1 The Two-Body Problem

In the previous example, we have considered a one-particle system and a central
external potential. In the following we are going to see how this configuration can
actually describe a two-particle system under an internal central potential i.e., the
gravitational (or Kepler's) potential. Consider a system formed by two point-like
masses moving under the influence of an internal central potential $U^{int} = U^{int}(r)$,
where r is the relative distance. We consider no other forces acting upon the system,
thus the energy conservation theorem in this case states

$$dE = dK + dU^{int} = 0\,. \tag{6.23}$$

Suppressing the int upper index and integrating, we obtain

$$E = \frac{1}{2}m_1 v_1^2 + \frac{1}{2}m_2 v_2^2 + U(r)\,, \tag{6.24}$$

where E stands for the total energy (which is a conserved quantity). As the potential
energy is central, the force acting between the two particles will be central, thus,
the angular momentum is also conserved. Again, due to angular momentum conservation, the motion takes place in a plane and, without loss of generality, we will
consider this plane to be (x, y). The expression for the total angular momentum reads

$$\mathbf{L} = m_1 \, \mathbf{r}_1 \times \mathbf{v}_1 + m_2 \, \mathbf{r}_2 \times \mathbf{v}_2 . \tag{6.25}$$

Let us now make use of Konig's two theorems and re-write the kinetic energy and the angular momentum in terms of the relative coordinates, velocity, reduced mass and the center of mass variables (as in Chap. 5). We obtain

$$E = \frac{1}{2} M \mathcal{V}^2 + \frac{1}{2} \mu \mathbf{v}^2 + U(r) ,$$
$$\mathbf{L} = M \, \mathcal{R} \times \mathcal{V} + \mu \, \mathbf{r} \times \mathbf{v} . \tag{6.26}$$

As there are no external forces acting on the system, the CM kinetic energy and CM angular momentum are also constants. We thus conclude that the relative quantities defined as

$$E_{rel} \equiv \frac{1}{2} \mu \mathbf{v}^2 + U(r) ,$$
$$\mathbf{L}_{rel} \equiv \mu \, \mathbf{r} \times \mathbf{v} , \tag{6.27}$$

are constants. This is a quite strong statement as it allows us to study the relative motion of the two masses as a one-dimensional problem.

Before moving on with this analysis, let us discuss the following limiting case. Consider that we are interested in the motion of the Earth around the Sun. As the Earth's mass m is much smaller that the Sun's mass M we have $m \ll M$ and so

$$\mu = \frac{Mm}{M+m} = \frac{m}{1+m/M} \approx m \left(1 - \frac{m}{M} \right) \approx m . \tag{6.28}$$

Thus, in terms of relative motion (ignoring the motion of the planetary system within the Universe, which would be described by the CM variables) we can describe the Earth's motion around the Sun (up to a good approximation) as point-like particle of mass m moving around a *static* Sun of mass M.

One naturally arising question should now be asked. Is it legitimate to consider planets as point-like particles as we are doing here? The answer is yes, also up to a good approximation. In Chap. 8 we will see that a spherical mass with a mass density distribution of the type $\rho(r)$ generates a gravitational field (and potential) outside its surface, that is equivalent to a point-like particle with the same mass, placed at the center of the sphere. As these suppositions are valid (the mass distributions of both the Earth and the Sun present spherical mass symmetry up to a good approximation i.e., their mass densities are approximately of the form $\rho(r)$, where r is the radial distance measured from the center of the sphere) we can consider that the system is formed by two point-like masses with relative distance given by the distance between their centres.

Having introduced the previous concepts, we will now move forward and analyse Kepler's potential (or the gravitational potential) and its peculiarities.

6.3 Kepler's Potential

Consider a system formed by two point-like masses that move under the influence of the mutual gravitational attraction i.e., with a potential energy given by

$$U(r) = -\frac{Gm_1m_2}{r} \equiv -\frac{k}{r}, \qquad k > 0, \tag{6.29}$$

as described in Chap. 4. In what follows we shall only analyse the relative motion, therefore we shall drop for simplicity the *rel* subindex for the energy and angular momentum. The total (relative) energy of the system will thus be given by

$$E = \frac{1}{2}\mu\dot{r}^2 + \frac{l^2}{2\mu r^2} - \frac{k}{r} = \frac{1}{2}\mu\dot{r}^2 + U_{eff}. \tag{6.30}$$

From the previous expression we can calculate the possible trajectories of the *virtual* mass μ, by finding the extrema of the effective potential energy. We have

$$\left.\frac{dU_{eff}}{dr}\right|_{r=r_c} = \frac{k}{r_c^2} - \frac{l^2}{\mu r_c^3} = 0. \tag{6.31}$$

Therefore, the extremum r_c is given by[5]

$$r_c = \frac{l^2}{k\mu}. \tag{6.32}$$

Let us calculate if this circular motion is stable or unstable i.e., if it corresponds to a minimum or a maximum. Taking the second derivative we obtain

$$\frac{d^2U_{eff}}{dr^2} = -\frac{2k}{r^3} + \frac{3l^2}{\mu r^4}. \tag{6.33}$$

For $r = r_c$, it reads

$$\left.\frac{d^2U_{eff}}{dr^2}\right|_{r=r_c} = \frac{k}{r_c^3} > 0, \tag{6.34}$$

which corresponds to a minimum and thus, to a stable circular motion. Note that the effective potential presents no other local extrema. The energy corresponding to the circular motion takes the simple form

---

[5]Note that we have called r_c because, as explained previously, an extremum corresponds to a circular trajectory.

Fig. 6.4 Kepler's effective potential energy

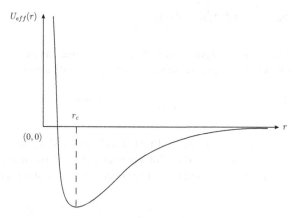

$$E(r_c) = U_{eff}(r_c) = \frac{l^2}{2\mu r_c^2} - \frac{k}{r_c} = \frac{l^2}{2\mu r_c^2} - \frac{l^2}{\mu r_c^2} = -\frac{l^2}{2\mu r_c^2}. \tag{6.35}$$

We observe that $E(r_c) < 0$. The behaviour at $0, +\infty$ is given by[6]

$$\lim_{r \to 0} U_{eff}(r) = +\infty, \qquad \lim_{r \to +\infty} U_{eff}(r) = 0. \tag{6.36}$$

Given the previous information we can approximately draw the shape of the potential. It is shown in Fig. 6.4.

As we have deduced previously, for $E = E_{\min} = E(r_c)$ the trajectory is circular and stable, thus bounded. From the previous figure we can deduce that the motion remains bounded for $E_{\min} < E < 0$. In this case the trajectory is elliptic. For $E > 0$ the motion is left-bounded (in terms of the (r, U_{eff}) graph), thus it only has one turning point and it corresponds to a hyperbola (as we shall see later on). The case of $E = 0$ corresponds to a parabolic motion. In this case μ will reach $r = \infty$ with no kinetic energy.

We are now going to solve the equations of motion for this potential and re-analyse the possible trajectories in terms of the total energy with greater detail. Newton's equations of motion in our case are given by

$$\mu \mathbf{a} = -\frac{dU}{dr} \mathbf{u}_r. \tag{6.37}$$

In polar coordinates

$$\mu(\ddot{r} - r\dot{\phi}^2)\mathbf{u}_r + (2\dot{r}\dot{\phi} + r\ddot{\phi})\mathbf{u}_\phi = -\frac{Gm_1 m_2}{r^2}\mathbf{u}_r. \tag{6.38}$$

---

[6]This is a subtle difference with respect to one dimensional potentials; as $r \geq 0$, the $(-\infty, 0)$ region is physically inaccessible.

We conclude that $(2\dot{r}\dot{\phi} + r\ddot{\phi}) = 0$ which is nothing but a direct consequence of angular momentum conservation. The angular momentum in polar coordinates can be written as

$$l = \mu r^2 \dot{\phi}, \tag{6.39}$$

and so

$$\frac{dl}{dt} = 0 = 2\mu r \dot{r}\dot{\phi} + \mu r^2 \ddot{\phi} = \mu r (2\dot{r}\dot{\phi} + r\ddot{\phi}), \tag{6.40}$$

which proves our previous statement.

Therefore, in order to obtain the equations of motion, we have to solve the following system of differential equations

$$\mu(\ddot{r} - r\dot{\phi}^2) = -\frac{k}{r^2} \tag{6.41}$$

$$\mu r^2 \dot{\phi} = l, \tag{6.42}$$

with l constant. We shall look for a solution of the type $r = r(\phi)$ instead of $r(t)$, as it turns out to be easier to obtain. We can thus express $\dot{r}$ as

$$\dot{r} = \frac{dr}{dt} = \frac{dr}{d\phi}\frac{d\phi}{dt} = \frac{dr}{d\phi}\dot{\phi}, \tag{6.43}$$

and so $\ddot{r}$ can simply be written as

$$\ddot{r} = \dot{\phi}\frac{d}{d\phi}\left(\dot{\phi}\frac{dr}{d\phi}\right). \tag{6.44}$$

Thus, the first equation (6.41) of the system can be re-expressed as

$$\dot{\phi}\frac{d}{d\phi}\left(\dot{\phi}\frac{dr}{d\phi}\right) - r\dot{\phi}^2 = -\frac{k}{\mu r^2}. \tag{6.45}$$

From the second equation, (6.42), we can straightforwardly obtain the explicit expression for $\dot{\phi}$

$$\dot{\phi} = \frac{l}{\mu r^2}. \tag{6.46}$$

By combining the two equations and simplifying, we obtain the following equivalent differential equation

$$\frac{d}{d\phi}\left(\frac{1}{r^2}\frac{dr}{d\phi}\right) - \frac{1}{r} = -\frac{k\mu}{l^2}. \tag{6.47}$$

Let us now perform a change of variable

$$\omega = \frac{1}{r} \quad \Rightarrow \quad \frac{d\omega}{d\phi} = -\frac{1}{r^2}\frac{dr}{d\phi}. \tag{6.48}$$

Our differential equation in terms of the new variable thus reads

$$\frac{d^2\omega}{d\phi^2} + \omega = \frac{k\mu}{l^2}. \tag{6.49}$$

The real solution can be straightforwardly obtained

$$\omega = \omega_0 \cos(\phi + \phi_0) + \frac{k\mu}{l^2}, \tag{6.50}$$

where ω_0 and ϕ_0 are two real constants. Let us now calculate the amplitude ω_0. From (6.30) we obtain

$$\dot{r}^2 = \frac{2E}{\mu} - \frac{l^2}{\mu^2 r^2} + \frac{2k}{\mu r}. \tag{6.51}$$

Introducing $\omega = 1/r$, we obtain $\dot{\omega}^2 = \dot{r}^2/r^4$ and so

$$\dot{\omega}^2 = \frac{1}{r^4}\left(\frac{2E}{\mu} - \frac{l^2}{\mu^2 r^2} + \frac{2k}{\mu r}\right). \tag{6.52}$$

On the other hand, using the explicit expression of ω we get

$$\dot{\omega}^2 = \omega_0^2 \sin^2(\phi + \phi_0)\dot{\phi}^2 = \omega_0^2 \sin^2(\phi + \phi_0)\frac{l^2}{\mu^2 r^4}. \tag{6.53}$$

From the last two equations we obtain

$$\omega_0^2 \sin^2(\phi + \phi_0) = \frac{2E\mu}{l^2} - \frac{1}{r^2} + \frac{2k\mu}{rl^2}. \tag{6.54}$$

If we now square the terms from (6.50) we get

$$\frac{1}{r^2} = \omega^2 = \omega_0^2 \cos^2(\phi + \phi_0) + \frac{\mu^2 k^2}{l^4} + 2\omega_0 \cos(\phi + \phi_0)\frac{\mu k}{l^2}. \tag{6.55}$$

Combining this expression with (6.54) and simplifying

$$\frac{2E\mu}{l^2} + \frac{2k\mu}{rl^2} = \omega_0^2 + \frac{\mu^2 k^2}{l^4} + 2\omega_0 \cos(\phi + \phi_0)\frac{\mu k}{l^2}. \tag{6.56}$$

Again, we can use the expression for ω to eliminate $\cos(\phi + \phi_0)$. We have

$$\cos(\phi + \phi_0) = \frac{1}{\omega_0}\left(\omega - \frac{\mu k}{l^2}\right) = \frac{1}{\omega_0}\left(\frac{1}{r} - \frac{\mu k}{l^2}\right). \tag{6.57}$$

By inserting this last equation back into (6.56) and simplifying we finally obtain

$$\omega_0 = \pm\sqrt{\frac{2E\mu}{l^2} + \frac{\mu^2 k^2}{l^4}}. \tag{6.58}$$

Without loss of generality we can choose the positive sign (as we still have the second integration constant ϕ_0, that can switch the sign of the expression). Finally the solution is given by the compact expression

$$\frac{1}{r} = \sqrt{\frac{2E\mu}{l^2} + \frac{\mu^2 k^2}{l^4}}\cos(\phi + \phi_0) + \frac{\mu k}{l^2}. \tag{6.59}$$

This equation corresponds to a generic elliptic curve with eccentricity

$$\epsilon = \frac{\omega_0}{(\mu k/l^2)}. \tag{6.60}$$

Depending on ϵ we can describe four types of elliptic curves:

- circle $\epsilon = 0$,
- ellipse $0 < \epsilon < 1$,

- parabola $\epsilon = 1$,
- hyperbola $\epsilon > 1$.

Having obtained the analytical solution, we can analyse the motion in terms of ω_0 (and implicitly in terms of the total energy and the eccentricity). For $\omega_0 = 0$, we have $E = E_{min}$ and

$$\frac{1}{r} = \frac{\mu k}{l^2}, \tag{6.61}$$

constant. As $\epsilon = 0$, we have a circular trajectory.

For the second case we have $\mu k/l^2 > \omega_0 > 0$, therefore $0 > E > E_{min}$ and $0 < \epsilon < 1$. The motion is elliptic in this case, and it is left for the reader as an exercise to check that the expression for r corresponds to that of an ellipse. One can check that $1/r$ reaches its maximum value for $\cos(\phi + \phi_0) = 1$ (thus r has a minimum value that we shall call pericenter r_p), and it reaches its minimum value for $\cos(\phi + \phi_0) = -1$ (thus r has a maximum value that we shall call apocenter r_a). These values are given by

$$\frac{1}{r_p} = \frac{\mu k}{l^2} + \omega_0, \tag{6.62}$$

Fig. 6.5 Elliptic motion for
$\mu k / l^2 > \omega_0 > 0$

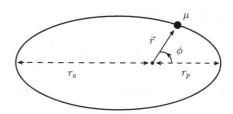

and

$$\frac{1}{r_a} = \frac{\mu k}{l^2} - \omega_0 . \tag{6.63}$$

If we choose $\phi_0 = 0$ for simplicity, as the initial condition, the corresponding motion is schematically shown in Fig. 6.5.

For the third case we have $\omega_0 = \mu k / l^2$ and so $E = 0$, thus $\epsilon = 1$. The pericenter is given by the condition $\cos \phi = 1$ (choosing $\phi_0 = 0$ as previously) and so

$$\frac{1}{r_p} = \frac{2\mu k}{l^2} . \tag{6.64}$$

In the limit $\phi \to \pi$ we have $1/r_a \to 0$ thus $r_a \to \infty$. The motion is thus left bounded (in terms of the (r, U_{eff}) graph) with one finite turning point r_p. From the energy conservation theorem, as the total energy has a constant value 0 and $U_{eff}(\infty) = 0$, we conclude that it reaches $r = \infty$ with $K = 0$. This motion is called parabolic.

Finally, the last case is given by $\omega_0 > \mu k / l^2$ and so $E > 0$ (with $\epsilon > 1$). The pericenter will be given by

$$\frac{1}{r_p} = \omega_0 + \frac{\mu k}{l^2} > \frac{2\mu k}{l^2} . \tag{6.65}$$

Note that the particle cannot reach the limit $\phi = \pi$ as r would turn negative. We thus have a left bounded motion (just as previously, with one finite turning point r_p) and additionally we have a limit angle (ϕ_l) given by the condition

$$\frac{1}{r_a} = \frac{1}{\infty} = 0 = \omega_0 \cos \phi_l + \frac{\mu k}{l^2} . \tag{6.66}$$

Again, we have considered $\phi_0 = 0$. Hence, the limit angle is given by

$$\phi_l = \arccos\left(-\frac{\mu k}{\omega_0 l^2}\right) . \tag{6.67}$$

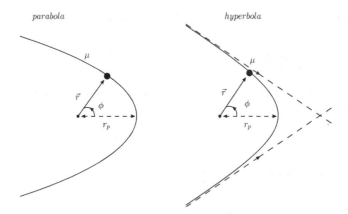

parabola   *hyperbola*

Fig. 6.6 Parabolic (left) and hyperbolic (right) motion

This motion is called hyperbolic. From the energy conservation theorem we deduce that the particle reaches $r = \infty$ with $K > 0$. Both the parabolic and hyperbolic trajectories are shown in Fig. 6.6.

Before Newton, Kepler had already discovered from direct observations that *the planets from the solar system describe elliptical orbits around the Sun*. This constitutes Kepler's first law. Kepler's second law states that *a straight line joining a planet and the Sun sweeps out equal areas in equal amounts of time*. This is of course, nothing but another way of stating the fact that the areolar velocity is constant (which is in general applicable to any central potential). Kepler's third law states that *the square of the period of a planet is proportional to the cube of the semimajor axis of its elliptical orbit*. We shall deduce this result as an exercise by using the expression of the areolar velocity and the expression of the area of an ellipse as a function of its axes. In the following, we shall also propose and solve a few exercises for one-dimensional potentials and two-dimensional central potentials.

6.4 Proposed Exercises

6.1 Study the one-dimensional potential energy function given by

$$V(x) = x^2(x-2)^2, \qquad (6.68)$$

where V is given in J (Joule) units. If the kinetic energy at $x = 1$ is 8 J, find the corresponding turning points.

Solution: Let us first find the maxima and the minima of the potential. We have

$$\frac{dV}{dx} = 4x^3 - 12x^2 + 8x = 4x(x^2 - 3x + 2) = 0. \tag{6.69}$$

Thus, one extremum is given by $x_1 = 0$. The other two extrema will be given by the solutions to the equation

$$x^2 - 3x + 2 = 0, \tag{6.70}$$

which are straightforwardly obtained

$$x_{2,3} = \frac{3 \pm \sqrt{9-8}}{2}. \tag{6.71}$$

Thus, $x_2 = 2$, $x_3 = 1$. By taking the second derivative we obtain

$$\frac{d^2V}{dx^2} = 12x^2 - 24x + 8, \tag{6.72}$$

and so

$$\frac{d^2V}{dx^2}\bigg|_{x=x_1} = 8 > 0 \quad \text{(minimum)},$$

$$\frac{d^2V}{dx^2}\bigg|_{x=x_2} = 8 > 0 \quad \text{(minimum)}, \tag{6.73}$$

$$\frac{d^2V}{dx^2}\bigg|_{x=x_3} = -4 < 0 \quad \text{(maximum)},$$

with

$$V(x_1) = V(x_2) = 0, \qquad V(x_3) = 1. \tag{6.74}$$

The behaviour in the $\pm\infty$ limit is also straightforward

$$\lim_{x \to \pm\infty} V(x) = +\infty. \tag{6.75}$$

    With all the previous information one can now easily draw the approximate shape of the potential. One should obtain a figure similar to Fig. 6.7. The turning points for the different energy level regions can be straightforwardly obtained.

    For the case $K = 8$ J at $x = 1$, we have $V(1) = 1$, and so the total energy will be given by

$$E = 8 + 1 = 9 \text{ J}. \tag{6.76}$$

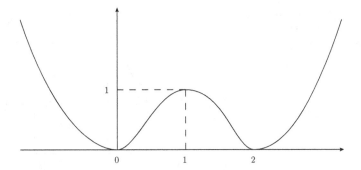

Fig. 6.7 Approximate shape of $V(x)$ and location of its local maximum and minima

From the figure one can observe that for $E = 9$ J, the motion has two turning points given by

$$V(x_R) = E = (x^R)^2(x^R - 2)^2 = 9.$$ (6.77)

We finally obtain $x_1^R = -1$ and $x_2^R = 3$.

6.2 Study the one-dimensional potential energy function given by

$$V(x) = -2x^6 + 2x^4 - \frac{1}{2}x^2 + \frac{1}{2},$$ (6.78)

where V is given in J (Joule) units.

Solution: Even if at first sight this potential might seem complicated, it is actually quite simple and just as straightforward to solve as the previous one. The extrema are given by

$$\frac{dV}{dx} = -12x^5 + 8x^3 - x = -x(12x^4 - 8x^2 + 1) = 0.$$ (6.79)

Thus, one extremum is given by $x_1 = 0$. The other four extrema will be given by the solutions to the equation

$$12(x^2)^2 - 8(x^2) + 1 = 0.$$ (6.80)

The solutions for x^2 is

$$x^2 = \frac{8 \pm \sqrt{64 - 48}}{24}.$$ (6.81)

Fig. 6.8 Approximate shape of $V(x)$ and location of its local maxima and minima

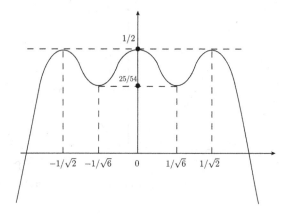

We thus obtain $x^2 = 1/2$ and $x^2 = 1/6$. The four remaining roots of (6.79) will therefore, be given by

$$x_2 = \frac{1}{\sqrt{2}}, \quad x_3 = -\frac{1}{\sqrt{2}}, \quad x_4 = \frac{1}{\sqrt{6}}, \quad x_5 = -\frac{1}{\sqrt{6}}. \tag{6.82}$$

By taking the second derivative we obtain

$$\frac{d^2V}{dx^2} = -60x^4 + 24x^2 - 1, \tag{6.83}$$

and so

$$\begin{aligned}
\frac{d^2V}{dx^2}\Big|_{x=x_1} &= -1 < 0 \qquad \text{(maximum)}, \\
\frac{d^2V}{dx^2}\Big|_{x=x_{2,3}} &= -4 < 0 \qquad \text{(maxima)}, \\
\frac{d^2V}{dx^2}\Big|_{x=x_{4,5}} &= \frac{4}{3} > 0 \qquad \text{(minima)}.
\end{aligned} \tag{6.84}$$

At these points V takes the following values

$$V(x_1) = V(x_2) = V(x_3) = \frac{1}{2}, \quad V(x_4) = V(x_5) = \frac{25}{54} \approx 0.46. \tag{6.85}$$

The behaviour in the $\pm\infty$ limit is

$$\lim_{x \to \pm\infty} V(x) = -\infty. \tag{6.86}$$

With all the previous information one can now draw the approximate shape of the potential. One should obtain a similar figure to the one shown in Fig. 6.8. The turning points for the different energy level regions are straightforward to obtain and will not be discussed.

6.3 Consider the elliptic motion of a reduced mass μ under Kepler's potential energy. Obtain the expression of the semimajor axis a as a function of ω_0 and $\mu k / l^2$.

Solution: We have previously found the expressions for the pericenter and apocenter

$$\frac{1}{r_p} = \frac{\mu k}{l^2} + \omega_0 \,,$$

$$\frac{1}{r_a} = \frac{\mu k}{l^2} - \omega_0 \,. \tag{6.87}$$

From the basic properties of an ellipse, we know that the semimajor axis is simply given by

$$a = \frac{r_p + r_a}{2} \,. \tag{6.88}$$

Using the previous expressions and simplifying we obtain

$$
\begin{aligned}
a &= \frac{1}{2} \frac{1}{\left(\dfrac{\mu k}{l^2} - \omega_0 \right)} + \frac{1}{2} \frac{1}{\left(\dfrac{\mu k}{l^2} + \omega_0 \right)} \\
&= \frac{\mu k / l^2}{\mu^2 k^2 / l^4 - \omega_0^2} \,.
\end{aligned} \tag{6.89}
$$

In the following we shall apply this result in proving Kepler's third law.

6.4 Using the expression for the areolar velocity

$$\frac{dA}{dt} = \frac{l}{2\mu} \,, \tag{6.90}$$

and the expression for the area of an ellipse

$$A = \pi a b \,, \tag{6.91}$$

where a and b are the semimajor and semiminor axes, prove Kepler's third law

$$T^2 = \frac{4\pi^2 \mu}{k} a^3 \,, \tag{6.92}$$

where T is the period, $k = G m_1 m_2$ and μ, the reduced mass.

Solution: On one hand the total area of the ellipse is proportional to the period i.e.,

$$A = \frac{l}{2\mu} \int_0^T dt = \frac{lT}{2\mu}. \tag{6.93}$$

On the other hand we can express the semiminor axis b in terms of the semimajor axis a and the eccentricity ϵ

$$b = a\sqrt{1 - \epsilon^2}. \tag{6.94}$$

By using (6.60) we can express b as

$$
\begin{aligned}
b &= a\sqrt{1 - \frac{\omega_0^2}{\mu^2 k^2 / l^4}} \\
&= a\sqrt{\frac{\mu^2 k^2 / l^4 - \omega_0^2}{\mu^2 k^2 / l^4}} \\
&= a\sqrt{\frac{l^2}{\mu k a}},
\end{aligned} \tag{6.95}
$$

where we have used the result from the previous exercise. Thus, we obtain the following expression

$$\frac{lT}{2\mu} = \pi a^2 \sqrt{\frac{l^2}{\mu k a}}, \tag{6.96}$$

and so the period T can be expressed as

$$T = \frac{2\pi\mu a^2}{l} \sqrt{\frac{l^2}{\mu k a}}. \tag{6.97}$$

Squaring the previous equation and simplifying we obtain the desired result

$$T^2 = \frac{4\pi^2 \mu}{k} a^3. \tag{6.98}$$

6.5 Find the possible trajectories (bounded or unbounded) for Kepler's potential as a function of the total energy E, by analysing the turning points of the effective potential. Calculate the total energy and the velocity v for the circular orbit.

Solution: Previously, in order to find the possible types of trajectories we have made use of the equation of motion $r(\phi)$. However, one does not have to solve the equations of motion in order to find the possible types of motion. One only has to calculate the tuning points of the effective potential energy. In our case we have

$$U_{eff} = \frac{l^2}{2\mu r^2} - \frac{k}{r}, \qquad (6.99)$$

with the turning points r_R given by

$$U_{eff}(r_R) = E. \qquad (6.100)$$

We can separately analyse the two possible cases $E \neq 0$ and $E = 0$. For the first case we have

$$\frac{l^2}{2\mu r_R^2} - \frac{k}{r_R} = E, \qquad (6.101)$$

and so

$$r_R^2 + \frac{k r_R}{E} - \frac{l^2}{2\mu E} = 0. \qquad (6.102)$$

Solving the previous equation we have

$$r_R = -\frac{k}{2E} \pm \frac{k}{2E}\sqrt{1 + \frac{2El^2}{k^2\mu}}. \qquad (6.103)$$

For $E > 0$ we have only one positive real solution and it corresponds to

$$r_R = -\frac{k}{2E} + \frac{k}{2E}\sqrt{1 + \frac{2El^2}{k^2\mu}}, \qquad (6.104)$$

and so in terms of the (r, U_{eff}) graph, we have a left bounded motion with one turning point. If $E < 0$ we shall re-express E as $E = -|E|$ in order to keep the negative terms strictly under control. We thus have two real positive solutions

$$r_R = \frac{k}{2|E|} \pm \frac{k}{2|E|}\sqrt{1 - \frac{2|E|l^2}{k^2\mu}}, \qquad (6.105)$$

(with $2|E|l^2 \leqslant k^2\mu$) and therefore the motion is bounded. For $2|E|l^2 < k^2\mu$ we have two non-degenerate solutions that correspond to the pericenter and apocenter. If $2|E|l^2 = k^2\mu$ we have one doubly degenerate solution which corresponds to the circular motion (the pericenter is equal to the apocenter).

The last case corresponds to $E = 0$ and so

$$\frac{l^2}{2\mu r_R^2} - \frac{k}{r_R} = 0,$$
(6.106)

which has a unique real positive solution

$$r_R = \frac{l^2}{2\mu k},$$
(6.107)

that corresponds to a left-bounded motion in terms of the (r, U_{eff}) graph.

We shall end this analysis with a few comments on the centripetal force and its role in the circular motion. As previously seen, for the circular motion we have one degenerate solution and the radius of the circle is given by

$$r_c = \frac{k}{2|E|} = -\frac{k}{2E}.$$
(6.108)

The energy then reads

$$E = -\frac{k}{2r_c}.$$
(6.109)

Again, as the motion is circular, $\dot{r} = 0$ and so, we can write down the following equation

$$E = -\frac{k}{2r_c} = \frac{1}{2}\mu v^2 - \frac{k}{r_c},$$
(6.110)

with $v = r\dot{\phi}$. We therefore obtain

$$v = \sqrt{\frac{k}{\mu r_c}},$$
(6.111)

(for $\dot{\phi} > 0$). Lets us now remember the expression for the centripetal force. In our case

$$F_{cc} = \frac{\mu v^2}{r_c} = \frac{k}{r_c^2} = \frac{Gm_1 m_2}{r_c^2} = \frac{GM\mu}{r_c^2},$$
(6.112)

where $M = m_1 + m_2$ is the total mass of the system. Thus, the gravitational force acts as the centripetal force in the case of the circular motion (as expected).

In the following we shall analyse central potentials of the type $\pm kr^n$ with n an integer and $k > 0$, a constant. We shall conclude that not for all integer values of n

there are solutions corresponding to closed bounded orbits. In fact one can check[7] that Kepler's potential energy $-k/r$, and kr^2 are the only two potential energies that give rise to such orbits.

6.6 Consider a particle of mass $m = 1$ kg and an external potential energy function given by

$$U(r) = -r^2/2, \qquad (6.113)$$

where U is given in Joule (J) units, with an angular momentum $l = 4$ kg m^2/s. Can $U(r)$ give rise to a circular orbit? If the total energy is 3 J obtain the pericenter and the apocenter. Is the region $E < 0$ physically accessible? How about $E = 0$?

Solution: In order to find out if a circular trajectory is physically allowed, we need to calculate the first derivative of the effective potential energy

$$\frac{dU_{eff}}{dr} = \frac{d}{dr}\left(\frac{l^2}{2mr^2} - \frac{r^2}{2}\right) = -\frac{l^2}{mr^3} - r = 0. \qquad (6.114)$$

We thus obtain

$$r_c^4 = -\frac{l^2}{m}, \qquad (6.115)$$

which has no positive real solution. Thus $U(r)$ cannot give rise to a circular orbit. Let us now calculate the turning points (for $E > 0$). We have

$$U_{eff}(r_R) = \frac{l^2}{2mr_R^2} - \frac{r_R^2}{2} = E. \qquad (6.116)$$

Manipulating the previous expression and simplifying we obtain

$$(r_R^2)^2 + 2\,r_R^2 E - \frac{l^2}{m} = 0, \qquad (6.117)$$

with the solution given by

$$r_R^2 = -E \pm \frac{1}{2}\sqrt{4E^2 + \frac{4l^2}{m}} = E\left(-1 \pm \sqrt{1 + \frac{l^2}{E^2 m}}\right). \qquad (6.118)$$

[7]Bertrand's theorem: Bertrand. J., "Théorème relatif au mouvement d'un point attiré vers un centre fixe" (1873), C. R. Acad. Sci. 77: 849–853.

As $\sqrt{1 + l^2/(E^2 m)} > 1$ we only have one real finite positive solution that corresponds to the pericenter. Inserting the given numerical values we obtain $r_p = \sqrt{2}$ m. The apocenter corresponds to $r_a = \infty$.

In order to find the answer to the next question, we shall re-write E as $E = -|E|$ and, the solution for r_R^2 becomes

$$r_R^2 = |E| \left(1 \pm \sqrt{1 + \frac{l^2}{|E|^2 m}} \right). \tag{6.119}$$

The previous expression presents one physically allowed solution (corresponding to the $+$ sign). One can trivially check that the motion for $E = 0$ is also physically allowed and it presents one finite turning point

$$r_R^2 = \sqrt{\frac{l^2}{m}}. \tag{6.120}$$

6.7 Repeat the previous exercise for

$$U(r) = r^2/2, \tag{6.121}$$

where U is given in J (joule) units, with a value for the total energy of $E = 5$ J.

Solution: Calculating the first derivative of the effective potential we obtain

$$\frac{dU_{eff}}{dr} = \frac{d}{dr} \left(\frac{l^2}{2mr^2} + \frac{r^2}{2} \right) = -\frac{l^2}{mr^3} + r = 0. \tag{6.122}$$

We thus obtain

$$r_c^4 = \frac{l^2}{m}, \tag{6.123}$$

and so the circular radius is given by

$$r_c = \sqrt[4]{\frac{l^2}{m}}. \tag{6.124}$$

Let us now find out if this orbit is stable. We obtain

$$\frac{d^2 U_{eff}}{dr^2} = \frac{3l^2}{mr^4} + 1 > 0, \tag{6.125}$$

therefore it corresponds to a minimum, and so the orbit is stable. Next, we are going to calculate the turning points (for $E > 0$)

$$U_{eff}(r_R) = \frac{l^2}{2mr_R^2} + \frac{r_R^2}{2} = E \,. \tag{6.126}$$

Manipulating the previous expression and simplifying we obtain

$$(r_R^2)^2 - 2r_R^2 E + \frac{l^2}{m} = 0 \,, \tag{6.127}$$

with the solution given by

$$r_R^2 = E \pm \frac{1}{2}\sqrt{4E^2 - \frac{4l^2}{m}} = E\left(1 \pm \sqrt{1 - \frac{l^2}{E^2 m}}\right) \,, \tag{6.128}$$

with $mE^2 \geqslant l^2$. We thus have two real positive solutions, or one doubly degenerated solution if $mE^2 = l^2$, which corresponds to the circular motion. Inserting the given numerical values we find $r_p = \sqrt{2}$ m and $r_a = 2\sqrt{2}$ m. One can straightforwardly check that the motion for $E \leqslant 0$ is physically inaccessible.

6.8 Consider a force field given by ($k > 0$)

$$\mathbf{F}(r) = \frac{k}{r^3}\mathbf{u}_r \,. \tag{6.129}$$

Obtain and analyse the potential energy function $U(r)$ that generates $\mathbf{F}(r)$, with the usual boundary condition $U(r \to \infty) = 0$.

Solution: Recall from Chap. 4 that for a central force $\mathbf{F}(r) = f(r)\mathbf{u}_r$, the potential energy function is given by

$$U(r) = -\int f(r)dr \,. \tag{6.130}$$

Therefore, in our case we have

$$U(r) = -k\int \frac{dr}{r^3} = \frac{k}{2r^2} + C \,, \tag{6.131}$$

with C a constant. From the boundary condition $C = 0$. The effective potential is then given by

$$U_{eff}(r) = \frac{l^2}{2mr^2} + \frac{k}{2r^2}. \tag{6.132}$$

Let us obtain the turning points. We have

$$E = \frac{l^2}{2mr_R^2} + \frac{k}{2r_R^2}. \tag{6.133}$$

This equation has a unique physically allowed solution, and that is

$$r_R = \sqrt{\frac{l^2}{2mE} + \frac{k}{2E}}, \tag{6.134}$$

with $E > 0$. The conclusions are straightforward. The reader is invited to analyse the previous potential for $k < 0$.

Further Reading

1. J.V. José, E.J. Saletan, *Classical Dynamics: A Contemporary Approach*. Cambridge University Press
2. S.T. Thornton, J.B. Marion, *Classical Dynamics of Particles and Systems*
3. H. Goldstein, C. Poole, J. Safko, *Classical Mechanics*, 3rd edn. Addison Wesley
4. J.R. Taylor, *Classical Mechanics*
5. D.T. Greenwood, *Classical Dynamics*. Prentice-Hall Inc.
6. D. Kleppner, R. Kolenkow, *An Introduction to Mechanics*
7. C. Lanczos, *The Variational Principles of Mechanics*. Dover Publications Inc.
8. W. Greiner, *Classical Mechanics: Systems of Particles and Hamiltonian Dynamics*. Springer
9. H.C. Corben, P. Stehle, *Classical Mechanics*, 2nd edn. Dover Publications Inc.
10. T.W.B. Kibble, F.H. Berkshire, *Classical Mechanics*. Imperial College Press
11. A.J. French, M.G. Ebison, *Introduction to Classical Mechanics*

Chapter 7
Non Relativistic Collisions

Abstract In this chapter we shall study both elastic and inelastic frontal (non-relativistic) collisions of point-like particles. We shall also introduce the notion of particle scattering by a hard (impenetrable sphere) and by a repulsive potential. As already usual in this book, the most generic aspects will be presented, and the reader will be invited to deduce additional theoretical aspects and apply them to some practical examples.

7.1 Frontal Collisions

In a frontal collision, the colliding particles move in the same direction but opposite sense, in some reference frame.[1] This is schematically shown in Fig. 7.1. If there are no external forces acting upon the system, the total momentum of the system is conserved. Therefore we can write down the following conservation law

$$\mathcal{P} = \mathcal{P}_{ini} \equiv \mathbf{p}_1 + \mathbf{p}_2 \equiv \mathcal{P}_{fin} = \mathbf{p}_1' + \mathbf{p}_2'. \tag{7.1}$$

Collisions can be categorized as elastic or inelastic. For elastic collisions the mechanical energy is conserved, whereas for inelastic ones it is not (however, the total momentum is conserved in both cases!). Neglecting mutual forces between the colliding particles, the energy conservation theorem for elastic collision states

$$K_1 + K_2 = K_1' + K_2'. \tag{7.2}$$

For inelastic collisions we have to add another term Q in the final state

$$K_1 + K_2 = K_1' + K_2' + Q, \tag{7.3}$$

---

[1] The comment *in some reference frame* refers to the fact that there are reference systems in which one of the particles can be at rest and therefore, the particles do not move in opposite sense in these cases.

© Springer Nature Switzerland AG 2020
V. Ilisie, *Lectures in Classical Mechanics*, Undergraduate Lecture
Notes in Physics, https://doi.org/10.1007/978-3-030-38585-9_7

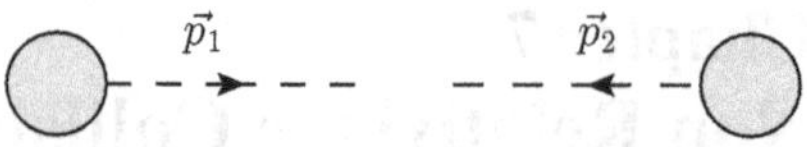

Fig. 7.1 Schematic
representation of a
two-particle frontal collision

Fig. 7.2 Two-particle
collision in the CM reference
frame

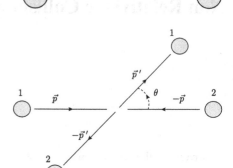

that accounts for the energy loss (or gain) after the collision.[2] If $Q > 0$ the reaction is
said to be exothermic and if $Q < 0$ it is said to be endothermic (which is more typical
in chemical reactions). Normally, two macroscopic particles that collide suffer an
energy loss ($Q > 0$) in the form of heat.

Additionally, if there are no external forces acting upon the system, **the angular momentum is conserved (for both elastic and inelastic collisions)**, hence the
collision is confined to a plane.

7.1.1   Elastic Collisions in the Center of Mass

In the center of mass (CM) reference frame we have the configuration shown in
Fig. 7.2. By definition, in this reference frame the total momentum of the system is
zero ($\mathcal{P} = \mathbf{0}$) and so

$$\mathbf{p} \equiv \mathbf{p}_1 = m_1 \mathbf{v}_1 = -\mathbf{p}_2 = -m_2 \mathbf{v}_2 , \qquad (7.4)$$

before the collision, and

$$\mathbf{p}' \equiv \mathbf{p}'_1 = m_1 \mathbf{v}'_1 = -\mathbf{p}'_2 = -m_2 \mathbf{v}'_2 , \qquad (7.5)$$

after the collision. It can be easily deduced that the cosine of the scattering angle is
given by the product

$$\cos \theta = \frac{\mathbf{p} \cdot \mathbf{p}'}{p \, p'} , \qquad (7.6)$$

---

[2]The previous expression is nothing but the energy conservation theorem (5.59) particularized to
our case.

with $\theta \in [0, \pi]$ and where $|\mathbf{p}| = p$, $|\mathbf{p}'| = p'$. The kinetic energies are given by

$$K_1 = \frac{p_1^2}{2m_1}, \qquad K_2 = \frac{p_2^2}{2m_2}, \qquad (7.7)$$

and

$$K_1' = \frac{p_1'^2}{2m_1}, \qquad K_2' = \frac{p_2'^2}{2m_2}, \qquad (7.8)$$

where we have introduced the magnitudes $|\mathbf{p}_i| = p_i$ and $|\mathbf{p}_i'| = p_i'$. It is left for the reader as an exercise to show that using the energy conservation law

$$K_1 + K_2 = K_1' + K_2', \qquad (7.9)$$

one obtains $p = p'$.

7.1.2 Elastic Collisions in the Laboratory Frame

By definition, in this reference frame, one of the particles is at rest. The momentum configuration and characteristic angles are shown in Fig. 7.3. Momentum conservation in this case states

$$\mathbf{p}_1^L = \mathbf{p}_1'^L + \mathbf{p}_2'^L, \qquad (7.10)$$

(with $\mathbf{p}_2^L = \mathbf{0}$). The energy conservation theorem in this case is given by

$$\frac{(p_1^L)^2}{2m_1} = \frac{(p_1'^L)^2}{2m_1} + \frac{(p_2'^L)^2}{2m_2}, \qquad (7.11)$$

Fig. 7.3 Two-particle collision in the Lab. reference frame

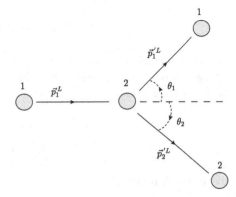

where $p_i^L = |\mathbf{p}_i^L|$ and $p_i'^L = |\mathbf{p}_i'^L|$. The angle θ_1 can be calculated similarly to the CM angle

$$\cos \theta_1 = \frac{\mathbf{p}_1^L \cdot \mathbf{p}_1'^L}{p_1^L \, p_1'^L} , \qquad (7.12)$$

and equally straightforward for θ_2.

7.1.3 Relating Both Frames

One can easily check that the collision plane in the Laboratory (Lab.) reference system is the same as the one that corresponds to the CM frame. Let us now relate the two reference frames. The velocity of the CM as seen from the Lab. reference frame will be given by

$$\mathcal{V} = \frac{1}{m_1 + m_2}(m_1 \mathbf{v}_1^L + m_2 \mathbf{v}_2^L) = \frac{m_1}{m_1 + m_2}\mathbf{v}_1^L . \qquad (7.13)$$

We can now relate the velocities measured in the Lab. system with the ones measured in the CM by means of a Galilean transformations i.e., the velocity of one of the particles measured in the Lab. frame is equal the CM velocity with respect to the Lab. frame plus the velocity of the particle with respect to the CM frame. This is true before the collision

$$\mathbf{v}_i^L = \mathcal{V} + \mathbf{v}_i , \qquad (7.14)$$

and after the collision

$$\mathbf{v}_i'^L = \mathcal{V} + \mathbf{v}_i' , \qquad (7.15)$$

with $i = 1, 2$. For the final state particle 1 we have

$$\mathbf{v}_1'^L = \mathcal{V} + \mathbf{v}_1' . \qquad (7.16)$$

Projecting the components of the previous vectors on the *collision axis* and, on the orthogonal axis (to the *collision axis*) that belongs to the *collision plane*, we obtain the following expressions

$$\begin{aligned} v_1'^L \cos \theta_1 &= \mathcal{V} + v_1' \cos \theta , \\ v_1'^L \sin \theta_1 &= v_1' \sin \theta , \end{aligned} \qquad (7.17)$$

and so, $\tan \theta_1$ is simply

$$\tan \theta_1 = \frac{v_1' \sin \theta}{V + v_1' \cos \theta} . \tag{7.18}$$

This is an expression that relates the Lab. angle θ_1 with the CM angle θ.[3] As we shall see in the following we can further simplify the previous expression in the case of elastic collisions. Let us write down the inverse transformation of (7.14) combined with (7.13). We obtain

$$\mathbf{v}_1 = \mathbf{v}_1^L - \mathbf{V} = \frac{m_1 + m_2}{m_1} \mathbf{V} - \mathbf{V} = \frac{m_2}{m_1} \mathbf{V} . \tag{7.19}$$

Considering the vector magnitudes in the previous equation and using $p = p'$ (for elastic collisions), we obtain the following relation in the CM frame

$$v_1 = \frac{1}{\rho} V = v_1' , \tag{7.20}$$

where we have defined the dimensionless term

$$\rho \equiv m_1 / m_2 . \tag{7.21}$$

Returning thus to the expression of $\tan \theta_1$ (7.18), it simplifies to

$$\tan \theta_1 = \frac{\sin \theta}{\rho + \cos \theta} , \tag{7.22}$$

(for elastic collisions). By differentiating the previous expression, one should be able to prove (it is left for the reader as an exercise) that there is a maximum θ_1 angle in the Lab. reference frame, given by

$$\tan \theta_1^{\max} = \frac{1}{\sqrt{\rho^2 - 1}} , \tag{7.23}$$

if $\rho \geq 1$. Next, we are going to analyse the most important characteristics of inelastic collisions and leave some open questions for the reader to solve.

7.1.4 Inelastic Collisions

In this section we shall briefly analyse the main characteristics of the inelastic collisions independently of the reference frame. Consider the initial colliding particles with velocities $\mathbf{v}_1^i$ and $\mathbf{v}_2^i$, and with final velocities $\mathbf{v}_1^f$ and $\mathbf{v}_2^f$ (after the collision) in

---

[3]Note, that we have only used momentum conservation in order to obtain the previous expression, therefore it is valid both for elastic and inelastic collisions.

an arbitrary reference frame. We define the restitution coefficient e as the magnitude

$$e \equiv \frac{|\mathbf{v}_2^f - \mathbf{v}_1^f|}{|\mathbf{v}_2^i - \mathbf{v}_1^i|} \equiv \frac{u_{21}^f}{u_{21}^i}. \tag{7.24}$$

It should be obvious that, as the previously defined coefficient only depends on relative magnitudes, it is independent on the reference frame. It is left for the reader as an exercise to check that

$$Q \equiv K_1^i + K_2^i - K_1^f - K_1^f = \frac{1}{2}\mu(u_{21}^i)^2(1 - e^2), \tag{7.25}$$

where $\mu = m_1 m_2/(m_1 + m_2)$ is the reduced mass, K_1^i and K_2^i are the initial state kinetic energies

$$K_1^i = \frac{1}{2}m_1(\mathbf{v}_1^i)^2, \qquad K_2^i = \frac{1}{2}m_2(\mathbf{v}_2^i)^2, \tag{7.26}$$

and, K_1^f and K_2^f the final state kinetic energies

$$K_1^f = \frac{1}{2}m_1(\mathbf{v}_1^f)^2, \qquad K_2^f = \frac{1}{2}m_2(\mathbf{v}_2^f)^2. \tag{7.27}$$

Depending on the value of e we can categorize the collisions as follows

- $e = 1$, elastic collision
- $e = 0$, completely inelastic collision
- $1 \geq e > 0$, exothermic collision (inelastic)
- $e < 1$, endothermic collision (inelastic).

Let us now analyse the CM and the Lab. frames. It left for the reader as an exercise to prove that in the CM reference frame

$$p^2 = p'^2 + 2\mu Q, \tag{7.28}$$

and

$$e = \frac{p}{p'}. \tag{7.29}$$

One should also be easily able to prove by using the relation (7.18) between the CM angle θ and the Lab. angle θ_1, that in the case of inelastic collisions there is a maximum allowed value for θ_1 given by

$$\sin \theta_1^{\max} = \frac{p'}{m_1 \mathcal{V}}. \tag{7.30}$$

7.2 Scattering by a Hard Sphere

Let us now define the parameters needed to describe the scattering by a hard sphere. We thus, need to introduce the concept of cross section. Consider for simplicity a uniform beam of particles, that are scattered by a hard sphere. As the beam is uniform it is sufficient to analyse two components of the beam separated by a distance ds, as shown in Fig. 7.4. The differential cross section (shown in the same figure) is nothing but the differential area of the disk

$$d\sigma = 2\pi s\, ds\,. \tag{7.31}$$

One can check that integrating in the $s \in [0, r]$ domain, one obtains the area of a disk of radius r.

Let us now define I as the incident particle density per unit surface and N as the density of scattered particles per unit solid angle. The total number of particles is conserved so, the incident and scattered number of particles must be the same. This translates into

$$I(\sigma)\, d\sigma = N(\Omega)\, d\Omega\,, \tag{7.32}$$

or equivalently

$$\frac{d\sigma}{d\Omega} = \frac{N(\Omega)}{I(\sigma)}\,, \tag{7.33}$$

where the right side of the previous expression represents the experimental measurement and the left side represents the prediction of the theoretical model.

We shall dedicate what is left of this chapter finding theoretical expressions for $d\sigma/d\Omega$ for different scattering elements such as a hard sphere or a (central) repulsive potential. Let us thus manipulate (7.31) in order to express it in the form $d\sigma/d\Omega$. Imagine that we do not integrate over the angle ϕ in between $[0, 2\pi]$. In this case we have a more general expression

$$d\sigma = s\, ds\, d\phi\,, \tag{7.34}$$

Fig. 7.4 Schematic representation of particle scattering by a hard sphere

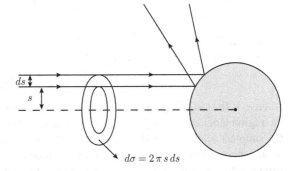

$$d\sigma = 2\pi s\, ds$$

Fig. 7.5 Relevant angles
and parameters for the
scattering by a hard sphere

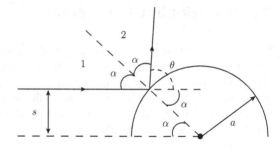

which will be valid even in the case of a non-uniform beam. Let us now remember
that in spherical coordinates the solid angle reads

$$d\Omega = \sin\theta \, d\theta \, d\phi. \tag{7.35}$$

By combining (7.34) and (7.35) we finally obtain the expression

$$\boxed{\frac{d\sigma}{d\Omega} = \frac{s}{\sin\theta}\left|\frac{ds}{d\theta}\right|}, \tag{7.36}$$

which is totally generic, as we have made no assumption on the scattering element.
We have only used the hard sphere example in order to introduce the basic concepts
regarding the cross section and the s parameter. The s parameter (also called **impact
parameter**), is usually defined for symmetric scattering elements (such as the sphere
or the repulsive central potential) as the distance from the symmetry axis (which
is parallel to the trajectory of the particle) to the line that describes the particle's
trajectory. Also note that we have introduced the modulus of $d\sigma/d\theta$ in order to
ensure that the cross section is always positive (a negative cross section has no
physical meaning).

Having introduced the previous generic concepts we shall now continue with the
hard sphere example. With the help of Fig. 7.5 the reader is invited to show that for
a hard sphere

$$s = a\cos\left(\frac{\theta}{2}\right), \tag{7.37}$$

and

$$\frac{d\sigma}{d\Omega} = \frac{a^2}{4}, \tag{7.38}$$

where a is the radius of the sphere. Some clarifying notes are needed. The angle α in
the region labelled as 1 is the same as the one in region 2 because we have assumed
the simplest scattering model, where the incident angle (with respect to the normal
of the spherical surface) is the same as the reflected angle. The remaining angles
shown in the figure can be deduced by means of geometric arguments.

7.3   Scattering by a Repulsive Potential

Consider the repulsive central potential energy

$$U(r) = \frac{k}{r}, \tag{7.39}$$

with $k > 0$. In the previous chapter we have found the solution for the equations of motion for Kepler's potential energy

$$U(r)_{Kep} = -\frac{k}{r}, \tag{7.40}$$

again, with $k > 0$. We can thus obtain the solution for the equations of motion for $U(r)$ directly from the solution of $U(r)_{Kep}$, by simply making the substitution $k \to -k$ and $\mu \to m$ (where m is the mass of the incident particle).[4] The solution for the equations of motion for $U(r)$ is therefore, given by

$$\frac{1}{r} = \sqrt{\frac{m^2 k^2}{l^4} + \frac{2 m E}{l^2}} \cos(\alpha - \alpha_0) - \frac{m k}{l^2}, \tag{7.41}$$

where, we have made the substitution $\phi \to \alpha$, in order to avoid any confusion with the ϕ angle from $d\Omega$. In order to simplify the notation we shall redefine $\alpha \to \alpha - \alpha_0$. Thus we finally obtain

$$\frac{1}{r} = \sqrt{\frac{m^2 k^2}{l^4} + \frac{2 m E}{l^2}} \cos \alpha - \frac{m k}{l^2}. \tag{7.42}$$

With the help of this equation the reader is invited to check that there is an asymptotic angle α_∞, equal in magnitude to $\alpha_{-\infty}$, corresponding to $r = \pm\infty$, as shown in Fig. 7.6 (where the black dot represents $r = 0$). From the same figure one can appreciate that

$$2\alpha_\infty + \theta = \pi \quad \Rightarrow \quad \alpha_\infty = \frac{\pi - \theta}{2}, \tag{7.43}$$

and so, we obtain

$$\cos \alpha_\infty = \sin \frac{\theta}{2}. \tag{7.44}$$

Finally, one can deduce as an exercise the following relations

---

[4]For Kepler's potential $k = Gm_1 m_2$ and μ is the reduced mass. In the present case we shall treat with a generic constant k and an incident particle of mass m.

Fig. 7.6 Schematic
representation of the
asymptotic angles and the s
parameter for a repulsive
potential

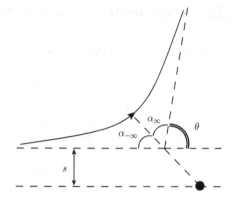

$$l = m v s = s\sqrt{2 m E} ,$$

$$s = \frac{k}{2E} \cot \frac{\theta}{2} , \tag{7.45}$$

where v is the total velocity of the particle (remember that $v \neq \dot{r}$), and the Rutherford
expression for the differential cross section

$$\frac{d\sigma}{d\Omega} = \frac{1}{4} \left(\frac{k}{2E} \right)^2 \frac{1}{\sin^4 \theta/2} . \tag{7.46}$$

7.4 Proposed Exercises

7.1 Show that in the CM reference frame the energy conservation theorem,
in the case of elastic collisions, implies that $p = p'$.

Solution: This exercise is very simple and it can be solved straightforwardly by
writing down explicitly the energy conservation theorem

$$
\begin{aligned}
K_1 + K_2 &= \frac{p_1^2}{2m_1^2} + \frac{p_2^2}{2m_2^2} = p^2 \left(\frac{1}{2m_1} + \frac{1}{2m_2} \right) \\
&= K_1' + K_2' = \frac{p_1'^{\,2}}{2m_1^2} + \frac{p_2'^{\,2}}{2m_2^2} \\
&= p'^2 \left(\frac{1}{2m_1} + \frac{1}{2m_2} \right) .
\end{aligned}
\tag{7.47}
$$

By comparing the first and the last line of the previous expression, one obtains the
desired result.

7.2 Show that the Lab. θ_1 angle is given by

$$\cos \theta_1 = \frac{\rho + \cos \theta}{\sqrt{1 + \rho^2 + 2\rho \cos \theta}}, \tag{7.48}$$

for elastic collisions.

Solution: Let us begin by re-introducing expression (7.16)

$$\mathbf{v}_1'^{L} = \boldsymbol{\mathcal{V}} + \mathbf{v}_1'. \tag{7.49}$$

By squaring the previous equation on both sides we obtain

$$(v_1'^{L})^2 = \mathcal{V}^2 + (v_1')^2 + 2\,\mathcal{V}\,v_1' \cos \theta. \tag{7.50}$$

Introducing (7.20) we obtain

$$(v_1'^{L})^2 = \rho^2 (v_1')^2 + (v_1')^2 + 2\,\rho\,(v_1')^2 \cos \theta, \tag{7.51}$$

whose solution is

$$v_1'^{L} = v_1' \sqrt{1 + \rho^2 + 2\rho \cos \theta}. \tag{7.52}$$

We have only considered the positive solution of the squared root because we are dealing with vector magnitudes, which are positive by definition. We can now introduce expression (7.17)

$$v_1'^{L} \cos \theta_1 = \mathcal{V} + v_1' \cos \theta. \tag{7.53}$$

By taking the quotient of the last two equations, and again making use of (7.20) we obtain the result we were looking for

$$\cos \theta_1 = \frac{\rho + \cos \theta}{\sqrt{1 + \rho^2 + 2\rho \cos \theta}}. \tag{7.54}$$

7.3 Using the expression

$$\tan \theta_1 = \frac{\sin \theta}{\rho + \cos \theta}, \tag{7.55}$$

valid for elastic collisions, check that

$$\tan \theta_1^{\max} = \frac{1}{\sqrt{\rho^2 - 1}},$$

(7.56)

if $\rho \geq 1$.

Solution: By differentiating $\tan \theta_1$ with respect to θ and setting the result to 0 we obtain the extrema of θ_1

$$\begin{aligned}
\frac{d \tan \theta_1}{d\theta} &= \frac{d}{d\theta} \left(\sin \theta (\rho + \cos \theta)^{-1} \right) \\
&= \cos \theta (\rho + \cos \theta)^{-1} + \sin^2 \theta (\rho + \cos \theta)^{-2} \\
&= \frac{\rho \cos \theta + 1}{(\rho + \cos \theta)^2} \\
&= 0.
\end{aligned}$$

(7.57)

Hence, $\cos \theta = -1/\rho$. From the previous expression we also conclude that the condition $\rho \geq 1$ must necessarily hold in order to have an extremum.

We have thus obtained that $\cos \theta = -1/\rho$ is an extremum if $\rho \geq 1$. We know that the minimum is 0 therefore it can only be a maximum or a saddle point. In our case (one can check by differentiating once again that) it is a maximum. Substituting this value into the expression of $\tan \theta_1$ we obtain

$$\tan \theta_1^{\max} = \frac{1}{\sqrt{\rho^2 - 1}}.$$

(7.58)

For $\rho \to 1$ we have $\cos \theta \to -1$ meaning that $\theta = \pi$, which corresponds to $\theta_1^{\max} = \pi/2$.

7.4 Check that for an elastic collision

$$\frac{K_1'^L}{K_1^L} = \frac{1 + 2\rho \cos \theta + \rho^2}{(1 + \rho)^2},$$

(7.59)

where

$$K_1'^L = \frac{(p_1'^L)^2}{2m_1}, \qquad K_1^L = \frac{(p_1^L)^2}{2m_1},$$

(7.60)

are the final and initial state kinetic energies of the incident particle in the Lab. frame.

Solution: First, we can simplify the quotient of the kinetic energies and obtain the following expression

$$\frac{K_1'^L}{K_1^L} = \frac{(v_1'^L)^2}{(v_1^L)^2}. \tag{7.61}$$

Let us now consider the relation between the magnitudes of the vectors $\mathbf{v}$ and $\mathbf{v}_1^L$ from (7.13). We obtain

$$\mathcal{V} = \frac{m_1}{m_1 + m_2} v_1^L, \tag{7.62}$$

or equivalently

$$v_1^L = \frac{m_1 + m_2}{m_1} \mathcal{V} = (1 + \rho^{-1}) \mathcal{V}. \tag{7.63}$$

By dividing the expression (7.51) by $(v_1^L)^2$ (with v_1^L given by the previous equation) we obtain

$$\begin{aligned}
\frac{(v_1'^L)^2}{(v_1^L)^2} &= \frac{(v_1')^2(1 + \rho^2 + 2\rho\cos\theta)}{(1 + \rho^{-1})^2 \mathcal{V}^2} \\
&= \frac{(v_1')^2(1 + \rho^2 + 2\rho\cos\theta)}{(1 + \rho^{-1})^2 \rho^2 (v_1')^2} \\
&= \frac{(1 + \rho^2 + 2\rho\cos\theta)}{(1 + \rho)^2},
\end{aligned} \tag{7.64}$$

which is the desired expression.

7.5 Consider the semicircular section of a spherical well, as shown in Fig. 7.7. A mass m_1 is released from a height h with respect to the bottom of the well. If the collision is totally inelastic, what are the velocities of both masses at the instants right before and right after the collision, and what is the maximum height they will both reach? If, on the contrary, the collision is totally elastic, what value should m_1 have (compared to m_2) in order for m_2 to reach a maximum height $2h$ after the collision? Consider no friction forces.

Solution: Right before the collision $v_2 = 0$, as no force is acting upon it besides gravity. The mass m_1 is released from a height h and, just before the collision its whole potential energy is transformed into kinetic energy. Thus

Fig. 7.7 Two-particle
collision in a spherical well

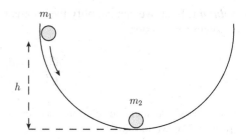

$$m_1 \, g \, h \;=\; \frac{1}{2} m_1 \, v_1^2 \quad \Rightarrow \quad v_1 = \sqrt{2 \, g \, h}\,. \qquad (7.65)$$

Right after the collision, both masses stick together and the momentum is conserved

$$m_1 v_1 \;=\; (m_1 + m_2) \, v\,, \qquad (7.66)$$

where v is the velocity of the system formed by the two masses $(m_1 + m_2)$. Therefore

$$v \;=\; \frac{m_1}{m_1 + m_2} \sqrt{2 \, g \, h}\,. \qquad (7.67)$$

The maximum height h' of the system $(m_1 + m_2)$ is reached when all the kinetic energy is transformed into potential energy

$$\frac{1}{2}(m_1 + m_2) v^2 \;=\; (m_1 + m_2) \, g \, h' \quad \Rightarrow \quad h' = \frac{m_1^2}{(m_1 + m_2)^2} \, h\,. \qquad (7.68)$$

As for the last question, we simply need to apply momentum and kinetic energy conservation. Momentum conservation (right before and right after the collision) reads

$$m_1 \, v_1 \;=\; -m_1 v_1' + m_2 \, v_2'\,, \qquad (7.69)$$

with $v_1, v_1', v_2' > 0$. As mentioned, m_2 has to reach $2h$, therefore we can re-express the previous equation in terms of h by using the potential energy. We obtain

$$m_1 \sqrt{2 \, g \, h} \;=\; -m_1 \, v_1' + m_2 \sqrt{4 \, g \, h}\,, \qquad (7.70)$$

and we can therefore express v_1' as

$$v_1' \;=\; -\sqrt{2 \, g \, h} + \frac{m_2}{m_1} \sqrt{4 \, g \, h}\,. \qquad (7.71)$$

The kinetic energy conservation can also be written in terms of h

$$\frac{1}{2} m_1 (2gh) = \frac{1}{2} m_1 (v_1')^2 + \frac{1}{2} m_2 (4gh). \tag{7.72}$$

Inserting the expression (7.71) into the previous one, we finally obtain

$$m_2 = m_1(\sqrt{2} - 1). \tag{7.73}$$

7.6 Consider an elastic collision in the Lab. frame. Knowing the values of $\theta_1 = 50°$, $v_1^L = 2$ m/s, $m_1 = 5$ kg and $m_2 = 8$ kg, obtain the magnitudes of the final state velocities for both final state particles ($v_1'^L$ and $v_2'^L$) and the angle θ_2.

Solution: For simplicity (in this exercise) we shall suppress the L upper index. Consider that the collision takes place in the (x, y) plane,[5] as shown in Fig. 7.8. The initial state is thus given by

$$\begin{aligned}
\mathbf{p}_1 &= m_1 (v_1, 0) = p_1 (1, 0), \\
\mathbf{p}_2 &= m_2 (0, 0) = (0, 0), \\
K &= \frac{1}{2} m_1 v_1^2 = \frac{p_1^2}{2m_1}.
\end{aligned} \tag{7.74}$$

As for the final state momenta we have

$$\begin{aligned}
\mathbf{p}_1' &= m_1 (v_1' \cos \theta_1, v_1' \sin \theta_1) = p_1' (\cos \theta_1, \sin \theta_1), \\
\mathbf{p}_2' &= m_2 (v_2' \cos \theta_2, v_2' \sin \theta_2) = p_2' (\cos \theta_2, \sin \theta_2),
\end{aligned} \tag{7.75}$$

with $\theta_1 > 0$ and $\theta_2 < 0$. The final total kinetic energy simply reads

$$K' = \frac{1}{2} m_1 (v_1')^2 + \frac{1}{2} m_2 (v_2')^2 = \frac{(p_1')^2}{2m_1} + \frac{(p_2')^2}{2m_2}. \tag{7.76}$$

From momentum and kinetic energy conservation we obtain the relations

$$\begin{aligned}
p_1 &= p_1' \cos \theta_1 + p_2' \cos \theta_2, \\
0 &= p_1' \sin \theta_1 + p_2' \sin \theta_2, \\
(p_1)^2 &= (p_1')^2 + \rho (p_2')^2,
\end{aligned} \tag{7.77}$$

---

[5]Note that this choice can always be made because the Lab. (or the CM) reference frame are not uniquely defined. They are defined up to rotations about the collision axis. Therefore we can always chose a reference frame where the collision plane is the (x, y) plane for example.

Fig. 7.8 Two-particle
collision in the Lab.
reference frame, where the
collision plane is the (x, y)
plane

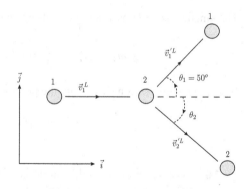

where, as usual $\rho = m_1/m_2$. Re-expressing the first two equations from the previous system as

$$p_1 - p_1' \cos \theta_1 = p_2' \cos \theta_2 \,,$$
$$-p_1' \sin \theta_1 = p_2' \sin \theta_2 \,, \qquad (7.78)$$

summing and squaring both expressions and simplifying, we obtain

$$(p_2')^2 = p_1^2 + (p_1')^2 - 2 p_1 p_1' \cos \theta_1 \,. \qquad (7.79)$$

Using the third expression, $(p_2')^2$ is given by

$$(p_2')^2 = \frac{1}{\rho}\left(p_1^2 - (p_1')^2\right). \qquad (7.80)$$

Combining (7.79) and (7.80) and simplifying we obtain

$$(p_1')^2(1 + \rho^{-1}) - 2 p_1 p_1' \cos \theta + p_1^2(1 - \rho^{-1}) = 0 \,. \qquad (7.81)$$

Solving the previous equation for p_1' and using the given numerical data we obtain the two solutions

$$p_1' = 7, 87 \text{ and } - 2.93 \ (\text{kg m/s}^2) \,. \qquad (7.82)$$

Discarding the negative value (as it makes no sense in this case) we obtain

$$v_1' = 1, 57 \, \text{m/s} \,, \qquad v_2' = 0.97 \, \text{m/s} \,, \qquad \theta_2 = 50.7° \,. \qquad (7.83)$$

7.7 A particle of mass m and velocity $\mathbf{v}$ explodes (due to internal forces) into two particles of masses $m_1 = 2m/3$ and $m_2 = m/3$. If the velocity of m_1 is orthogonal to $\mathbf{v}$, as shown in Fig. 7.9, calculate $\mathbf{v}_2$, ϕ and Q as functions of $v = |\mathbf{v}|$, $v_1 = |\mathbf{v}_1|$ and the masses.

Fig. 7.9 A particle of mass m and velocity **v**, that explodes into two fragments of masses $2m/3$ and $m/3$

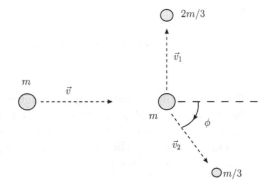

Solution: As in the previous exercise, without loss of generality we will consider that the reaction takes place in the (x, y) plane. The initial state configuration is given by

$$\mathbf{p} = m\,\mathbf{v} = m\,v\,(1, 0)\,,$$
$$K = \frac{1}{2}m\,v^2\,. \tag{7.84}$$

After the explosion the momenta and kinetic energy read (with $\phi < 0$)

$$\mathbf{p_1} = m_1\,\mathbf{v_1} = m_1\,v_1\,(0, 1)\,,$$
$$\mathbf{p_2} = m_2\,\mathbf{v_2} = m_2\,v_2\,(\cos\phi, \sin\phi) = m_2\,(v_{2,x}, v_{2,y})\,, \tag{7.85}$$
$$K' = \frac{1}{2}m_1\,v_1^2 + \frac{1}{2}m_2(v_{2,x}^2 + v_{2,y}^2)\,.$$

From momentum conservation we obtain

$$m\,v = m_2\,v_{2,x} \quad \Rightarrow \quad v_{2,x} = \frac{m}{m_2}v = 3v\,, \tag{7.86}$$

and

$$0 = m_1\,v_1 + m_2\,v_{2,y} \quad \Rightarrow \quad v_{2,y} = -\frac{m_1}{m_2}v_1 = -2v_1\,. \tag{7.87}$$

The angle ϕ will thus given by

$$\tan\phi = -2v_1/3v\,. \tag{7.88}$$

The final state kinetic energy can therefore be written, as a function of v and v_1, as

$$K' = m\,v_1^2 + \frac{3}{2}m\,v^2\,, \tag{7.89}$$

and so

$$Q = K - K' = -m\,(v_1^2 + v^2) \;<\; 0. \qquad (7.90)$$

7.8 Consider the system of simple pendulums shown in Fig. 7.10. If at some instant m_1 is released and collides frontally with m_2, calculate the height that each mass will reach after the collision (h_1 and h_2) as functions of m_1, m_2 and the restitution coefficient e (with $0 \le e \le 1$). If $m_1 = 5$ kg, $m_2 = 2$ kg, calculate h_1 and h_2 for $e = 0,\ 0.5,\ 1$.

Solution: From the energy conservation theorem, right before the collision

$$m_1\,g\,d \;=\; \frac{1}{2} m_1\,v_1^2 , \qquad (7.91)$$

and so, v_1 is simply given by

$$v_1 \;=\; \sqrt{2\,g\,d} . \qquad (7.92)$$

Right after the collision momentum conservation reads

$$m_1\,v_1 \;=\; -m_1\,v_1' + m_2\,v_2' , \qquad (7.93)$$

with $v_1, v_1', v_2' > 0$. Note that we cannot use the energy conservation theorem for the kinetic energy right before and right after the collision because we have assumed a generic collision, which can be inelastic. Let us now introduce the restitution coefficient

$$e \;=\; \frac{|\mathbf{v}_2' - \mathbf{v}_1'|}{|\mathbf{v}_2 - \mathbf{v}_1|} , \qquad (7.94)$$

which, in our case translates into

$$e \;=\; \frac{v_2' + v_1'}{v_1} , \qquad (7.95)$$

where the relative positive sign in between v_2' and v_1' in the numerator, is due to the fact that right after the collision, we have $\mathbf{v}_1' = -v_1'\,\mathbf{i}$ (with $\mathbf{i}$, the horizontal axis). We thus obtain

$$v_1' \;=\; e\,v_1 - v_2' . \qquad (7.96)$$

Fig. 7.10 A system formed
by a pendulum of mass m_1
that collides frontally with a
pendulum of mass m_2,
initially at rest

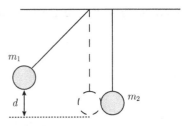

Introducing this expression into (7.93) we obtain

$$v_2' = \frac{v_1(1+e)}{1 + m_2/m_1}.$$ (7.97)

Using energy conservation (the kinetic energy right after the collision of the second
particle that transforms into potential energy) we can express the velocity v_2' in terms
of the height h_2

$$v_2' = \sqrt{2g h_2},$$ (7.98)

where h_2 is the final (maximum) height of the particle of mass m_2 after the collision.
By using (7.92) we finally obtain

$$h_2 = \frac{d\, m_1^2(1+e)^2}{(m_1+m_2)^2}.$$ (7.99)

Similarly we obtain

$$h_1 = \frac{d(m_1 - e\, m_2)^2}{(m_1+m_2)^2}.$$ (7.100)

Note that if $e = 0$ (totally inelastic collision) we obtain $h_1 = h_2$ which is the correct
result as both masses stick together after the collision, and reach the same final height
together. The last task that corresponds to the numerical analysis is left for the reader.

7.9 With the help of Fig. 7.5 show that for a hard sphere

$$s = a \cos\left(\frac{\theta}{2}\right) \quad \text{and} \quad \frac{d\sigma}{d\Omega} = \frac{a^2}{4},$$ (7.101)

where a is the radius of the sphere. Calculate the total cross section (integrating
for all solid angles). Calculate also, the cross section for the region $0 < \theta < \pi/6$.

Solution: From the previously mentioned figure one can conclude that

$$\theta = \pi - 2\alpha = \pi - 2\arcsin\frac{s}{a} \quad \Rightarrow \quad \arcsin\frac{s}{a} = \frac{\pi - \theta}{2} \tag{7.102}$$

and so

$$s = a\sin\frac{\pi - \theta}{2} = a\cos\frac{\theta}{2}. \tag{7.103}$$

We can now calculate the differential cross section. Using

$$2\sin x\cos x = \sin 2x, \tag{7.104}$$

we obtain

$$\frac{d\sigma}{d\Omega} = \frac{s}{\sin\theta}\left|\frac{ds}{d\theta}\right| = \frac{a^2\cos(\theta/2)\sin(\theta/2)}{2\sin\theta} = \frac{a^2}{4}. \tag{7.105}$$

The total cross section (integrating for all solid angles) is simply given by

$$\sigma = \pi a^2. \tag{7.106}$$

For the restricted region $0 < \theta < \pi/6$ we obtain

$$\begin{aligned}
\sigma &= \frac{a^2}{4}\int_0^{2\pi} d\phi \int_0^{\pi/6}\sin\theta d\theta \\
&= \frac{a^2}{4}2\pi\int_{\sqrt{3}/2}^{1} d\cos\theta \\
&= \frac{a^2\pi}{2}(1 - \sqrt{3}/2).
\end{aligned} \tag{7.107}$$

7.10 Using the equation of motion for a repulsive Coulomb-type potential energy $U(r) = k/r$ with $(k > 0)$

$$\frac{1}{r} = \sqrt{\frac{m^2k^2}{l^4} + \frac{2mE}{l^2}}\cos\alpha - \frac{mk}{l^2}, \tag{7.108}$$

obtain the asymptotic angle α_∞. Also obtain the following relations

$$l = mvs = s\sqrt{2mE},$$

$$s = \frac{k}{2E}\cot\frac{\theta}{2}, \tag{7.109}$$

(where v is the incident velocity) and the Rutherford expression for the differential cross section

$$\frac{d\sigma}{d\Omega} = \frac{1}{4}\left(\frac{k}{2E}\right)^2 \frac{1}{\sin^4 \theta/2} \, . \tag{7.110}$$

Calculate the pericenter.

Solution: From the expression of $1/r$, it should result obvious that there is a limiting angle $\alpha_{\pm\infty}$ with $|\alpha_{-\infty}| = |\alpha_\infty|$ and it is given by $r \to \pm\infty$

$$\sqrt{\frac{m^2k^2}{l^4} + \frac{2m E}{l^2}}\cos\alpha_\infty - \frac{m k}{l^2} = 0 \, , \tag{7.111}$$

which translates into

$$\cos\alpha_\infty = \frac{1}{\sqrt{1 + \frac{2El^2}{mk^2}}} \, . \tag{7.112}$$

The total energy at $r \to \pm\infty$ is given by the kinetic energy

$$E = \frac{1}{2}m v^2 \, , \tag{7.113}$$

where v is the total velocity of the incident particle (without making the separation into the radial and angular part). Therefore

$$v = \sqrt{\frac{2E}{m}} \, . \tag{7.114}$$

The angular momentum at $r \to \pm\infty$, considering (without loss of generality) the (x, y) plane as the plane of motion

$$\mathbf{l} = m(\pm\infty\mathbf{i} + s\mathbf{j}) \times v\mathbf{i} \, , \tag{7.115}$$

and so

$$l = m v s = s\sqrt{2m E} \, . \tag{7.116}$$

Using the previously deduced relations we obtain

$$\frac{1}{\cos\alpha_\infty} = \sqrt{1 + \left(\frac{2E}{k}\right)^2 s^2} \, , \tag{7.117}$$

which can be re-expressed as

$$s = \frac{k}{2E} \tan \alpha_\infty = \frac{k}{2E} \cot \frac{\theta}{2}, \qquad (7.118)$$

where we have used (7.44). Using the previous expression we can calculate the differential cross section

$$\begin{aligned} \frac{d\sigma}{d\Omega} &= \frac{s}{\sin\theta} \left| \frac{ds}{d\theta} \right| \\ &= \frac{k}{2E} \frac{\cot(\theta/2)}{\sin\theta} \left| \frac{ds}{d\theta} \right| \\ &= \left(\frac{k}{2E}\right)^2 \frac{\cot(\theta/2)}{\sin\theta} \frac{1}{2\sin^2(\theta/2)} \\ &= \frac{1}{4}\left(\frac{k}{2E}\right)^2 \frac{1}{\sin^4(\theta/2)} . \qquad (7.119) \end{aligned}$$

Finally, using the expression for the effective potential we can calculate the pericenter

$$E = U_{eff}(r_P) = \frac{l^2}{2mr_P} + \frac{k}{r_P}, \qquad (7.120)$$

or equivalently by using the solution for the equations of motion given by the expression of $1/r$.

Further Reading

1. J.V. José, E.J. Saletan, *Classical Dynamics: A Contemporary Approach*. Cambridge University Press
2. S.T. Thornton, J.B. Marion, *Classical Dynamics of Particles and Systems*
3. H. Goldstein, C. Poole, J. Safko, *Classical Mechanics*, 3rd edn. Addison Wesley
4. J.R. Taylor, *Classical Mechanics*
5. D.T. Greenwood, *Classical Dynamics*. Prentice-Hall Inc.
6. D. Kleppner, R. Kolenkow, *An Introduction to Mechanics*
7. C. Lanczos, *The Variational Principles of Mechanics*. Dover Publications Inc.
8. W. Greiner, *Classical Mechanics: Systems of Particles and Hamiltonian Dynamics*. Springer
9. H.C. Corben, P. Stehle, *Classical Mechanics*, 2nd edn. Dover Publications Inc.
10. T.W.B. Kibble, F.H. Berkshire, *Classical Mechanics*. Imperial College Press
11. M.G. Calkin, *Lagrangian and Hamiltonian Mechanics*
12. A.J. French, M.G. Ebison, *Introduction to Classical Mechanics*

Chapter 8
Continuous Mass Distributions.
Gravitational Potential and Field

Abstract We shall dedicate this chapter to the study of the gravitational potential
and field, originated by continuous mass distributions. We will apply the obtained
results to different geometries and mass density distributions. Special emphasis will
made on potentials with spherical symmetry. Other additional exercises regarding
satellites, stationary orbits and escape velocities are also proposed.

8.1 Introduction

Consider a system of N point-like particles with coordinates $\mathbf{r}_i$ in a given reference
frame, as shown in Fig. 8.1 (top). Consider a point in space P with coordinates given
by $\mathbf{r}$ in the same reference frame. The gravitational potential at P, generated by the
presence of the masses m_i, will be given by the sum of all individual contributions
of each mass

$$\phi(\mathbf{r}) = -G \sum_i \frac{m_i}{|\mathbf{r} - \mathbf{r}_i|}. \tag{8.1}$$

Let us now generalize the previous expression for the continuum case. Consider
that the distances among a given mass and the surrounding particles is infinitesimal.
Then, in order to describe the position of a differential mass element dm, we can no
longer use the $\mathbf{r}_i$ coordinates. We will need a continuous function which we shall call
$\mathbf{r}'$. This is shown in Fig. 8.1 (bottom). In this case, expressing dm as a function of
the mass density ρ and the differential volume one obtains the following expression

$$\phi(\mathbf{r}) = -G \iiint_V dV' \frac{\rho(\mathbf{r}')}{|\mathbf{r} - \mathbf{r}'|}. \tag{8.2}$$

Note that we have explicitly introduced the *primed* notation for the integration variable $\mathbf{r}'$, in order to distinguish it from $\mathbf{r}$, which is the position where the gravitational
potential is *measured*. The previous expression looks rather simple, however, analytical solutions are only known for highly symmetrical systems. Here we shall analyse
a few examples.

© Springer Nature Switzerland AG 2020

V. Ilisie, *Lectures in Classical Mechanics*, Undergraduate Lecture
Notes in Physics, https://doi.org/10.1007/978-3-030-38585-9_8

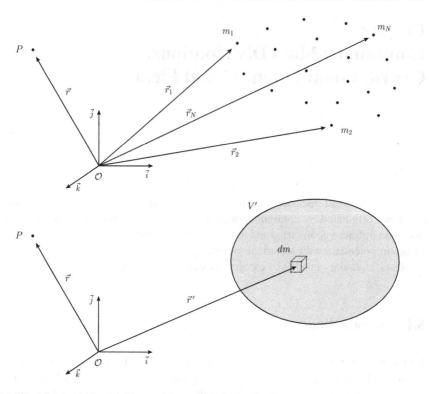

Fig. 8.1 System of N point-like particles, and the point P where the gravitational potential is *measured* (top) together with continuum generalization $N \to \infty$ (bottom)

8.2   Potentials with Spherical Symmetry

For this analysis let us consider that our volume is a sphere and place the $\{\mathcal{O}, \mathbf{i}, \mathbf{j}, \mathbf{k}\}$ reference system at the center of the sphere. We shall also consider that the mass density distribution is a function that only depends on the radial distance r i.e., $\rho(\mathbf{r}') = \rho(r')$ (where $r' = |\mathbf{r}'|$). Without further complications, one can demonstrate that in this particular case, the gravitational potential will be of the form $\phi(\mathbf{r}) = \phi(r)$ (where $r = |\mathbf{r}|$). Thus, due to the spherical symmetry of our configuration, for the calculation of the potential, P can be placed along the $\mathbf{k}$ axis i.e., with coordinates $\mathbf{r} = r\mathbf{k}$, as shown Fig. 8.2. The final result will be valid for any point placed at this given distance.

Writing $\mathbf{r}'$ in spherical coordinates, one obtains a simple expression for $|\mathbf{r} - \mathbf{r}'|$

$$
\begin{aligned}
|\mathbf{r} - \mathbf{r}'| &= \left(r'^2 \sin^2 \theta \cos^2 \phi + r'^2 \sin^2 \theta \sin^2 \phi + (r - r' \cos \theta)^2\right)^{1/2} \\
&= (r^2 + r'^2 - 2rr' \cos \theta)^{1/2} .
\end{aligned}
\tag{8.3}
$$

Fig. 8.2 Spherical mass distribution with the reference system placed at the center of the sphere. The point P is placed along the **k** axis

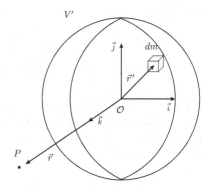

As there is no possible confusion, in order to maintain a simple notation, we have not used the primed notation for the angles. A differential volume element in spherical coordinates (as seen in Chap. 2) is given by the expression

$$dV' = r'^2\, dr'\, \sin\theta\, d\theta\, d\phi \equiv r'^2\, dr'\, d\Omega\,, \tag{8.4}$$

where we have introduced the differential solid angle

$$d\Omega = \sin\theta\, d\theta\, d\phi\,. \tag{8.5}$$

As there is no ϕ-dependence within the integrand (8.2) in our particular case, one can integrate over this variable (and obtain 2π). Performing the usual change of variable

$$\int_0^\pi \sin\theta\, d\theta = \int_{-1}^1 d\cos\theta\,, \tag{8.6}$$

will further simplify our integration. One obtains the gravitational potential in spherical coordinates, given by the expression

$$\phi(r) = -2\pi G \int_0^R dr' \int_{-1}^1 d\cos\theta\, r'^2\, \rho(r')\, (r^2 + r'^2 - 2rr' \cos\theta)^{-1/2}\,, \tag{8.7}$$

where R is the radius of the sphere. Let us now perform the integration over the variable θ

$$
\begin{aligned}
I &\equiv \int_{-1}^1 d\cos\theta (r^2 + r'^2 - 2rr' \cos\theta)^{-1/2} \\
&= -\frac{(r^2 + r'^2 - 2rr' \cos\theta)^{1/2}}{rr'} \bigg|_{\cos\theta=-1}^{\cos\theta=1} \\
&= \frac{1}{rr'} \left(-\sqrt{(r - r')^2} + \sqrt{(r + r')^2} \right).
\end{aligned}
\tag{8.8}
$$

We can distinguish two cases:

- $r > r' \quad \Rightarrow \quad \sqrt{(r - r')^2} = r - r' \quad \Rightarrow \quad I = 2/r,$ (8.9)
- $r < r' \quad \Rightarrow \quad \sqrt{(r - r')^2} = r' - r \quad \Rightarrow \quad I = 2/r',$ (8.10)

where we have made use of $\sqrt{x^2} = |x|$. In the case $r = r'$ both expressions are identical. Introducing the Heaviside theta function ($\theta(x) = 1$ for $x > 0$ and $\theta(x) = 0$ for $x < 0$; for $x = 0$ it is usually defined as $\theta(0) = 1/2$) we can compactly write down the previous result

$$I = \frac{2}{r} \theta(r - r') + \frac{2}{r'} \theta(r' - r).$$ (8.11)

We finally obtain the expression for the gravitational potential for a spherical mass distribution in compact form

$$\phi(r) = -4\pi G \int_0^R dr' \, \rho(r') \left(\frac{r'^2}{r} \theta(r - r') + r' \theta(r' - r) \right).$$ (8.12)

We can observe that, the expression for the potential outside the sphere will only receive one contribution, meanwhile if the point P is placed inside the sphere, both terms will contribute to the integral. This will be better understood later on, in the Proposed Exercises section.

8.3   Gravitational Field and Gauss's Law

The gravitational field of a finite mass distribution, just as for a point-like particle, can be obtained by simply taking the gradient

$$\mathbf{g}(\mathbf{r}) = -\nabla \phi(\mathbf{r}).$$ (8.13)

Using the expression (8.1) and

$$\nabla \left(\frac{1}{|\mathbf{r} - \mathbf{r}'|} \right) = -\frac{\mathbf{r} - \mathbf{r}'}{|\mathbf{r} - \mathbf{r}'|^3},$$ (8.14)

as in shown in (4.16), we obtain the expression for the gravitational field $\mathbf{g}$

$$\mathbf{g}(\mathbf{r}) = -G \sum_i \frac{m_i}{|\mathbf{r} - \mathbf{r}_i|^3} (\mathbf{r} - \mathbf{r}_i).$$ (8.15)

The generalization in the continuum limit is straightforward

$$\mathbf{g}(\mathbf{r}) = -G \iiint_V dV' \frac{\rho(\mathbf{r}')}{|\mathbf{r} - \mathbf{r}'|^3} (\mathbf{r} - \mathbf{r}') . \qquad (8.16)$$

In the following we shall deduce Gauss's Law for the gravitational field. Consider the following unitary vector

$$\mathbf{u}_{r_i} \equiv \frac{\mathbf{r} - \mathbf{r}_i}{|\mathbf{r} - \mathbf{r}_i|} . \qquad (8.17)$$

Thus, the gravitational field of a point-like particle with mass m_i can be expressed as

$$\mathbf{g}(\mathbf{r}) = -G \frac{m_i}{|\mathbf{r} - \mathbf{r}_i|^2} \mathbf{u}_{r_i} . \qquad (8.18)$$

Let us now consider an arbitrary convex closed surface S. The expression for $d\mathbf{s}$ in spherical coordinates is given by (2.44)

$$d\mathbf{s} = r^2 \sin\theta \, d\theta \, d\phi \, \mathbf{u}_r + \cdots , \qquad (8.19)$$

where the dots stand for the rest of components, that are of no relevance, as they will not contribute to our case. If we place the coordinate origin at $\mathbf{r}_i$, with respect to which we calculate $d\mathbf{s}$, then we obtain the following expression

$$d\mathbf{s} = |\mathbf{r} - \mathbf{r}_i|^2 \, d\Omega \, \mathbf{u}_{r_i} + \cdots = ds_{\theta\phi} \, \mathbf{u}_{r_i} + \cdots . \qquad (8.20)$$

Let us now take a look at the following integral in spherical coordinates:

$$\oiint_S \mathbf{g} \cdot d\mathbf{s} \qquad (8.21)$$

If the mass is placed inside or on the surface then

$$\begin{aligned}
\oiint_S \mathbf{g} \cdot d\mathbf{s} &= -G \oiint_S \left(\frac{m_i}{|\mathbf{r} - \mathbf{r}_i|^2} \mathbf{u}_{r_i} \right) \left(|\mathbf{r} - \mathbf{r}_i|^2 \, d\Omega \, \mathbf{u}_{r_i} + \cdots \right) \\
&= -G \, m_i \int_\Omega d\Omega \\
&= -4\pi G m_i . \qquad (8.22)
\end{aligned}$$

We observe that the result does not depend on the relative distance $(\mathbf{r} - \mathbf{r}_i)$. If, on the other hand, the mass is placed outside the surface then we have two contributions to the surface integral, as schematically shown in Fig. 8.3

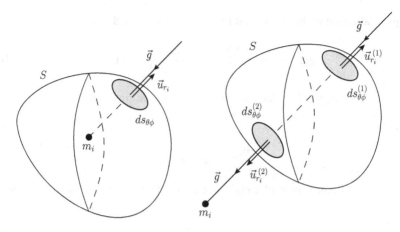

Fig. 8.3 Schematic representation of a closed convex surface S together with the field's direction and sense for an enclosed mass (left), and for a mass placed outside the surface (right). The components $ds_{\theta\phi}, ds_{\theta\phi}^{(1,2)}$ and their normal vectors $\mathbf{u}_{r_i}, \mathbf{u}_{r_i}^{(1,2)}$ are also shown

$$\oiint_S \mathbf{g} \cdot (d\mathbf{s}^{(1)} + d\mathbf{s}^{(2)}) = -G\,m_i \int_\Omega d\Omega \left(\mathbf{u}_{r_i} \cdot (\mathbf{u}_{r_i}^{(1)} + \mathbf{u}_{r_i}^{(2)})\right)$$

$$= -G\,m_i \int_\Omega d\Omega\,(1-1) = 0 . \qquad (8.23)$$

We can see that, due to the fact that $\mathbf{u}_{r_i}^{(1)}$ and $\mathbf{u}_{r_i}^{(2)}$ have opposite senses, the two contributions cancel each other.

One should note that, for the case of the mass placed inside the surface, we have not made any assumption on the exact position of the mass inside the surface. By placing another mass m_j somewhere inside the surface S, we obtain another contribution $-4\pi G m_j$ to the integral. Thus the following result is completely generic

$$\oiint_S \mathbf{g} \cdot d\mathbf{s} = -4\pi G \sum_i m_i , \qquad (8.24)$$

where $\sum_i m_i$ is the sum of all the masses enclosed by, or on the surface S. This is called Gauss's Law. Its generalization to the continuum case is straightforward

$$\oiint_S \mathbf{g} \cdot d\mathbf{s} = -4\pi G \iiint_V \rho(\mathbf{r}')dV' = -4\pi G \int_V dm = -4\pi G M_S , \qquad (8.25)$$

where V is the volume enclosed by the surface S, and where we have defined M_S as the mass enclosed by the surface S. Note that the beauty of this law consists in the fact that we have made no assumption on the shape of the surface[1] or the mass distribution in its interior. The law states that the flux of the gravitational field through an arbitrary convex surface is proportional to the mass enclosed within the surface.

Using Gauss's theorem we can additionally obtain Gauss's law in differential form

$$-4\pi G \iiint_V \rho(\mathbf{r}')dV' = \iiint_V \boldsymbol{\nabla} \cdot \mathbf{g}\, dV' . \tag{8.26}$$

Considering a differential volume we get

$$\boldsymbol{\nabla} \cdot \mathbf{g} = -4\pi G\rho . \tag{8.27}$$

However, we shall not make use of this result in our calculations.

8.4 Gauss's Law for Spherical Mass Distributions

Let us now turn back to the expression (8.25). We can use it to calculate $\mathbf{g}$ for the simple case of a spherical mass distribution (and we shall also use it for some other cases in the Proposed Exercises section). In our particular case (spherical distribution) we can express $\mathbf{g}$ as

$$\mathbf{g}(\mathbf{r}) = g(r)\,\mathbf{u}_r . \tag{8.28}$$

Therefore, if we consider a spherical surface with its center placed at the coordinate origin, then

$$\oiint_S \mathbf{g} \cdot d\mathbf{s} = 4\pi r^2 g(r) , \tag{8.29}$$

where r is the radius of the sphere. By comparing this result to (8.25), we obtain the expression for $g(r)$ for a spherical mass distribution

$$g(r) = -\frac{G\,M_r}{r^2} , \tag{8.30}$$

where M_r is the mass enclosed by the sphere of radius r. This simple configuration, as we shall see in the following, will help us solve a great deal of interesting cases and get an good approximate idea about the gravitational potential and field generated by planets.

---

[1] We have only supposed that is convex (any straight line passes through the surface mostly twice) and differentiable.

Let us now address another question that turns out to be crucial. Under what circumstances the gravitational field will have the form $\mathbf{g}(\mathbf{r}) = g(r)\,\mathbf{u}_r$. We know that the most generic expression for $\mathbf{g}$ is given by (8.16). Thus the answer to this question depends on the function $\rho(\mathbf{r}')$. Recall the expression (8.12). From this result we can conclude that *if $\rho(\mathbf{r}')$ does not have any angular dependence i.e., is of the form $\rho(r')$ then the field (and also the potential) will be central.* We shall check this result explicitly with a few proposed exercises. The implications of this statement are pretty strong. It means that if the Earth, or any other planet in general, has a mass distribution disposed in spherical shells, which each shell having a uniform density, the gravitational field outside the surface will be a central field. For the Earth (and in many other cases) this is approximately true, and as consequence, the gravitational field is approximately constant everywhere on the surface of the Earth.

8.5 Proposed Exercises

8.1 Calculate the mass of a sphere of radius R for the following cases: ρ constant, $\rho(r') = k/r'$, $\rho(r') = kr'$, and for a star with a density $\rho(r') = ke^{-r'/r_0}$, with k and r_0 positive constants.

Solution: In order to obtain the mass, the integral we need to solve in all cases is

$$M = \iiint_V \rho(r')\,dV' = \int_\Omega d\Omega \int_0^R r'^2 \rho(r')\,dr'. \tag{8.31}$$

As ρ has no angular dependence we can integrate the solid angle, and obtain

$$M = 4\pi \int_0^R r'^2 \rho(r')\,dr'. \tag{8.32}$$

For the first configuration (constant ρ) we simply obtain

$$M_{(1)} = \frac{4}{3}\pi R^3 \rho. \tag{8.33}$$

For the second case, the mass will be given by

$$M_{(2)} = 4\pi \int_0^R r'^2 \left(\frac{k}{r'}\right) dr' = 2\pi k R^2. \tag{8.34}$$

For the third case we have

$$M_{(3)} = 4\pi \int_0^R r'^2 (kr')\,dr' = \pi k R^4. \tag{8.35}$$

As for the last mass distribution, we obtain

$$
M_{(4)} = 4\pi \int_0^R r'^2 (ke^{-r'/r_0}) \, dr',
$$

$$
= 4\pi k \left(2r_0^3 - r_0 e^{-\frac{R}{r_0}} \left(R^2 + 2Rr_0 + 2r_0^2 \right) \right). \tag{8.36}
$$

In the following exercises we shall analyse the gravitational potential and field for some of the previous mass distributions.

8.2 Consider a spherical mass distribution with a constant mass density $\rho(r') = \rho$. Calculate the potential $\phi(r)$ for an external point $r > R$ and for a point inside the sphere $r < R$. Check that $\phi(r)$ is continuous at $r = R$. Obtain the potential at $r = 0$. Calculate the gravitational field for $r > R$ and $r < R$ and at $r = 0$, first by directly taking the gradient $\nabla \phi$ and second, by using Gauss's law. Check the continuity of $\mathbf{g}(r)$ at $r = R$.

Solution: The gravitational potential (8.12) for a constant mass density reads

$$
\phi(r) = -4\pi G\rho \int_0^R dr' \left(\frac{r'^2}{r} \theta(r - r') + r' \, \theta(r' - r) \right). \tag{8.37}
$$

For $r > R$, we only have one contribution (as $r' \leq R$, $\theta(r - r') = 1$). Thus, the previous expression simplifies to

$$
\phi(r > R) = -4\pi G\rho \int_0^R dr' \left(\frac{r'^2}{r} \right) = -\frac{4\pi G\rho R^3}{3r}. \tag{8.38}
$$

The mass of a uniform sphere (with constant mass density) is $M = \rho 4\pi R^3/3$, thus $\phi(r > R)$ expressed in terms of the mass M takes the form

$$
\phi(r > R) = -\frac{GM}{r}. \tag{8.39}
$$

From the previous expression we can conclude that the potential at a point outside a uniform sphere (with a constant mass density) is identical to the potential corresponding to a point-like particle with the same mass placed at the centre of the sphere. Later on, we shall explicitly check that this is true even for $\rho \neq$ constant. This is an interesting result with strong implications that brings enormous simplifications to many problems. As already seen in Chap. 6, we can study planetary motion by considering planets as point-like particles with all its mass concentrated in its center of mass.

Fig. 8.4 A spherical mass distribution of radius R. For a point inside the sphere at a distance r from the center, the two regions ($r' > r$ and $r < r$) that contribute to the integral are schematically shown

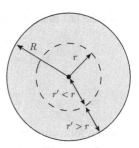

Returning to our exercise, could we now directly take the limit $r \to 0$ in order to obtain the gravitational field at the center of the sphere? The answer is no, as the previous expression is only valid outside the sphere (for $r > R$). In order to properly obtain the potential at $r = 0$ we first have to obtain the expression for $\phi(r < R)$. In this case, we have two contributions to the integral and the expression (8.37) can be written as

$$\phi(r < R) = -4\pi G\rho \left(\int_0^r dr' \frac{r'^2}{r} + \int_r^R dr' r' \right) . \tag{8.40}$$

In order to better understand the two terms from the previous integral one can take a look at Fig. 8.4. The result is

$$\phi(r < R) = -4\pi G\rho \left(\frac{1}{3}r^2 + \frac{1}{2}(R^2 - r^2) \right) . \tag{8.41}$$

Rearranging terms and introducing the expression for the mass (obtained in the previous exercise) we obtain

$$\phi(r < R) = -GM \left(\frac{3}{2R} - \frac{r^2}{2R^3} \right) . \tag{8.42}$$

Note that the potential is continuous at $r = R$ as both expressions give the same result

$$\phi(r = R) = -\frac{GM}{R^2} . \tag{8.43}$$

Given the expression for $\phi(r < R)$, we can take the limit $r \to 0$ and obtain

$$\phi(r = 0) = -GM \left(\frac{3}{2R} \right) . \tag{8.44}$$

Note that the previous expression presents no singularity at $r = 0$. This is due to the non-zero volume of the mass distribution. Nonetheless we can recover the divergent behaviour of a point-like mass at $r = 0$, by shrinking the volume to a point i.e., by taking the limit $R \to 0$.

Lets us now calculate the gravitational field. First we shall calculate it by taking the gradient of the potential function. For $r > R$ we obtain

$$\mathbf{g}(r > R) = GM \, \nabla \left(\frac{1}{r} \right) = -\frac{GM\mathbf{r}}{r^3} = -\frac{GM}{r^2}\mathbf{u}_r \,, \tag{8.45}$$

which is obviously the expected result. For $r < R$ expression for the field reads

$$\mathbf{g}(r < R) = -\frac{GM}{2R^3} \, \nabla r^2 = -\frac{GM}{R^3}\mathbf{r} \,. \tag{8.46}$$

Note that the field is also continuous at $r = R$,

$$\mathbf{g}(r = R) = -\frac{GM}{R^2}\mathbf{u}_r \,. \tag{8.47}$$

The field at $r = 0$ can be obtained from (8.46)

$$\mathbf{g}(r = 0) = \mathbf{0} \,. \tag{8.48}$$

Finally let us check that we obtain the same result by applying Gauss's law (8.30). For the field outside the surface we simply have

$$g(r > R) = -\frac{GM}{r^2} \,, \tag{8.49}$$

as the whole mass is contained within the surface. For the second case

$$
\begin{aligned}
g(r < R) &= -\frac{GM_r}{r^2} \\
&= -\frac{G}{r^2} \int_0^r \rho r'^2 \int_\Omega d\Omega \\
&= -G\frac{4\pi}{3}\rho r \\
&= -\frac{GM}{R^3}r \,,
\end{aligned}
\tag{8.50}
$$

which again, corresponds to the previously obtained result.

8.3 Repeat the previous exercise for the spherical mass distribution shown in Fig. 8.5.

Solution: Let us start by calculating the potential inside the inner sphere. As inside the inner sphere the density is $\rho = 0$, the only contribution will be given by the outer

Fig. 8.5 A spherical mass
distribution of radius b and
density $\rho = k/r'$ that
contains a spherical hole
($\rho = 0$) with the same center
and radius a

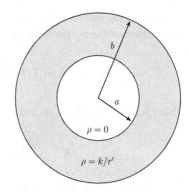

shell (for r in between a and b)

$$\phi(r < a) = -4\pi G \int_a^b dr' \left(\frac{k}{r'}\right) r' = -4\pi k G(b - a). \qquad (8.51)$$

As the potential is constant, the field will be zero. Using Gauss's law, one obtains
the same result, as the mass contained within the sphere of radius r is null. Again,
one should observe, that due to the non-zero volume of the mass, the result is finite
at $r = 0$.

The second region we are going to analyse is $a < r < b$. From $0 < r < a$ we
have no contribution as $\rho = 0$. Thus, similar to the previous exercise, we will have
two contributions to the integral (corresponding to $r' \in (a, r)$ and $r' \in (r, b)$). We
obtain

$$\phi(a < r < b) = -4\pi G \int_a^r dr' \left(\frac{k}{r'}\right) \frac{r'^2}{r} - 4\pi G \int_r^b dr' \left(\frac{k}{r'}\right) r'$$

$$= -4\pi k G \frac{(r^2 - a^2)}{2r} - 4\pi k G(b - r), \qquad (8.52)$$

which simplifies to

$$\phi(a < r < b) = -2\pi k G(2b - r - a^2/r). \qquad (8.53)$$

In order to calculate the gradient we will use the following results

$$\nabla r = \frac{\mathbf{r}}{r}, \qquad \nabla \left(\frac{1}{r}\right) = -\frac{\mathbf{r}}{r^3}. \qquad (8.54)$$

The expression for the gravitational field is therefore given by

$$\mathbf{g}(a < r < b) = -\nabla \phi(a < r < b) = 2k\pi G \left(\frac{a^2}{r^3} - \frac{1}{r}\right) \mathbf{r}. \qquad (8.55)$$

We can observe that both the potential and the field are continuous at $r = a$. We shall now apply Gauss's law to calculate $g(a < r < b)$. The result is straightforward

$$
\begin{aligned}
g(a < r < b) &= -\frac{G}{r^2} \int_a^r \left(\frac{k}{r'}\right) r'^2 \int_\Omega d\Omega \\
&= -\frac{-4\pi k G}{r^2} \frac{1}{2}(r^2 - a^2) \\
&= 2\pi k G \left(\frac{a^2}{r^2} - 1\right),
\end{aligned}
\tag{8.56}
$$

which is the same result as the one obtained previously. Note that even if the density is divergent for $r' \to 0$ both the potential and the field are finite in the limit $a \to 0$.

Let us now move on and analyse the last region with $r > b$. We only have one contribution, namely

$$
\phi(r > b) = -4\pi G \int_a^b dr' \left(\frac{k}{r'}\right) \frac{r'^2}{r} = -2\pi k G \frac{(b^2 - a^2)}{r}.
\tag{8.57}
$$

The gravitational field is simply given by

$$
\mathbf{g}(r > b) = -2\pi k G \frac{(b^2 - a^2)}{r^3} \mathbf{r}.
\tag{8.58}
$$

Calculating the enclosed mass, we obtain

$$
M = 2\pi k (b^2 - a^2),
\tag{8.59}
$$

and therefore

$$
\mathbf{g}(r > b) = -\frac{GM}{r^2} \mathbf{u}_r.
\tag{8.60}
$$

Again, both the field and the potential are continuous at $r = b$.

8.4 Repeat the previous exercise for $\rho = kr'$.

Solution: The calculation is similar to the previous exercise and we shall directly present the final results. For the first region we have

$$
\phi(r < a) = -\frac{4}{3}\pi k G (b^3 - a^3),
$$
$$
\mathbf{g}(r < a) = \mathbf{0}.
\tag{8.61}
$$

For the second region

$$\phi(a < r < b) = \pi kG \left(\frac{r^3}{3} + \frac{a^4}{r} - \frac{4b^3}{3} \right) ,$$

$$\mathbf{g}(a < r < b) = -\pi kG \left(r - \frac{a^4}{r^3} \right) \mathbf{r} , \qquad (8.62)$$

where we have used $\nabla r^3 = 3r \, \mathbf{r}$. As for the last region we obtain

$$\phi(r > b) = -\pi kG \frac{(b^4 - a^4)}{r} ,$$

$$\mathbf{g}(r > b) = -\pi kG \frac{(b^4 - a^4)}{r^3} \mathbf{r} . \qquad (8.63)$$

8.5 Consider an object of volume V and mass density ρ_0 (constant or not). Consider a sub-volume $V' \subset V$. Given a point P with coordinates $\mathbf{r}$ (with respect to some arbitrary reference frame), obtain the variation of $\mathbf{g}(\mathbf{r})$ at P when replacing the mass density $\rho_0 \to \rho$ in V', as schematically shown in Fig. 8.6.

Solution: Initially V' has a mass density given by ρ_0. Thus all the volume V has the same mass density ρ_0. The gravitational field at P can then be virtually separated in two contributions

$$\mathbf{g}(\mathbf{r}, \rho, V) = \mathbf{g}(\mathbf{r}, \rho_0, V - V') + \mathbf{g}(\mathbf{r}, \rho_0, V') . \qquad (8.64)$$

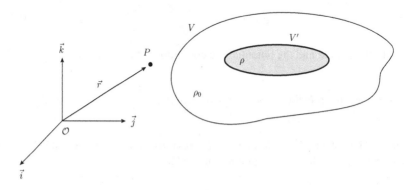

Fig. 8.6 A body with volume V and mass density ρ_0 except a sub-region $V' \subset V$ with a different mass density ρ

After performing the substitution $\rho_0 \rightarrow \rho$ in V', the gravitational field measured at P will be given by the following expression

$$\mathbf{g}'(\mathbf{r}, \rho, \rho_0, V) = \mathbf{g}(\mathbf{r}, \rho_0, V - V') + \mathbf{g}(\mathbf{r}, \rho, V'). \qquad (8.65)$$

Therefore the variation of $\mathbf{g}$ due to the replacement $\rho_0 \rightarrow \rho$ in V' (called gravitation anomaly), will be simply given by

$$\begin{aligned}
\Delta\mathbf{g}(\mathbf{r}, \rho, \rho_0, V') &= \mathbf{g}(\mathbf{r}, \rho, V) - \mathbf{g}'(\mathbf{r}, \rho, \rho_0, V) \\
&= \mathbf{g}(\mathbf{r}, \rho_0, V') - \mathbf{g}(\mathbf{r}, \rho, V') \qquad (8.66) \\
&= \mathbf{g}(\mathbf{r}, \rho_0 - \rho, V'),
\end{aligned}$$

which can be compactly expressed as

$$\Delta\mathbf{g}(\mathbf{r}, \rho, \rho_0, V') = \mathbf{g}(\mathbf{r}, \Delta\rho, V'), \qquad (8.67)$$

where we have introduced $\Delta\rho = \rho_0 - \rho$.

This is a quite strong result with great implications. It allows us to calculate gravitational anomalies on the Earth's surface due to the presence of bodies with known geometries and different densities. Or, on the contrary, having measured the gravitational anomaly, one can estimate the shape and density of a buried mass. We shall apply this result in the following exercises, to some simple geometries.

8.6 Calculate the gravitational anomaly on the surface of the Earth due to a buried spherical mass of radius R and constant density ρ, at a point P placed as shown in Fig. 8.7. Consider the Earth's mass density ρ constant.

Solution: This problem can be straightforwardly solved by making use of the previous result. The gravitational anomaly will be given by the expression (8.67). As both ρ and ρ_0 are constants, we only need to calculate the gravitational field at $\mathbf{r}$ due to the spherical mass, and perform the substitution $\rho \rightarrow \Delta\rho$. As learned previously, the gravitational field outside the surface of a sphere with constant density is equivalent to the gravitational field of a point-like particle with the same mass, placed at the center of the sphere. Thus the expression for the anomaly is simply

$$\Delta\mathbf{g}(\mathbf{r}) = -G\frac{4\pi R^3 \Delta\rho}{3(h^2 + x^2)^{3/2}} (h\mathbf{j} + x\mathbf{i}), \qquad (8.68)$$

where we have considered the vertical axis given by $\mathbf{j}$. There is one important thing one should note from the previous result. Due to the fact the buried sphere is placed arbitrarily within the Earth's volume, the expression does not describe a central field. Therefore we do not only have a variation of the strength of the field along the vertical, we also have a horizontal deflection.

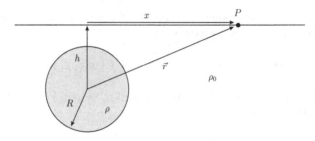

Fig. 8.7 A buried spherical mass of radius R and constant mass density ρ, and the point P where the gravitational field is measured

Fig. 8.8 A massive rod of constant density λ placed along the **i** axis and with finite length l

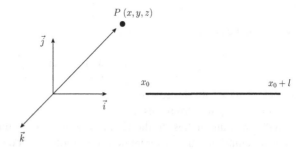

8.7 Calculate the potential and gravitational field at a point P generated by a rod of length l and constant linear mass density λ, placed along the **i** axis (as shown in Fig. 8.8). Obtain the gravitational field of an infinitely large rod at a point P, placed along the **j** axis. Can the result easily be generalized for infinite rod placed along an arbitrary axis and an arbitrary point P?

Solution: For the particular case of a linear mass density, the expression of the potential is given by

$$\phi(\mathbf{r}) = -G \int_l \frac{\lambda(\mathbf{r})\, dr'}{|\mathbf{r} - \mathbf{r}'|} . \tag{8.69}$$

In our case this expression simplifies to

$$\phi(\mathbf{r}) = -G\lambda \int_{x_0}^{x_1} \frac{dx'}{\sqrt{(x - x')^2 + y^2 + z^2}} , \tag{8.70}$$

where we have defined $x_1 \equiv x_0 + l$. The result of this integral is straightforward

$$\phi(\mathbf{r}) = -G\lambda \left[-\ln\left(x - x' + \sqrt{(x - x')^2 + y^2 + z^2} \right) \right]_{x_0}^{x_1}$$

$$= -G\lambda \ln\left(\frac{x - x_0 + \sqrt{(x - x_0)^2 + y^2 + z^2}}{x - x_1 + \sqrt{(x - x_1)^2 + y^2 + z^2}} \right) . \tag{8.71}$$

The gravitational field can be easily obtained as usual by taking the gradient. We obtain $\mathbf{g} = g_x\mathbf{i} + g_y\mathbf{j} + g_z\mathbf{k}$ with

$$g_x = G\lambda\left(\frac{1}{r_0} - \frac{1}{r_1}\right), \tag{8.72}$$

where we have introduced

$$r_i = \sqrt{(x - x_i)^2 + y^2 + z^2}, \tag{8.73}$$

and with $i = 0, 1$. The other two components are

$$g_y = -G\lambda\,y\,\mathcal{F}(x, x_0, x_1), \tag{8.74}$$
$$g_z = -G\lambda z\,\mathcal{F}(x, x_0, x_1), \tag{8.75}$$

with the function $\mathcal{F}(x, x_0, x_1)$ given by

$$\mathcal{F}(x, x_0, x_1) = \frac{x_0^2 - r_0x_0 + x_1(r_1 - x_1) + x(r_0 - r_1 - 2x_0 + 2x_1)}{r_0r_1(r_0 + x - x_0)(r_1 + x - x_1)}. \tag{8.76}$$

Let us now particularize the previous result for the point P with coordinates $x = z = 0$ and $y \neq 0$. In this case we only have two components for the field, namely g_x and g_y, and the expressions for r_0 and r_1 read

$$r_0 = \sqrt{x_0^2 + y^2}, \qquad r_1 = \sqrt{x_1^2 + y^2}. \tag{8.77}$$

In the infinitely large rod limit $l \to \infty$, we have $x_0 \to -\infty$ and $x_1 \to \infty$. For r_0 and r_1 we obtain

$$\lim_{x_0 \to -\infty} r_0 = \infty, \qquad \lim_{x_1 \to \infty} r_1 = \infty. \tag{8.78}$$

Thus, in this limit $g_x = 0$, which is a pretty intuitive result. It can be interpreted the following way. As the rod is infinite along the $\mathbf{i}$ axis, the gravitational pull from the *left part* of the rod (with respect to the projection of the point P on the $\mathbf{i}$ axis) is always compensated by the gravitational pull from *right part*, and the net result is zero. For g_y with $x = z = 0$ we have

$$g_y = -G\lambda\,y\left(-\frac{x_0}{r_0r_1(r_1 - x_1)} - \frac{x_1}{r_0r_1(r_0 - x_0)}\right). \tag{8.79}$$

Let us analyse the first term in the limit $x_0 \to -\infty$ and $x_1 \to \infty$:

$$\lim_{x_0 \to -\infty}\lim_{x_1 \to \infty} -\frac{x_0}{r_0r_1(r_1 - x_1)}. \tag{8.80}$$

If we make the substitution $x_0 \rightarrow -l/2$ and $x_1 \rightarrow l/2$, the previous expression is equivalent to

$$\lim_{l \to \infty} \frac{l\left(l + \sqrt{l^2 + 4y}\right)}{y^2(l^2 + 4y)} = \frac{2}{y^2}. \tag{8.81}$$

Similarly, the second term of (8.79) vanishes. Therefore in the infinite rod limit we obtain the following simple expression for the gravitational field

$$\mathbf{g} = -G\frac{2\lambda}{y}\mathbf{j}. \tag{8.82}$$

Note that translating the point P along the $\mathbf{i}$ axis, the previous expression will not change. It should also result obvious that the gravitational field only depends on the distance to the rod thus, if we displace P along the $\mathbf{k}$ axis the previous expression will transform into

$$\mathbf{g} = -G\frac{2\lambda}{\sqrt{y^2 + z^2}}(y\mathbf{j} + z\mathbf{k}). \tag{8.83}$$

From the previous result we can deduce that the field generated by a rod placed along some arbitrary axis $\mathbf{u}$ will generate a field at an arbitrary point P

$$\mathbf{g} = -G\frac{2\lambda}{d}\mathbf{u}_d, \tag{8.84}$$

where d is the distance from the point to the rod and $\mathbf{u}_d$ is the unitary vector with origin in P, and pointing towards the rod, with the property $\mathbf{u}_d \perp \mathbf{u}$.

8.8 Using Gauss's law, calculate the gravitational anomaly on the surface of the Earth due to a buried cylinder with constant density ρ_M, infinite length and radius R (as in Fig. 8.9). Hint: use polar coordinates.

Solution: Let us first calculate the gravitational field outside the surface of an infinitely large cylinder placed along the $\mathbf{u}_z$ axis in polar coordinates. For this case we will use the generic Gauss's law (8.25). The Gauss's surface that we can use is the surface of the cylinder of radius ρ, that encloses the cylinder of radius R. It can be separated into two contributions. The first one will be given by

$$\rho\, d\rho\, d\phi\, \mathbf{u}_z + d\rho\, dz\, \mathbf{u}_\phi, \tag{8.85}$$

for the two ends of the cylinder (for the *rest* of the cylinder ρ is constant). However, as the two ends are placed at infinity, the previous contribution can be neglected.

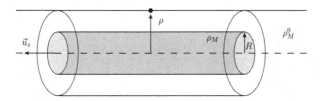

Fig. 8.9 A buried cylinder of constant mass density ρ_M, infinite length and radius R. The Earth's constant mass density in this case is given by ρ_M^0. The Gauss cylinder of radius ρ (used for applying Gauss's law) is also shown

Therefore the only non-negligible contribution will be given by

$$\rho \, d\phi \, dz \, \mathbf{u}_\rho \, . \tag{8.86}$$

Let us now calculate the gravitational field. As in the case of the infinite rod, the gravitational pull along the $\mathbf{u}_z$ axis will be null. Using similar symmetry arguments, the component g_ϕ will also be zero. Therefore, the only surviving term will be g_ρ. Thus, applying Gauss's law we obtain

$$\lim_{l\to\infty} g_\rho \, \rho \int_0^{2\pi} d\phi \int_{-l}^{l} dz = \lim_{l\to\infty} (-4\pi G \rho_M) \int_0^{2\pi} d\phi' \int_{-l}^{l} dz' \int_0^{R} \rho' d\rho' . \tag{8.87}$$

Simplifying the two *infinite* integrals from both sides of the previous expression we obtain

$$\mathbf{g} = g_\rho(\rho) \, \mathbf{u}_\rho = -\frac{2\pi G \rho_M R^2}{\rho} \mathbf{u}_\rho , \tag{8.88}$$

which is the expression for a gravitational field outside a cylinder of infinite length and radius R at a distance ρ from its axis. By calculating the gravitational field of a finite length cylinder and taking the limit $l \to \infty$ one would obtain exactly the same result. Note also that, due to the axial symmetry of the cylinder, the previous result corresponds to an infinite rod of linear mass density $\lambda = \pi R^2 \rho_M$, placed along the symmetry axis of the cylinder.

Finally the gravitational anomaly will be simply given by

$$\Delta g_\rho = -\frac{2\pi G \Delta \rho_M R^2}{\rho} . \tag{8.89}$$

Again, as in the previous case (the infinite rod), the anomaly has both vertical and horizontal components.

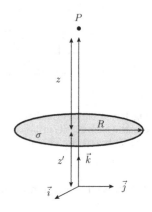

Fig. 8.10 A two-dimensional disk of constant mass density σ and radius R. The gravitational field is measured at a point P placed at a distance z from the center of the disk

8.9 Calculate the gravitational field generated by an uniform disk of superficial mass density σ at a point P, placed along the **k** axis at a distance z from the center of the disk, as shown in Fig. 8.10. Using the generic result, obtain the gravitational field in the two cases: $R \to \infty$ and $z \gg R$. Comment upon these results. Check that the expression in the limit $R \to \infty$ can be obtained applying Gauss's law and symmetry arguments. Obtain the gravitational anomaly along the vertical axis generated by a buried disk (of a very large radius).

Solution: Let us first obtain the field by direct integration. It is given by

$$\mathbf{g}(\mathbf{r}) = -G\sigma \iint_S \frac{ds'}{|\mathbf{r} - \mathbf{r}'|^3} (\mathbf{r} - \mathbf{r}') , \tag{8.90}$$

where $ds' = |d\mathbf{s}'|$. In polar coordinates we have

$$\mathbf{r} = (z + z')\mathbf{k} , \qquad \mathbf{r}' = \rho'\mathbf{u}'_\rho + z'\mathbf{u}_z . \tag{8.91}$$

Here we shall use **k** and $\mathbf{u}_z$ indistinctly (also note that $\mathbf{u}_z = \mathbf{u}'_z$). Therefore we simply obtain

$$\mathbf{r} - \mathbf{r}' = z\mathbf{u}_z - \rho'\mathbf{u}'_\rho , \qquad |\mathbf{r} - \mathbf{r}'|^3 = \left(\rho'^2 + z^2\right)^{3/2} . \tag{8.92}$$

Let us now turn our attention to ds'. As $dz' = 0$ we obtain a simple expression in polar coordinates: $d\mathbf{s}' = \rho' d\rho' d\phi \, \mathbf{u}_z$. Therefore $ds' = \rho' d\rho' d\phi$. The expression of gravitational field transforms into

$$\mathbf{g}(\mathbf{r}) = -G\sigma \int_0^R \int_0^{2\pi} \frac{\rho' d\rho' d\phi}{\left(\rho'^2 + z^2\right)^{3/2}} (z\mathbf{u}_z - \rho'\mathbf{u}'_\rho) . \tag{8.93}$$

The previous equation simplifies when integrating over ϕ. We have

$$\int_0^{2\pi} d\phi\, \mathbf{u}'_\rho = \int_0^{2\pi} d\phi\, (\cos\phi\, \mathbf{i} + \sin\phi\, \mathbf{j}) = 0, \qquad (8.94)$$

and therefore

$$\mathbf{g}(\mathbf{r}) = -2\pi G\sigma z \int_0^R \frac{\rho'\, d\rho'}{\left(\rho'^2 + z^2\right)^{3/2}}\, \mathbf{u}_z. \qquad (8.95)$$

Integrating we obtain

$$\mathbf{g}(\mathbf{r}) = -2\pi G\sigma z\, \mathbf{u}_z \left(\frac{1}{z} - \frac{1}{(z^2 + R^2)^{1/2}}\right). \qquad (8.96)$$

We shall now analyse the two limiting cases. For $R \to \infty$ we obtain

$$\mathbf{g}(\mathbf{r}) = -2\pi G\sigma\, \mathbf{u}_z, \qquad (8.97)$$

which is an expression that does not depend on the distance of the point P with respect to the infinite plane. This is a peculiar result and it can be understood intuitively. Roughly speaking, as one gets closer to the surface, the contributions from far away regions become less intense, and this effect is compensated *exactly* by regions that *are getting closer*. The inverse phenomena takes place when moving farther from the plane. The net effect (of the field variation) in both cases is zero.

For the second limiting case, we must perform a Taylor expansion

$$\begin{aligned}
\mathbf{g}(\mathbf{r}) &= -2\pi G\sigma z\, \mathbf{u}_z \left(\frac{1}{z} - \frac{1}{z(1 + R^2/z^2)^{1/2}}\right) \\
&= -2\pi G\sigma z\, \mathbf{u}_z \left(\frac{1}{z} - \frac{1}{z} + \frac{1}{2}\frac{R^2}{z^2} + \cdots\right),
\end{aligned} \qquad (8.98)$$

therefore, except for higher order terms we obtain

$$\mathbf{g}(\mathbf{r}) \simeq -G\frac{\pi R^2 \sigma}{z}\, \mathbf{u}_z, \qquad (8.99)$$

which is the behaviour of a point-like mass (of mass $\pi R^2\sigma$). This result is also very intuitive (from far away a disk *looks* like a point).

Let us now calculate the limiting case, $R \to \infty$, using Gauss's law. It should be obvious from symmetry arguments, that the components of the field along any direction, except the one given by the $\mathbf{k}$ axis, will be null. The only surviving component will be the z component. The expression for the field will be

Fig. 8.11 A cylindrical
Gauss surface for calculating
the gravitational field of a
disk of infinite radius and
constant mass density σ

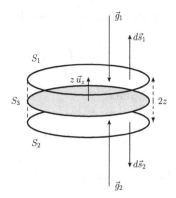

$$\mathbf{g} = \mathbf{g}_1 = -g(|z|)\,\mathbf{u}_z\,, \tag{8.100}$$

above the disk and

$$\mathbf{g} = \mathbf{g}_2 = g(|z|)\,\mathbf{u}_z\,, \tag{8.101}$$

below the disk. In order to apply Gauss's theorem we define a cylindrical surface as in Fig. 8.11, of height $2z$. The surface will be given by S_1, S_2 and S_3. The lateral surface S_3 as it will be placed at infinity, its contribution can be neglected. We therefore obtain

$$\oiint_S \mathbf{g}\cdot ds = \iint_{S_1} \mathbf{g}_1\cdot ds_1 + \iint_{S_2} \mathbf{g}_2\cdot ds_2 \tag{8.102}$$

Without loss of generality, in order to simplify our calculation, we shall place the center of the disk at the origin of coordinates. We thus obtain

$$\oiint_S \mathbf{g}\cdot ds = -2g(|z|)\int_0^{2\pi} d\phi \int_0^{\infty} \rho\,d\rho\,. \tag{8.103}$$

Gauss's law then states

$$-2g(|z|)\int_0^{2\pi} d\phi \int_0^{\infty} \rho\,d\rho = -4\pi G\sigma \int_0^{2\pi} d\phi \int_0^{\infty} \rho\,d\rho\,. \tag{8.104}$$

Simplifying we obtain

$$g(|z|) = 2\pi G\sigma\,, \tag{8.105}$$

which corresponds to the previously obtained result.

As for the gravitational anomaly, we obtain

$$\Delta g(|z|) = 2\pi G \Delta \sigma, \tag{8.106}$$

which is independent of the depth at which the disk is buried.

Its is worth mentioning, as a conclusion regarding the last few exercises, that having calculated the gravitational field of extensive objects, we have also indirectly learned a great deal of concepts about electrostatics. The electric field of a set of point-like electrical charges (in the vacuum) is given by the following expression

$$\mathbf{E(r)} = \frac{1}{4\pi\epsilon_0} \sum_i \frac{q_i}{|\mathbf{r} - \mathbf{r}_i|^3} (\mathbf{r} - \mathbf{r}_i), \tag{8.107}$$

where ϵ_0 is the vacuum permittivity constant. By making the substitutions $(4\pi\epsilon_0)^{-1} \rightarrow -G$ and $q_i \rightarrow m_i$ one obtains the gravitational field. The generalization the continuum case is

$$\mathbf{E(r)} = \frac{1}{4\pi\epsilon_0} \iiint_V dV' \frac{\rho(\mathbf{r}')}{|\mathbf{r} - \mathbf{r}_i|^3} (\mathbf{r} - \mathbf{r}_i), \tag{8.108}$$

where $\rho(\mathbf{r}')$ is the charge density. Therefore, making the proper substitutions one can go from electrostatics to gravitation and vice-versa.

Having calculated the potential energy of an extensive spherical mass, we can straightforwardly apply these results for describing the motion of satellites moving around the Earth and their escape velocity. This will be done in the remaining proposed exercises.

8.10 The escape velocity of an object moving under the influence of a gravitational field, is defined as the minimum velocity needed to escape the gravitational pull. In terms of kinetic and potential energy, it corresponds to the initial kinetic energy, needed to reach $r = \infty$ with zero kinetic energy. Using the energy conservation theorem calculate the escape velocity of a rocket launched from the surface of the Earth.

Solution: The generic energy conservation theorem (5.59) for a system of particles states

$$dE = dK + dU^{ext} + dU^{int} = dW_{nc}^{ext} + dW_{nc}^{int}. \tag{8.109}$$

In our particular case the are no non-conservative forces. There is no external potential term either, therefore the previous expression simplifies to (dropping the *int* upper index)

$$dE = dK + dU^{int} = 0, \tag{8.110}$$

and so, the total energy is conserved. Integrating the previous expression we obtain

$$K_i + U_i = K_f + U_f, \tag{8.111}$$

where i stands for the initial configuration and f for the final one. The gravitational potential on the surface of the Earth is given by

$$\phi = -G\frac{M}{R}, \tag{8.112}$$

and therefore the potential energy (where m is the mass of the rocket) will take the form

$$U = -G\frac{Mm}{R}. \tag{8.113}$$

The energy conservation law then states

$$\frac{1}{2}mv_i^2 - G\frac{Mm}{R} = \frac{1}{2}mv_f^2 - G\frac{Mm}{R+h}, \tag{8.114}$$

where h is the distance from the surface of the Earth to the rocket. By definition the escape velocity corresponds to $h \rightarrow \infty$ and $v_f = 0$, thus we obtain

$$\frac{1}{2}mv_{esc}^2 - G\frac{Mm}{R} = 0, \tag{8.115}$$

which translates into

$$v_{esc} = \sqrt{\frac{2GM}{R}}. \tag{8.116}$$

Note that the previous result is independent of the rocket's mass.

8.11 Suppose the Earth had a constant mass density ρ but twice its diameter. Argue if the following statements are true or false.

- The mass of the Earth would be double.
- The intensity of the gravitational field on its surface would be double.
- The escape velocity for a rocket launched from the surface of the Earth would be double.

Solution: The mass of a spherical object of radius R with constant density ρ is given by

$$M = \rho V = \rho \frac{4}{3} \pi R^3 . \tag{8.117}$$

If we now make the substitution $R \rightarrow R' = 2R$ we obtain

$$M' = \rho \frac{4}{3} \pi (2R)^3 = 8 M , \tag{8.118}$$

and therefore the first statement is false. Let us now move to the second statement. The gravitational field on the surface of the Earth is given by

$$\mathbf{g(r)} = -G \frac{M}{R^2} \mathbf{u}_r = -G \frac{4}{3} \pi \rho \frac{R^3}{R^2} \mathbf{u}_r = -\frac{4}{3} G \rho \pi R \, \mathbf{u}_r . \tag{8.119}$$

Performing the substitution $R \rightarrow R' = 2R$ we obtain $\mathbf{g}' = 2\mathbf{g}$ and so, the second statement is true. The remaining task is to analyse the last statement. In the previous exercise we have calculated the escape velocity. Expressing the Earth's mass in terms of its density we obtain

$$v_{esc} = \sqrt{\frac{2GM}{R}} = \sqrt{2G \frac{4}{3} \pi \rho \frac{R^3}{R}} = R \sqrt{\frac{8}{3} \pi \rho \, G} , \tag{8.120}$$

and so the last statement is also true.

8.12 A satellite is placed on a geostationary platform at a distance D from the surface of the Earth, as shown in Fig. 8.12. Calculate the velocity that the satellite needs to be launched with, in order to describe a circular motion around the Earth. How about if we want the satellite to describe an elliptic orbit? Relate the previous answer to the corresponding escape velocity (of the object placed on the platform).

Solution: The fact that the platform is geostationary means that it has no velocity relative to the Earth. The gravitational potential energy is given by Kepler's potential energy with $k = GMm$, where M is the Earth's mass and m the mass of the satellite. Therefore, as seen in Chap. 6, the object will describe a circular orbit (of radius r_c) if

$$r_c = \frac{l^2}{k m} = \frac{l^2}{GM m^2} , \tag{8.121}$$

where we have supposed $m \ll M$ (thus $\mu \simeq m$) and with

$$l^2 = m^2 r_c^4 \omega^2 = m^2 r_c^2 v_\theta^2 . \tag{8.122}$$

Fig. 8.12 A satellite of mass m placed at a geostationary platform at a distance D from the surface of the Earth

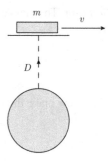

Combining the previous expression we obtain

$$v_\theta = \sqrt{\frac{GM}{r_c}}. \tag{8.123}$$

In our case $r_c = R + D$, where R is the Earth's radius, therefore the velocity v of the satellite (in order to describe a circular trajectory) will be simply given by

$$v = v_\theta = \sqrt{\frac{GM}{R + D}}. \tag{8.124}$$

For the second part of the exercise, let us first calculate the escape velocity of the satellite from the platform. Just as in the previous exercises

$$\frac{1}{2}mv_{esc}^2 - G\frac{Mm}{R + D} = 0 \tag{8.125}$$

therefore

$$v_{esc} = \sqrt{\frac{2GM}{R + D}}. \tag{8.126}$$

As explained in Chap. 6, the orbit will be elliptic if the total energy is negative i.e.,

$$\frac{1}{2}mv^2 - G\frac{Mm}{R + D} < 0, \tag{8.127}$$

which translates into

$$v < \sqrt{\frac{2GM}{R + D}} = v_{esc}. \tag{8.128}$$

Note that the following condition must also hold: $v > v_\theta$. For $v < v_\theta$ the object would end up falling towards the surface of the Earth.

Further Reading

1. J.V. José, E.J. Saletan, *Classical Dynamics: A Contemporary Approach*. Cambridge University Press
2. S.T. Thornton, J.B. Marion, *Classical Dynamics of Particles and Systems*
3. H. Goldstein, C. Poole, J. Safko, *Classical Mechanics*, 3rd edn. Addison Wesley
4. J.R. Taylor, *Classical Mechanics*
5. D.T. Greenwood, *Classical Dynamics*. Prentice-Hall Inc.
6. D. Kleppner, R. Kolenkow, *An Introduction to Mechanics*
7. C. Lanczos, *The Variational Principles of Mechanics*. Dover Publications Inc.
8. W. Greiner, *Classical Mechanics: Systems of Particles and Hamiltonian Dynamics*. Springer
9. H.C. Corben, P. Stehle, *Classical Mechanics*, 2nd edn. Dover Publications Inc.
10. T.W.B. Kibble, F.H. Berkshire, *Classical Mechanics*. Imperial College Press
11. M.G. Calkin, *Lagrangian and Hamiltonian Mechanics*
12. A.J. French, M.G. Ebison, *Introduction to Classical Mechanics*
13. B.C. Consuelo, G.F. Antonio, R.D. Marcelo, *Campos electromagnéticos*. Editorial Universidad de Sevilla
14. J.D. Jackson, *Classical Electrodynamics*, 3rd edn.

Chapter 9
Non-inertial Reference Systems

Abstract So far we have stated that, up to a good approximation, the Earth can be considered an inertial reference system. Strictly speaking, it is obviously not an inertial reference frame as it rotates about its axis, orbits around the Sun, etc. In this chapter we shall deduce the expression for the fictitious forces that appear in non-inertial reference systems by using two approaches, one of them based in Euler's angles (introduced in Appendix F) and a second one, which can be considered more heuristic. We shall then apply these generic results to a set of simplified problems regarding the trajectory and motion of physical systems on the surface of the Earth. We will also analyse in detail the effects of both the Coriolis and the centrifugal force, avoiding any possible confusion about the magnitude of both effects.

9.1   Velocity and Angular Velocity

Let us start by considering an inertial reference frame $\{\mathcal{O}, \mathbf{i}, \mathbf{j}, \mathbf{k}\}$ and a second reference system $\{\mathcal{O}', \mathbf{i}', \mathbf{j}', \mathbf{k}'\}$, that is obtained from the initial one by a rotation and a translation i.e., the coordinates $\mathbf{r}$ of a point P in $\mathcal{O}$, and the ones in $\mathcal{O}'$ (given by $\mathbf{r}'$), are related through

$$\mathbf{r} = \mathcal{R} + \mathbf{r}', \tag{9.1}$$

which can explicitly be written in matrix form as

$$(\mathbf{i}, \mathbf{j}, \mathbf{k}) \begin{pmatrix} x \\ y \\ z \end{pmatrix} = (\mathbf{i}, \mathbf{j}, \mathbf{k}) \begin{pmatrix} \mathcal{R}_x \\ \mathcal{R}_y \\ \mathcal{R}_z \end{pmatrix} + (\mathbf{i}', \mathbf{j}', \mathbf{k}') \begin{pmatrix} x' \\ y' \\ x' \end{pmatrix}, \tag{9.2}$$

(in this chapter, in contrast with other chapters, we shall not employ $\mathcal{R}$, $\mathcal{V}$ and $\mathcal{A}$ as referring to the CM coordinates, velocity and acceleration). The rotation matrix[1] R is given by (Fig. 9.1)

---

[1]Do not confuse it with the translation $\mathcal{R}$.

© Springer Nature Switzerland AG 2020
V. Ilisie, *Lectures in Classical Mechanics*, Undergraduate Lecture
Notes in Physics, https://doi.org/10.1007/978-3-030-38585-9_9

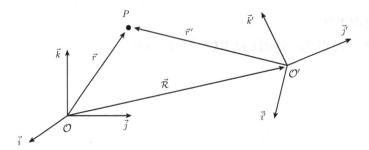

Fig. 9.1 A reference frame $\{\mathcal{O}, \mathbf{i}, \mathbf{j}, \mathbf{k}\}$ and a primed reference frame $\{\mathcal{O}', \mathbf{i}', \mathbf{j}', \mathbf{k}'\}$, that is obtained from the original one through a rotation and a translation

$$
\begin{pmatrix} \mathbf{i}' \\ \mathbf{j}' \\ \mathbf{k}' \end{pmatrix} = R \begin{pmatrix} \mathbf{i} \\ \mathbf{j} \\ \mathbf{k} \end{pmatrix} .
\tag{9.3}
$$

Equivalently, by transposing the previous expression we obtain

$$
(\mathbf{i}', \mathbf{j}', \mathbf{k}') = (\mathbf{i}, \mathbf{j}, \mathbf{k}) \, R^T ,
\tag{9.4}
$$

and so (9.2) becomes

$$
(\mathbf{i}, \mathbf{j}, \mathbf{k}) \begin{pmatrix} x \\ y \\ z \end{pmatrix} = (\mathbf{i}, \mathbf{j}, \mathbf{k}) \begin{pmatrix} \mathcal{R}_x \\ \mathcal{R}_y \\ \mathcal{R}_z \end{pmatrix} + (\mathbf{i}, \mathbf{j}, \mathbf{k}) \, R^T \begin{pmatrix} x' \\ y' \\ x' \end{pmatrix} .
\tag{9.5}
$$

Let us now consider that the rotation matrix R is expressed in terms of Euler's angles, as explained in Appendix F. Let us also consider that Euler's angles, change with time. In Appendix F, we have deduced that the time variation of R was given by

$$
\frac{dR}{dt} = \Omega \, R ,
\tag{9.6}
$$

where Ω is given by the matrix

$$
\Omega = \begin{pmatrix} 0 & \omega_3' & -\omega_2' \\ -\omega_3' & 0 & \omega_1' \\ \omega_2' & -\omega_1' & 0 \end{pmatrix} .
\tag{9.7}
$$

The quantities ω_1', ω_2' and ω_3' are the components of the angular velocity in the primed basis

$$
\boldsymbol{\omega} = \omega_1' \mathbf{i}' + \omega_2' \mathbf{j}' + \omega_3' \mathbf{k}' ,
\tag{9.8}
$$

which can be expressed in terms of Euler's angles and their time variation as given in (F.9). By transposing (9.6) we obtain

$$\frac{dR^T}{dt} = R^T \Omega^T . \tag{9.9}$$

Having obtained the previous results we can relate the velocity of a point-like particle in both reference frames

$$\frac{d\mathbf{r}}{dt} = \frac{d\mathcal{R}}{dt} + \frac{d\mathbf{r}'}{dt} . \tag{9.10}$$

Explicitly introducing the rotation matrix and its time variation, the previous expression turns into

$$(\mathbf{i}, \mathbf{j}, \mathbf{k}) \begin{pmatrix} dx/dt \\ dy/dt \\ dz/dt \end{pmatrix} = (\mathbf{i}, \mathbf{j}, \mathbf{k}) \begin{pmatrix} d\mathcal{R}_x/dt \\ d\mathcal{R}_y/dt \\ d\mathcal{R}_z/dt \end{pmatrix} + (\mathbf{i}, \mathbf{j}, \mathbf{k}) R^T \begin{pmatrix} dx'/dt \\ dy'/dt \\ dz'/dt \end{pmatrix}$$
$$+ (\mathbf{i}, \mathbf{j}, \mathbf{k}) \frac{dR^T}{dt} \begin{pmatrix} x' \\ y' \\ z' \end{pmatrix} , \tag{9.11}$$

or equivalently

$$(\mathbf{i}, \mathbf{j}, \mathbf{k}) \begin{pmatrix} v_x \\ v_y \\ v_z \end{pmatrix} = (\mathbf{i}, \mathbf{j}, \mathbf{k}) \begin{pmatrix} \mathcal{V}_x \\ \mathcal{V}_y \\ \mathcal{V}_z \end{pmatrix} + (\mathbf{i}', \mathbf{j}', \mathbf{k}') \begin{pmatrix} v_x' \\ v_y' \\ v_z' \end{pmatrix}$$
$$+ (\mathbf{i}', \mathbf{j}', \mathbf{k}') \Omega^T \begin{pmatrix} x' \\ y' \\ z' \end{pmatrix} . \tag{9.12}$$

Let us now analyse the last term of the previous expression and explicitly perform the matrix product

$$(\mathbf{i}', \mathbf{j}', \mathbf{k}') \begin{pmatrix} 0 & -\omega_3' & \omega_2' \\ \omega_3' & 0 & -\omega_1' \\ -\omega_2' & \omega_1' & 0 \end{pmatrix} \begin{pmatrix} x' \\ y' \\ z' \end{pmatrix} = \mathbf{i}'(\omega_2' z' - \omega_3' y') + \mathbf{j}'(\omega_3' x' - \omega_1' z')$$
$$+ \mathbf{k}'(\omega_1' y' - \omega_2' x') , \tag{9.13}$$

which is equivalent to the vector product $\boldsymbol{\omega} \times \mathbf{r}'$ i.e.,

$$(\mathbf{i}', \mathbf{j}', \mathbf{k}') \Omega^T \begin{pmatrix} x' \\ y' \\ z' \end{pmatrix} = \boldsymbol{\omega} \times \mathbf{r}' . \tag{9.14}$$

We are left with the following simple expression

$$\mathbf{v} = \boldsymbol{\mathcal{V}} + \mathbf{v}' + \boldsymbol{\omega} \times \mathbf{r}'. \tag{9.15}$$

Calculating the second time derivative, we obtain the acceleration

$$\frac{d\mathbf{v}}{dt} = \frac{d\boldsymbol{\mathcal{V}}}{dt} + \frac{d\mathbf{v}'}{dt} + \frac{d\boldsymbol{\omega}}{dt} \times \mathbf{r}' + \boldsymbol{\omega} \times \frac{d\mathbf{r}'}{dt}. \tag{9.16}$$

By applying the same technique we get to the following expression

$$\mathbf{a} = \boldsymbol{\mathcal{A}} + \mathbf{a}' + \boldsymbol{\omega} \times \mathbf{v}' + \boldsymbol{\alpha} \times \mathbf{r}' + (\boldsymbol{\omega} \times \boldsymbol{\omega}) \times \mathbf{r}'$$
$$+ \boldsymbol{\omega} \times \mathbf{v}' + \boldsymbol{\omega} \times (\boldsymbol{\omega} \times \mathbf{r}'), \tag{9.17}$$

where we have defined the angular acceleration as

$$\boldsymbol{\alpha} = \frac{d\omega_1'}{dt}\mathbf{i}' + \frac{d\omega_2'}{dt}\mathbf{j}' + \frac{d\omega_3'}{dt}\mathbf{k}'. \tag{9.18}$$

As $\boldsymbol{\omega} \times \boldsymbol{\omega} = \mathbf{0}$, the final expression simplifies to

$$\mathbf{a} = \boldsymbol{\mathcal{A}} + \mathbf{a}' + 2\boldsymbol{\omega} \times \mathbf{v}' + \boldsymbol{\alpha} \times \mathbf{r}' + \boldsymbol{\omega} \times (\boldsymbol{\omega} \times \mathbf{r}'). \tag{9.19}$$

Multiplying by the mass m on both sides and isolating $m\mathbf{a}'$ we obtain

$$m\mathbf{a}' = m\mathbf{a} - m\boldsymbol{\mathcal{A}} - 2m\boldsymbol{\omega} \times \mathbf{v}' - m\boldsymbol{\alpha} \times \mathbf{r}' - m\boldsymbol{\omega} \times (\boldsymbol{\omega} \times \mathbf{r}'). \tag{9.20}$$

We can now identify (or *label*) all the terms from the previous equation:

- $\mathbf{F}' = m\mathbf{a}'$ is the total force, over the particle of mass m, measured in the non-inertial reference system.
- $\mathbf{F} = m\mathbf{a}$ is the force, over the particle of mass m, measured in the inertial reference system.
- $\mathbf{F} = -m\mathbf{A}$ is the fictitious force due to the translational acceleration.
- $\mathbf{F}_{cor} = -2m\boldsymbol{\omega} \times \mathbf{v}'$ is called the Coriolis force.
- $\mathbf{F}_{\alpha} = -m\boldsymbol{\alpha} \times \mathbf{r}'$ is the fictitious force due to the angular acceleration of the reference system.
- $\mathbf{F}_{centrif} = -m\boldsymbol{\omega} \times (\boldsymbol{\omega} \times \mathbf{r}')$ is the so-called centrifugal force.

Even if not explicitly stated, it should be obvious that both the Coriolis and the centrifugal forces are also fictitious.

Before moving on to the next section we shall introduce a second method, which is rather heuristic, which is however very oftenly used to obtain the previous results without making use of Euler's angles. The advantage is that the reader does not have to previously know anything about Euler's angles and rigid solid motion, however, the physical interpretation of the angular velocity might result a bit unclear. This is due to the fact that the reader does not know at first sight, how to interpret ω in terms of the rotation angles.

9.1.1 The Heuristic Approach

Starting from the same initial expression

$$(\mathbf{i}, \mathbf{j}, \mathbf{k}) \begin{pmatrix} x \\ y \\ z \end{pmatrix} = (\mathbf{i}, \mathbf{j}, \mathbf{k}) \begin{pmatrix} \mathcal{R}_x \\ \mathcal{R}_y \\ \mathcal{R}_z \end{pmatrix} + (\mathbf{i}', \mathbf{j}', \mathbf{k}') \begin{pmatrix} x' \\ y' \\ x' \end{pmatrix}, \qquad (9.21)$$

we differentiate on both sides as follows

$$(\mathbf{i}, \mathbf{j}, \mathbf{k}) \begin{pmatrix} dx/dt \\ dy/dt \\ dz/dt \end{pmatrix} = (\mathbf{i}, \mathbf{j}, \mathbf{k}) \begin{pmatrix} d\mathcal{R}_x/dt \\ d\mathcal{R}_y/dt \\ d\mathcal{R}_z/dt \end{pmatrix} + (\mathbf{i}', \mathbf{j}', \mathbf{k}') \begin{pmatrix} dx'/dt \\ dy'/dt \\ dz'/dt \end{pmatrix}$$
$$+ \left(\frac{d\mathbf{i}'}{dt}, \frac{d\mathbf{j}'}{dt}, \frac{d\mathbf{k}'}{dt} \right) \begin{pmatrix} x' \\ y' \\ z' \end{pmatrix}. \qquad (9.22)$$

Note that this expression is almost identical to the one obtained with the first approach, the only difference is that we have not explicitly introduced the matrix R. We now assume that the variation of the components of the basis can be written as some linear combination of the basis components themselves i.e.,

$$\frac{d\mathbf{i}'}{dt} = \alpha_{11}\mathbf{i}' + \alpha_{12}\mathbf{j}' + \alpha_{13}\mathbf{k}',$$
$$\frac{d\mathbf{j}'}{dt} = \alpha_{21}\mathbf{i}' + \alpha_{22}\mathbf{j}' + \alpha_{23}\mathbf{k}', \qquad (9.23)$$
$$\frac{d\mathbf{k}'}{dt} = \alpha_{31}\mathbf{i}' + \alpha_{32}\mathbf{j}' + \alpha_{33}\mathbf{k}'.$$

Let us now introduce the orthonormality properties

$$\mathbf{i}' \cdot \mathbf{i}' = \mathbf{j}' \cdot \mathbf{j}' = \mathbf{k}' \cdot \mathbf{k}' = 1,$$
$$\mathbf{i}' \cdot \mathbf{j}' = \mathbf{i}' \cdot \mathbf{k}' = \mathbf{j}' \cdot \mathbf{k}' = 0. \qquad (9.24)$$

Differentiating the previous expressions we obtain the following properties of α_{ij}

$$\alpha_{ij} = -\alpha_{ji}, \qquad \alpha_{ii} = 0, \qquad (9.25)$$

$\forall i, j$. Thus, if we define the quantity

$$\boldsymbol{\omega} = \alpha_{23}\mathbf{i'} + \alpha_{31}\mathbf{j'} + \alpha_{12}\mathbf{k'} \equiv \omega_1'\mathbf{i'} + \omega_2'\mathbf{j'} + \omega_3'\mathbf{k'}, \qquad (9.26)$$

we can re-express (9.23) as

$$\frac{d\mathbf{i'}}{dt} = \boldsymbol{\omega} \times \mathbf{i'}, \qquad \frac{d\mathbf{j'}}{dt} = \boldsymbol{\omega} \times \mathbf{j'}, \qquad \frac{d\mathbf{k'}}{dt} = \boldsymbol{\omega} \times \mathbf{k'}. \qquad (9.27)$$

We therefore obtain the same result as previously, having introduced some quantity that has the properties of an angular velocity, without being able to infer, as mentioned previously, its physical interpretation in terms of the rotation angles.

9.2 Motion over the Earth's Surface

Let us now apply al the previously deduced results for the motion over the surface of the Earth, by employing some simple approximations. We shall consider that the Earth is spherical and that it rotates about the *inertial* **k** axis as shown in Fig. 9.2. Consider an inertial reference frame $\{\mathbf{i}, \mathbf{j}, \mathbf{k}\}$ with origin at the center of the Earth. Consider that the Earth rotates from west (W) to east (E) with angular velocity $\boldsymbol{\omega} = \omega\,\mathbf{k}$ (expressed in the inertial reference frame). Finally, also consider a point P over the surface of the Earth (at rest with respect to the Earth) in the northern hemisphere, and a non-inertial reference frame $\{\mathbf{i'}, \mathbf{j'}, \mathbf{k'}\}$ with origin in P. The coordinates of P can be easily expressed in the non-inertial reference frame as $\boldsymbol{\mathcal{R}} = \mathcal{R}\,\mathbf{k'}$ (with $\mathcal{R}$ the Earth's radius).

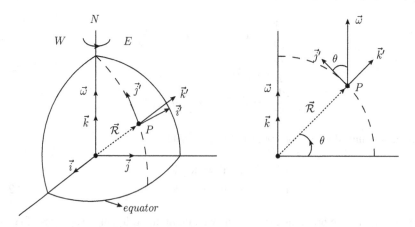

Fig. 9.2 An inertial reference frame with origin at the center of the Earth, together with a non-inertial (primed) reference system with origin P, placed over the Earth's surface, at rest with the Earth. Only the northern hemisphere is shown

For this configuration we must be careful about one subtle detail. In the previous section, when deriving the equations of motion in a non-inertial reference system we have always expressed $\mathcal{R}$ in the inertial basis. Here we have expressed it (because it was straightforward) in the primed reference system. We should thus re-express it in terms of the inertial basis and calculate $\mathcal{V}$ and $\mathcal{A}$ (which are non zero because, even if the magnitude of $\mathcal{R}$ is constant, it changes its direction with respect to the inertial reference frame). There is, however, a second alternative. If we express $\mathcal{R}$ from the beginning in the non-inertial frame i.e.,

$$\mathcal{R} = \mathcal{R}'_x \mathbf{i}' + \mathcal{R}'_y \mathbf{j}' + \mathcal{R}'_z \mathbf{k}', \tag{9.28}$$

then

$$\frac{d\mathcal{R}}{dt} = \mathcal{V} + \omega \times \mathcal{R}, \tag{9.29}$$

and

$$\frac{d^2\mathcal{R}}{dt} = \mathcal{A} + 2\omega \times \mathcal{V} + \alpha \times \mathcal{R} + \omega \times (\omega \times \mathcal{R}), \tag{9.30}$$

with

$$\mathcal{V} = \frac{d\mathcal{R}'_x}{dt}\mathbf{i}' + \frac{d\mathcal{R}'_y}{dt}\mathbf{j}' + \frac{d\mathcal{R}'_z}{dt}\mathbf{k}', \tag{9.31}$$

and

$$\mathcal{A} = \frac{d\mathcal{V}'_x}{dt}\mathbf{i}' + \frac{d\mathcal{V}'_y}{dt}\mathbf{j}' + \frac{d\mathcal{V}'_z}{dt}\mathbf{k}'. \tag{9.32}$$

We would therefore have to make the following substitutions

$$\mathcal{V} \to \mathcal{V} + \omega \times \mathcal{R}, \tag{9.33}$$
$$\mathcal{A} \to \mathcal{A} + 2\omega \times \mathcal{V} + \alpha \times \mathcal{R} + \omega \times (\omega \times \mathcal{R}), \tag{9.34}$$

in (9.15) and (9.19). By doing so, and by considering that, the magnitude of $\mathcal{R}$ is constant, and also that, up to a good approximation ω can also be considered constant, we are left with the following simplified equation of motion for the acceleration

$$\mathbf{a}' = \mathbf{a} - 2\omega \times \mathbf{v}' - \omega \times \left(\omega \times (\mathbf{r}' + \mathcal{R})\right). \tag{9.35}$$

If we only consider small distances from the origin P, when compared to Earth's radius $|\mathbf{r}'| \ll |\mathcal{R}|$, we can further simplify the previous expression to

$$\mathbf{a}' = \mathbf{a} - 2\omega \times \mathbf{v}' - \omega \times (\omega \times \mathcal{R}). \tag{9.36}$$

We can conclude from the previous equation that we have two dominant fictitious accelerations in the Earth's non-inertial reference system, the Coriolis acceleration and the centrifugal acceleration.

Let us now express the angular velocity in the non-inertial basis and explicitly calculate the previous scalar products. For this purpose we shall make use of the angle θ we have introduced in Fig. 9.2, which is called latitude. Latitude is always measured with respect to the equator. Colatitude in change, can be expressed as $\pi/2 - \theta$. However, here we shall always work with the latitude. In terms of θ we thus obtain the following

$$\boldsymbol{\omega} = \omega \cos\theta\,\mathbf{j}' + \omega\sin\theta\,\mathbf{k}',$$ (9.37)

and so the vector products can be straightforwardly calculated. We obtain

$$2\boldsymbol{\omega} \times \mathbf{v}' = 2\omega(v_z'\cos\theta - v_y'\sin\theta)\mathbf{i}' + 2\omega\,v_x'\sin\theta\mathbf{j}' - 2\omega v_x'\cos\theta\mathbf{k}',$$ (9.38)

and

$$\boldsymbol{\omega} \times (\boldsymbol{\omega} \times \mathcal{R}) = \mathcal{R}\omega^2\sin\theta\cos\theta\mathbf{j}' - \mathcal{R}\omega^2\cos^2\theta\mathbf{k}'.$$ (9.39)

As we will consider small distances from the origin (compared to the Earth's radius), in the previous equations we can consider the latitude θ to be constant.

Note that we have performed the previous calculations for a reference system placed in the northern hemisphere. A natural question is, how are the equations going to change for the southern hemisphere. In the southern hemisphere, we will have the configuration shown in Fig. 9.3. Thus if we follow the previously mentioned convention (the latitude is measured with respect to the equator) then θ in the southern hemisphere is measured counter-clockwise and so it is negative. One should thus realize, that $\boldsymbol{\omega}$ can be written using the same expression (9.37).

In conclusion, all the previous expressions are valid in both hemispheres, with $\theta > 0$ in the northern hemisphere, $\theta < 0$ in the southern hemisphere and $\theta = 0$ along the equator.

Fig. 9.3 The reference system configuration in the southern hemisphere together with the latitude θ (which is measured counter-clockwise, thus it is negative)

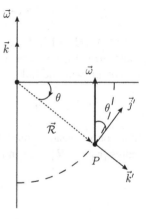

9.3 Free Fall

We shall now re-analyse the free fall problem in the Earth's reference system by introducing non-inertial effects. We will also introduce a perturbative method for solving the equations of motion, as there is no generic known analytical solution. The *inertial* gravitation acceleration is given by $\mathbf{g}$, and it can be easily expressed in the primed basis as

$$\mathbf{g} = -g\,\mathbf{k}' \tag{9.40}$$

with $g > 0$. We must now put special emphasis on the centrifugal acceleration. Many authors omit to give a clear explanation of its effects and they center their attention on the Coriolis effect only. The reader might even find affirmations regarding the smallness of the deflection due to the centrifugal acceleration when compared to the Coriolis acceleration. This affirmations are obviously misleading and it depends on how one defines the vertical axis. It shall be explained in the following.

Note that the centrifugal acceleration does not depend on the velocity of the object, and so if we set $\mathbf{v}'$ to zero we obtain the expression for the acceleration term, in the non-inertial reference frame to be $\mathbf{g} - \boldsymbol{\omega} \times (\boldsymbol{\omega} \times \mathcal{R})$. We shall call this term, for obvious reasons, the effective gravitational acceleration:

$$\begin{aligned}
\mathbf{g}_{eff} &= \mathbf{g} - \boldsymbol{\omega} \times (\boldsymbol{\omega} \times \mathcal{R}), \\
&= -(g - \mathcal{R}\omega^2 \cos^2\theta)\,\mathbf{k}' - \mathcal{R}\omega^2 \sin\theta \cos\theta \mathbf{j}' \\
&\equiv g'_{z,eff}\,\mathbf{k}' + g'_{y,eff}\,\mathbf{j}'.
\end{aligned} \tag{9.41}$$

Thus, it is really $\mathbf{g}_{eff}$ the acceleration that we experience on the Earth's surface. Two things should be noted. First, that $\mathbf{g}_{eff}$ has a component along the $\mathbf{j}'$ axis.[2] Second, this term is anything but small (due to the presence of $\mathcal{R}$), when compared to the deviation due to the Coriolis effect. We shall see that the deflection can actually be one order of magnitude larger than the Coriolis force. The misunderstanding comes from the fact that this *leading effect* is reabsorbed by redefining the vertical axis. One, by convention, rotates the primed reference frame about the $\mathbf{i}'$ axis in order to align the vertical with the direction of $\mathbf{g}_{eff}$, which is the *natural* vertical axis for a reference system at rest with the Earth.

Let us therefore find the proper orthogonal rotation matrix (that we shall call $\mathbb{R}_\phi$) that rotates the primed basis into a double primed basis, where $\mathbf{g}_{eff}$ is aligned with the vertical, as shown in Fig. 9.4 (for the N hemisphere). The matrix acting on the vector components of $\mathbf{g}_{eff}$ reads

---

[2]which in the N hemisphere corresponds to a southerly deflection and, in the S hemisphere corresponds to a northerly deflection.

Fig. 9.4 $\mathbf{g}_{eff}$ in the N
hemisphere and the double
primed basis where $\mathbf{g}_{eff}$ is
aligned with the vertical $\mathbf{k}''$

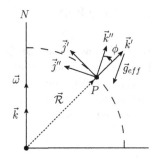

$$\mathbb{R}_\phi \begin{pmatrix} 0 \\ g'_{y,eff} \\ g'_{z,eff} \end{pmatrix} \equiv \begin{pmatrix} 1 & 0 & 0 \\ 0 & \cos\phi & \sin\phi \\ 0 & -\sin\phi & \cos\phi \end{pmatrix} \begin{pmatrix} 0 \\ g'_{y,eff} \\ g'_{z,eff} \end{pmatrix} = \begin{pmatrix} 0 \\ 0 \\ g''_{z,eff} \end{pmatrix}. \qquad (9.42)$$

Therefore we obtain the two following equations

$$\cos\phi\, g'_{y,eff} + \sin\phi\, g'_{z,eff} = 0,$$
$$-\sin\phi\, g'_{y,eff} + \cos\phi\, g'_{z,eff} = g''_{z,eff}. \qquad (9.43)$$

From the first equation we can straightforwardly deduce ϕ

$$\tan\phi = -\frac{g'_{y,eff}}{g'_{z,eff}} = \frac{\mathcal{R}\omega^2 \sin\theta \cos\theta}{\mathcal{R}\omega^2 \cos^2\theta - g}. \qquad (9.44)$$

Using the previous expression let us now calculate the order of magnitude of ϕ. The Earth's angular velocity is $\omega = 2\pi/T \approx 7.27 \times 10^{-5}$ rad s^{-1}, the Earth's radius is $\mathcal{R} \approx 6.37 \times 10^6$ m and $g \approx 9.81$. Therefore

$$|\phi| \sim \arctan(0.0034) \sim 0.0034\,\text{rad} \sim 0.2°. \qquad (9.45)$$

We can thus conclude that the angle is small. As $\mathcal{R}\omega^2 \sim 0.033 \ll g$ we can make the following Taylor expansion

$$\begin{aligned}
\tan\phi &= -\frac{\mathcal{R}\omega^2 \sin\theta \cos\theta}{g} \left(\frac{1}{1 - \mathcal{R}\omega^2 \cos^2\theta/g} \right) \\
&= -\frac{\mathcal{R}\omega^2 \sin\theta \cos\theta}{g} \left(1 + \mathcal{R}\omega^2 \cos^2\theta/g + \mathcal{O}(\omega^4) \right), \qquad (9.46) \\
&= -\frac{\mathcal{R}\omega^2 \sin\theta \cos\theta}{g} + \mathcal{O}(\omega^4).
\end{aligned}$$

Remember also, that for small angles

$$\tan \phi \approx \phi \approx \sin \phi, \tag{9.47}$$

and so, up to $\mathcal{O}(\omega^2)$ we have

$$\sin \phi \approx \phi \approx -\frac{\mathcal{R}\omega^2 \sin\theta \cos\theta}{g},$$
$$\cos \phi \approx 1. \tag{9.48}$$

The negative sign only indicates the sense of the rotation of the angle (clockwise, when looking at the (y', z') plane from the positive region of the $\mathbf{i}'$ axis). We can observe that this deviation reaches its maximum (absolute) value for $\theta = \pm\pi/4$, for which $|\phi| \approx 0.1° \approx 0.0017$ rad. For a freely falling object from 100 m the deviation of the plumb would be approximately $100 \times 0.0017 = 0.17$ m, which is a quite large contribution.

We can now calculate $g''_{z,eff}$. Up to order ω^2 we obtain

$$\begin{aligned}
g''_{z,eff} &= -\sin\phi\, g'_{y,eff} + \cos\phi\, g'_{z,eff} \\
&= g'_{z,eff} + \mathcal{O}(\omega^4) \\
&= -g + \mathcal{R}\omega^2 \cos^2\theta + \mathcal{O}(\omega^4).
\end{aligned} \tag{9.49}$$

So, up to $\mathcal{O}(\omega^2)$ which is the order we are mostly interested in, the rotation of the axes does not affect the z component of the effective gravitational acceleration.

Let us now discuss what is the effect of rotating the coordinate axis on the Coriolis term. This term already contains one power of ω therefore the rotation would introduce additional contributions of $\mathcal{O}(\omega^3)$, which are smaller than $\mathcal{O}(\omega^2)$, and therefore will be neglected in our calculations.

Having estimated all the previous contributions, one can check that in the **double primed** basis where $\mathbf{g}_{eff}$ is aligned with the vertical (**dropping the** eff **label** from $g''_{z,eff}$ in order to simplify the notation), the equations of motion (9.36) form a coupled system of differential equations

$$\begin{aligned}
\ddot{x}'' &= 2\omega \sin\theta\, \dot{y}'' - 2\omega \cos\theta\, \dot{z}'', \\
\ddot{y}'' &= -2\omega \sin\theta\, \dot{x}'', \\
\ddot{z}'' &= -g''_z + 2\omega \cos\theta\, \dot{x}''.
\end{aligned} \tag{9.50}$$

This is, what one can find in the majority of textbooks, unfortunately without explicitly going through the previous calculations.

The previous system in general, has no analytical solution. However, due to the fact that $\omega \ll 1$, we can solve the previous system perturbatively up to the desired order of ω as we shall see in the following.

Let us consider a body placed at rest over some platform (that is at rest with the Earth) at some height z''_0. This object is released at some instant $t_0 = 0$ and falls to the ground. In order to describe its motion we have to solve (9.50) perturbatively.

The first step in applying perturbation theory is to set $\omega = 0$. We thus obtain

$$\ddot{x}'' = 0, \qquad \ddot{y}'' = 0, \qquad \ddot{z}'' = -g_z'', \tag{9.51}$$

which is a non-coupled system that can be trivially integrated. Considering the initial conditions $t_0 = 0$ and $\dot{x}_0'' = \dot{y}_0'' = \dot{z}_0'' = 0$, we simply obtain

$$
\begin{aligned}
\dot{x}'' &= 0, \\
\dot{y}'' &= 0, \\
\dot{z}'' &= -g_z'' t.
\end{aligned} \tag{9.52}
$$

The second step is to introduce the previous velocities back into (9.50). We obtain, at first order in perturbation theory

$$
\begin{aligned}
\ddot{x}'' &= 2\omega \cos\theta \, g_z'' t, \\
\ddot{y}'' &= 0, \\
\ddot{z}'' &= -g_z''.
\end{aligned} \tag{9.53}
$$

Integrating,

$$
\begin{aligned}
\dot{x}'' &= \omega \cos\theta \, g_z'' t^2, \\
\dot{y}'' &= 0, \\
\dot{z}'' &= -g_z'' t.
\end{aligned} \tag{9.54}
$$

If one wishes to obtain a higher order approximations, one can re-insert the previous results back into (9.50). For now we are only interested in the first order terms therefore, we directly integrate the previous expressions and get

$$
\begin{aligned}
x'' &= \frac{1}{3}\omega \cos\theta \, g_z'' t^3, \\
y'' &= 0 \\
z'' &= -\frac{1}{2} g_z'' t^2 + z_0''.
\end{aligned} \tag{9.55}
$$

Note that there is a positive deviation along the $\mathbf{i}'(=\mathbf{i}'')$ axis (eastern deflection) in both hemispheres, that is due to the Coriolis acceleration (called the Coriolis effect). We can further eliminate t by using the expression for z''. If t is the time that it takes the object to reach the ground, having started its free fall at z_0'', then $z''(t) = 0$ and so

$$t = \sqrt{\frac{2z_0''}{g_z''}}. \tag{9.56}$$

Therefore, we finally obtain

$$x'' = \frac{2\omega z_0''}{3}\sqrt{\frac{2z_0''}{g}}\cos\theta. \tag{9.57}$$

Let us now numerically analyse the previous results in order to estimate the order of magnitude of the deviation. For $\theta = \pi/4$ (45° northern latitude), $\omega = 2\pi/T = 7.27 \times 10^{-5}$ s^{-1}, $\mathcal{R} = 6.37 \times 10^6$ m, $g = 9.81$ m s^{-2} and $z_0'' = 100$ m, we obtain

$$x'' = 0.015\,\text{m}. \tag{9.58}$$

Having introduced al the previous concepts and the method to solve the system of equations, in the final section of this chapter we shall propose and solve a few interesting exercises related to deflections due to the Coriolis effect, for objects or projectiles that are launched from the surface of the Earth (some results may even seem rather counter-intuitive). Before this, we shall end the theoretical section by briefly analysing Foucault's pendulum.

9.4 Foucault's Pendulum

Consider a simple pendulum of length l and mass m, in some arbitrary inertial reference system, as shown in Fig. 9.5. If we consider small oscillations $\alpha \ll$, then we can consider $z \simeq$ constant. The oscillations in the (x, y) plane are given by

$$m\,\mathbf{a} = -m\,g\sin\alpha\,\mathbf{u}_r, \tag{9.59}$$

where $\mathbf{a} = \ddot{x}\mathbf{i} + \ddot{y}\mathbf{j}$, $\mathbf{u}_r = \mathbf{r}/r$, $r = |\mathbf{r}|$ and $\mathbf{r} = x\mathbf{i} + y\mathbf{j}$. Expressing $\sin\alpha$ as $\sin\alpha = r/l$ and cancelling m from both sides of the previous expression we obtain

$$\mathbf{a} = -\frac{g}{l}\mathbf{r}. \tag{9.60}$$

Let us now consider the same pendulum in the (previously introduced) double primed non-inertial reference frame (where $\mathbf{g}_{eff}$ is aligned with the vertical). Up to $\mathcal{O}(\omega^2)$ the equations of motion are simply

$$\mathbf{a}' = \mathbf{a} - 2\boldsymbol{\omega} \times \mathbf{v}', \tag{9.61}$$

where we haven't written down explicitly the centrifugal terms as (in the double primed basis) it is absorbed into $\mathbf{g}_{eff}$ i.e., $\mathbf{a} = (g_z''/l)\mathbf{r}$. Considering $z'' \simeq$ constant, the equations of motion in the double primed non-inertial reference frame read

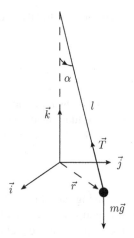

$$\ddot{x}'' = 2\,\omega \sin\theta\,\dot{y}'' - \frac{g_z''}{l}\,x'' , \tag{9.62}$$

and

$$\ddot{y}'' = -2\,\omega \sin\theta\,\dot{x}'' - \frac{g_z''}{l}\,y'' . \tag{9.63}$$

This is a set of coupled differential equations and it is beyond the goal of this book to describe the methods for solving such equations, or the analysis of the corresponding type of oscillations.[3] We shall thus briefly present the solution. If one sums to the first equation, the second equation multiplied by the complex number i and simplifies, one obtains

$$\ddot{\beta} + 2\,i\,\omega \sin\theta\dot{\beta} + \frac{g_z''}{l}\beta = 0 , \tag{9.64}$$

where we have defined $\beta \equiv x'' + i\,y''$. One can check that the solution to this equation is given by

$$\beta(t) = \beta_0\,e^{-i\omega \sin\theta t}\cos\left(\sqrt{\omega^2 \sin^2\theta + \frac{g_z''}{l}}\,t + \phi_0\right) , \tag{9.65}$$

with β_0 and ϕ_0 two real constants. As $\omega^2 \ll g_z''/l$, up to a good approximation we can eliminate ω^2 from the squared root. We can thus write the solution in terms of x'' and y'' as

---

[3]The previous expressions correspond to the equations that describe a damped oscillator.

$$x''(t) \simeq \beta_0 \cos(\omega \sin \theta \, t) \cos\left(\sqrt{\frac{g_z''}{l}} t + \phi_0\right), \tag{9.66}$$

and

$$y''(t) \simeq -\beta_0 \sin(\omega \sin \theta \, t) \cos\left(\sqrt{\frac{g_z''}{l}} t + \phi_0\right). \tag{9.67}$$

Note that the plane of oscillation of the pendulum rotates periodically, due to the Coriolis force, with its period given by

$$T_{rot} = \frac{2\pi}{\omega |\sin \theta|}, \tag{9.68}$$

meaning that at the equator $\theta = 0$ it does not rotate ($T_{rot} \to \infty$) and at the poles $T_{rot} \simeq$ 24 h, which is the Earth's rotation period. Note also that the period of oscillation is that of a simple pendulum (with g replaced by g_z''), as seen in Chap. 4.

9.5 Proposed Exercises

9.1 Towards which cardinal direction is a projectile deviated, due to the Coriolis force, launched under the following circumstances?

- Towards N in the northern/southern hemisphere.
- Towards S in the northern/southern hemisphere.
- Towards E in the northern/southern hemisphere.
- Towards W in the northern/southern hemisphere.
- Downwards/upwards in the northern/southern hemisphere.

Solution: In order to answer these questions we have to remember the expression for the acceleration in the non-inertial reference system. Considering that we are in the double primed reference system i.e., ignoring as a first order approximation the centrifugal force, we are left with

$$\mathbf{a}' = \mathbf{a} - 2\boldsymbol{\omega} \times \mathbf{v}', \tag{9.69}$$

where

$$-2\boldsymbol{\omega} \times \mathbf{v}' = -2\omega(v_z' \cos \theta - v_y' \sin \theta)\mathbf{i}' - 2\omega \, v_x' \sin \theta \mathbf{j}' + 2\omega v_x' \cos \theta \mathbf{k}'. \tag{9.70}$$

With the previous expression we can easily answer all the above questions.

For the first case we have $v'_x = 0$ and $v'_y > 0$. If we consider that $v'_z \ll v'_y$ and we are in the northern hemisphere $\sin \theta > 0$, then

$$-2\boldsymbol{\omega} \times \mathbf{v}' = 2\,\omega\, v'_y \sin \theta\, \mathbf{i}' \quad > \quad 0\,, \tag{9.71}$$

which corresponds to an eastern deflection. In the southern hemisphere it would correspond to a western deflection. As for the second case, it can be easily obtained by simply inverting the sign of v'_y.

For the third case we have $v'_x > 0$ and $v'_y = 0$ in the northern hemisphere ($\sin \theta > 0$). If $v'_z \ll v'_x$ we have

$$-2\boldsymbol{\omega} \times \mathbf{v}' = -2\,\omega\, v'_x \sin \theta \mathbf{j}' + 2\,\omega\, v'_x \cos \theta \mathbf{k}'\,, \tag{9.72}$$

which corresponds to an southern-upward deflection. In the southern hemisphere it would correspond to a northern-upward deflection. The fourth case can also be easily obtained from the previous, by simply inverting the sign of v'_x.

Last, if it is launched upwards or downwards the expression is

$$-2\boldsymbol{\omega} \times \mathbf{v}' = -2\,\omega\, v'_z \cos \theta \mathbf{i}'\,, \tag{9.73}$$

with $v'_z > 0$ or $v'_z < 0$ therefore with an westerly or easterly deflection (independently of the hemisphere). Remember that we have already studied the easterly deflection for a free-falling body.

In the following we shall analyse in great detail the behaviour of a body that is launched from the surface of the Earth, reaches some final height h and afterwards undergoes free-fall. Hence, it will first suffer a westerly deflection on its way up, and an easterly deflection on its way down, and the net result will not be zero.

9.2 Consider the equations of motion (9.50) in the double primed reference system. Consider an object of mass m that is thrown upwards with initial velocity $\mathbf{v}_0 = v_0\, \mathbf{z}''$. Calculate the deflection, due to the Coriolis force, when it reaches the maximum height. Calculate the total deflection when it reaches the ground. Compare the result with the free-fall case.

Solution: In order to make it easier for the reader to follow the calculation we shall write down again the system corresponding to the equations of motion

$$\begin{aligned}
\ddot{x}'' &= 2\,\omega \sin \theta\, \dot{y}'' - 2\,\omega \cos \theta\, \dot{z}''\,, \\
\ddot{y}'' &= -2\,\omega \sin \theta\, \dot{x}''\,, \\
\ddot{z}'' &= -g''_z + 2\,\omega \cos \theta\, \dot{x}''\,.
\end{aligned} \tag{9.74}$$

By setting $\omega = 0$, the previous system reads

$$\ddot{x}'' = 0,$$
$$\ddot{y}'' = 0, \tag{9.75}$$
$$\ddot{z}'' = -g_z''.$$

Integrating, we obtain

$$\dot{x}'' = 0,$$
$$\dot{y}'' = 0, \tag{9.76}$$
$$\dot{z}'' = -g_z''t + v_0.$$

Inserting these expressions back into the initial system of equations we get

$$\ddot{x}'' = 2\omega\cos\theta\,(g_z''t - v_0),$$
$$\ddot{y}'' = 0, \tag{9.77}$$
$$\ddot{z}'' = -g_z''.$$

By integrating once we obtain the components of the velocity as functions of time at $\mathcal{O}(\omega)$

$$\dot{x}'' = \omega\cos\theta\,(g_z''t^2 - 2v_0t),$$
$$\dot{z}'' = -g_z''t + v_0, \tag{9.78}$$

and $\dot{y}'' = 0$. Integrating once again we obtain the expressions for the coordinates as functions of time

$$x''(t) = \omega\cos\theta\left(\frac{1}{3}g_z''t^3 - v_0t^2\right),$$
$$z''(t) = -\frac{1}{2}g_z''t^2 + v_0t, \tag{9.79}$$

with $y''(t) = 0$. The time t_{up} (considering $t_0 = 0$) that it takes for the object to reach its maximum height is simply given by $t_{up} = v_0/g_z''$ i.e. for $\dot{z}''(t_{up}) = 0$. Having calculated this parameter, we can now account for the initial conditions for the fall. First of all, note that it will have an initial velocity along the $\mathbf{i}''$ axis. It is given by

$$\bar{v}_0 \equiv \dot{x}''(t_{up}) = -\omega\cos\theta\,\frac{v_0^2}{g_z''}. \tag{9.80}$$

The maximum height that it will reach, it is given by

$$h_z \equiv z''(t_{up}) = \frac{1}{2} \frac{v_0^2}{g_z''}, \qquad (9.81)$$

and the deflection along the $\mathbf{i}''$ axis is simply

$$d_x \equiv x''(t_{up}) = -\frac{2}{3} \frac{v_0^3}{g_z''^2} \omega \cos\theta. \qquad (9.82)$$

It corresponds to a westerly deflection, in contrast with the easterly deflection corresponding to the free fall case.

The final step is to solve equations (9.74) for the fall. We shall ignore the y'' coordinate as it will bring no contribution. By setting $\omega = 0$ we obtain (9.75). By integrating once, the result reads

$$\dot{x}'' = \bar{v}_0,$$
$$\dot{z}'' = -g_z'' t,$$

where again, for simplicity we have set $t_0 = 0$. By inserting these expressions back into (9.74) and keeping only terms of $\mathcal{O}(\omega)$ we obtain

$$\ddot{x}'' = 2\omega \cos\theta \, g_z'' t,$$
$$\ddot{z}'' = -g_z''.$$

Integrating once we get

$$\dot{x}'' = \omega \cos\theta \, g_z'' t^2 + \bar{v}_0,$$
$$\dot{z}'' = -g_z'' t.$$

Integrating one last time, we finally obtain

$$x''(t) = \frac{1}{3}\omega \cos\theta \, g_z'' t^3 + \bar{v}_0 t + d_x,$$
$$z''(t) = -\frac{1}{2} g_z'' t^2 + h_z.$$

Knowing that it takes the same amount of time for the object to reach the maximum height and to fall back to the ground, the total deflection will be given by

$$x''(t_{up}) = -\frac{4}{3}\omega \cos\theta \frac{v_0^3}{g_z''^2}, \qquad (9.83)$$

which corresponds to a westerly deflection. Note that we could have directly deduced this result from (9.79), by calculating $x''(2t_{up})$.

Fig. 9.6 Direction and sense
of the launched projectile
with initial velocity v_0

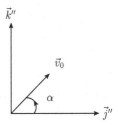

9.3 A projectile is launched from the surface of the Earth, from a point P
of colatitude θ, with an initial velocity $\mathbf{v}_0$ forming an angle α ($\alpha \in (0, \pi/2)$)
with the $\mathbf{j}''$ axis, as shown in Fig. 9.6. Calculate the deflection along the $\mathbf{i}''$ axis
when it reaches the ground. Is it positive or negative?

Solution: Just as previously, by setting $\omega = 0$ and integrating we obtain

$$
\begin{aligned}
\dot{x}'' &= 0, \\
\dot{y}'' &= v_0 \cos \alpha, \\
\dot{z}'' &= -g_z'' t + v_0 \sin \alpha,
\end{aligned}
\tag{9.84}
$$

where $t_0 = 0$. Inserting these expressions back into the original equations of motion
and integrating twice we obtain the following result

$$
\begin{aligned}
x''(t) &= \omega \left(v_0 t^2 (\cos \alpha \sin \theta - \sin \alpha \cos \theta) + \frac{1}{3} g_z'' t^3 \cos \theta \right), \\
y''(t) &= v_0 t \cos \alpha, \\
z''(t) &= -\frac{1}{2} g_z'' t^2 + v_0 t \sin \alpha.
\end{aligned}
\tag{9.85}
$$

As explained in the previous exercise, the deflection can be simply calculated as
$x''(2t_{up})$. Let us therefore calculate t_{up}. If we call the maximum height h, then

$$
z''(t_{up}) = h = -\frac{1}{2} g_z'' t_{up}^2 + v_0 t_{up} \sin \alpha.
\tag{9.86}
$$

Rearranging and simplifying we obtain the following equation for t_{up}

$$
t_{up}^2 - \frac{2 v_0}{g_z''} \sin \alpha \, t_{up} + \frac{2h}{g_z''} = 0,
\tag{9.87}
$$

whose solution is

$$t_{up} = \frac{v_0}{g_z''} \sin \alpha \pm \frac{1}{2} \sqrt{\frac{4v_0^2 \sin^2 \alpha}{g_z''^2} - \frac{8h}{g_z''}}. \tag{9.88}$$

Using energy conservation we obtain the relation between v_0 and h

$$\frac{1}{2} m v_0^2 \sin^2 \alpha = m g_z'' h, \tag{9.89}$$

and so t_{up} is simply given by

$$t_{up} = \frac{v_0}{g_z''} \sin \alpha = \sqrt{\frac{2h}{g_z''}}. \tag{9.90}$$

Just as previously $t_{up} = t_{fall}$ so $t_{up} + t_{fall} = 2t_{up}$ and so the total deflection will be given by

$$x''(2t_{up}) = \frac{4}{3} \omega \frac{v_0^3}{g_z''^2} \sin^2 \alpha (3 \cos \alpha \sin \theta - \sin \alpha \cos \theta). \tag{9.91}$$

It is left for the reader to analyse under what conditions the previous deflection is positive or negative.

9.4 A projectile is launched from the surface of the Earth, from a point P of colatitude θ, with an initial velocity $\mathbf{v_0}$ forming an angle α ($\alpha \in (0, \pi/2)$) with the $\mathbf{i}''$ axis as shown in Fig. 9.7. Calculate the deflection when it reaches the ground.

Solution: This is a similar exercise to the previous one, however it presents a few additional subtleties. Setting $\omega = 0$ and integrating we obtain

$$\begin{aligned}
\dot{x}'' &= v_0 \cos \alpha, \\
\dot{y}'' &= 0, \\
\dot{z}'' &= -g_z'' t + v_0 \sin \alpha.
\end{aligned} \tag{9.92}$$

Inserting these expressions back into the original equations we obtain

$$\begin{aligned}
\ddot{x}'' &= 2\omega(g_z'' t - v_0 \sin \alpha) \cos \theta, \\
\ddot{y}'' &= -2\omega v_0 \cos \alpha \sin \theta, \\
\ddot{z}'' &= -g_z'' + 2\omega v_0 \cos \alpha \cos \theta.
\end{aligned} \tag{9.93}$$

Fig. 9.7 Direction and sense
of the launched projectile
with initial velocity v_0

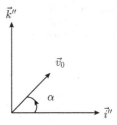

Integrating we obtain

$$
\begin{aligned}
\dot{x}'' &= \omega\, g_z'' t^2 \cos\theta - 2\,\omega\, v_0\, t\, \sin\alpha \cos\theta + v_0 \cos\alpha\,, \\
\dot{y}'' &= -2\,\omega\, v_0\, t\, \cos\alpha \sin\theta\,, \\
\dot{z}'' &= -g_z'' t + 2\,\omega\, v_0\, t \cos\alpha \cos\theta + v_0 \sin\alpha\,.
\end{aligned}
\tag{9.94}
$$

Integrating one last time we finally get

$$
\begin{aligned}
x''(t) &= \frac{1}{3}\omega\, g_z'' t^3 \cos\theta - \omega\, v_0\, t^2\, \sin\alpha \cos\theta + v_0\, t \cos\alpha\,, \\
y''(t) &= -\omega\, v_0\, t^2\, \cos\alpha \sin\theta\,, \\
z''(t) &= -\frac{1}{2} g_z'' t^2 + \omega\, v_0\, t^2 \cos\alpha \cos\theta + v_0\, t \sin\alpha\,.
\end{aligned}
\tag{9.95}
$$

If t is of the order of seconds or a even minutes, the $\omega\, v_0\, t^2$ term is subdominant (when compared to $v_0 t$) and we can safely ignore it. Thus for $t = t_{up}$ we have the same equation as in the previous exercise i.e.,

$$
t_{up} = \frac{v_0}{g_z''} \sin\alpha = \sqrt{\frac{2h}{g_z''}}\,.
\tag{9.96}
$$

Also note, that we have deflections along both axes $\mathbf{i}''$ and $\mathbf{j}''$ in this case. Dropping the $v_0\, t \cos\alpha$ term (from $x''(t)$) which does not correspond to a deflection, we obtain

$$
\begin{aligned}
x''(2t_{up}) &= \frac{1}{3}\omega\, g_z'' t_{up}^3 \cos\theta - \omega\, v_0\, t_{up}^2\, \sin\alpha \cos\theta \\
&= -\frac{4}{3}\omega\, \frac{v_0^3}{g^2} \cos\theta \sin^3\alpha\,,
\end{aligned}
\tag{9.97}
$$

$$
\begin{aligned}
y''(2t_{up}) &= -\omega\, v_0\, t_{up}^2\, \cos\alpha \sin\theta \\
&= -4\,\omega\, \frac{v_0^3}{g^2} \sin\theta \sin^2\alpha \cos\alpha\,.
\end{aligned}
\tag{9.98}
$$

Fig. 9.8 Clockwise rotation
of the oscillation plane in the
northern hemisphere

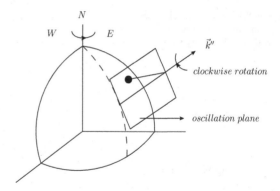

We shall end this chapter by briefly analysing some basic aspects of Foucault's
pendulum.

9.5 Consider Foucault's pendulum with a string of negligible mass and length
of 10 m that is placed at 45° of colatitude. Calculate the effective gravity in the
double primed basis (where $\mathbf{g}_{eff}$ is aligned with the vertical)

$$\mathbf{g}_{eff} = g_z'' \, \mathbf{k}'' . \qquad (9.99)$$

Consider $g_z'' = -9.83$ m/s^2 at the north pole, and the Earth's radius $\mathcal{R} =$
6378 km. Calculate also, its oscillation and rotation period. Obtain the sense
of rotation of the oscillation plane.

Solution: Remember that the expression of the effective gravitational pull is

$$g_z'' = -g + \mathcal{R}\omega^2 \cos^2 \theta , \qquad (9.100)$$

and so, at the poles (where $\cos \theta = 0$) presents its maximum absolute value (which
corresponds to the *inertial* value g). For $\theta = \pi/4$ we obtain $g_z'' \approx -9.81$ m/s^2. The
oscillation frequency is given by

$$\Omega = \sqrt{\frac{g_z''}{l}} , \qquad (9.101)$$

and so the oscillation period is simply

$$T_{osc} = \frac{2\pi}{\Omega} = 2\pi \sqrt{\frac{l}{g_z''}} , \qquad (9.102)$$

which, for $\theta = \pi/4$ is $T_{osc} \approx 6.344$ s. At the poles this period would be $T_{osc} \approx 6.337$ s. The rotation period of the plane of oscillation can be also straightforwardly calculated. We obtain

$$T_{rot} = \frac{24\,\text{h}}{|\sin(\pi/4)|} \approx 33.9\,\text{h}. \tag{9.103}$$

In order to give an answer to the last question we need to focus on the following terms of the solution of the equations of motion

$$x''(t) \sim \beta_0 \cos(\omega t \sin \theta),$$
$$y''(t) \sim -\beta_0 \sin(\omega t \sin \theta). \tag{9.104}$$

By choosing $\beta_0 > 0$ and making a Taylor expansion of $\cos(\omega \sin \theta t)$ and $\sin(\omega \sin \theta t)$ for small t, we obtain

$$x''(t) \sim 1 - \frac{\epsilon^2}{2} + \mathcal{O}(\epsilon^3),$$
$$y''(t) \sim -\epsilon + \mathcal{O}(\epsilon^3), \tag{9.105}$$

where we have considered $\epsilon = \omega t \sin \theta$. From the previous expressions we conclude that in the southern hemisphere ($\sin \theta < 0$) it rotates counter-clockwise and, in the northern hemisphere it rotates clockwise, as shown in Fig. 9.8.

Further Reading

1. J.V. José, E.J. Saletan, *Classical Dynamics: A Contemporary Approach*. Cambridge University Press
2. S.T. Thornton, J.B. Marion, *Classical Dynamics of Particles and Systems*
3. H. Goldstein, C. Poole, J. Safko, *Classical Mechanics*, 3rd edn. Addison Wesley
4. J.R. Taylor, *Classical Mechanics*
5. D.T. Greenwood, *Classical Dynamics*. Prentice-Hall Inc.
6. D. Kleppner, R. Kolenkow, *An Introduction to Mechanics*
7. C. Lanczos, *The Variational Principles of Mechanics*. Dover Publications Inc.
8. W. Greiner, *Classical Mechanics: Systems of Particles and Hamiltonian Dynamics*. Springer
9. H.C. Corben, P. Stehle, *Classical Mechanics*, 2nd edn. Dover Publications Inc.
10. T.W.B. Kibble, F.H. Berkshire, *Classical Mechanics*. Imperial College Press
11. M.G. Calkin, *Lagrangian and Hamiltonian Mechanics*
12. A.J. French, M.G. Ebison, *Introduction to Classical Mechanics*

Chapter 10
Rigid Body Dynamics

Abstract This chapter introduces from basic principles the equations that describe rigid solid dynamics, making use of the Euler angles introduced in Appendix F (also previously discussed in detail in Chap. 9). The reader is thus highly encouraged to read at least Appendix F before starting the lecture of this chapter. New concepts such as the inertia tensor, will be studied in detail for both the discrete and the continuum case. All relevant equations will be deduced transparently from simple expressions i.e., the kinetic energy and the angular momentum. The Euler equations of motion will also be deduced, analysed and solved for a few relevant cases, along with the analysis of motion stability for a rigid body.

10.1 Discrete Case

Consider an N-particle system whose components maintain a constant relative distance i.e.,

$$|\mathbf{r}_\alpha - \mathbf{r}_\beta| = C_{\alpha\beta} \tag{10.1}$$

for any pair α, β, where $C_{\alpha\beta}$ are non-negative constants and where $\mathbf{r}_{\alpha,\beta}$ are the coordinates of the particles in some reference system. This corresponds to a discrete rigid solid. The definition can be straightforwardly extended to a continuous object by considering $\mathbf{r}_\alpha$ and $\mathbf{r}_\beta$ the coordinates of any two arbitrary points of the object.

Let us now consider an arbitrary reference frame $\{\mathcal{O}, \mathbf{i}, \mathbf{j}, \mathbf{k}\}$ and a second reference frame $\{\mathcal{O}', \mathbf{i}', \mathbf{j}', \mathbf{k}'\}$ at rest with the system that forms a discrete rigid solid, as shown in Fig. 10.1. The two coordinate systems are related through the transformation

$$\mathbf{r}_\alpha = \mathbf{c} + \mathbf{r}'_\alpha, \tag{10.2}$$

© Springer Nature Switzerland AG 2020
V. Ilisie, *Lectures in Classical Mechanics*, Undergraduate Lecture
Notes in Physics, https://doi.org/10.1007/978-3-030-38585-9_10

Fig. 10.1 A reference frame $\{\mathcal{O}, \mathbf{i}, \mathbf{j}, \mathbf{k}\}$ and a primed reference frame $\{\mathcal{O}', \mathbf{i}', \mathbf{j}', \mathbf{k}'\}$, that is at rest with the discrete rigid solid (and which can be obtained through a rotation and a translation of the non-primed system)

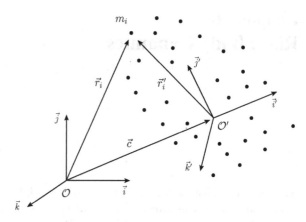

with $\alpha = 1, \ldots, N$. The previous expression can be explicitly written in matrix form as

$$(\mathbf{i}, \mathbf{j}, \mathbf{k}) \begin{pmatrix} x_\alpha \\ y_\alpha \\ z_\alpha \end{pmatrix} = (\mathbf{i}, \mathbf{j}, \mathbf{k}) \begin{pmatrix} c_x \\ c_x \\ c_z \end{pmatrix} + (\mathbf{i}', \mathbf{j}', \mathbf{k}') \begin{pmatrix} x'_\alpha \\ y'_\alpha \\ x'_\alpha \end{pmatrix}. \tag{10.3}$$

Let us now consider the rotation matrix R that relates the two bases

$$\begin{pmatrix} \mathbf{i}' \\ \mathbf{j}' \\ \mathbf{k}' \end{pmatrix} = R \begin{pmatrix} \mathbf{i} \\ \mathbf{j} \\ \mathbf{k} \end{pmatrix}. \tag{10.4}$$

This matrix can be expressed in terms of Euler's angles, as in Appendix F, i.e., $R = R_{\psi\theta\phi}$.

As in the previous chapter[1] we consider that Euler's angles change with time (see Appendix F). The time variation of R can be expressed in terms of the matrix product

$$dR/dt = \Omega R, \tag{10.5}$$

or equivalently

$$dR^T/dt = R^T \Omega^T, \tag{10.6}$$

where Ω is explicitly given by

---

[1] We repeat the same arguments as in the previous chapter for completeness, for the possible reader that has not yet studied non-inertial reference frames.

$$\Omega = \begin{pmatrix} 0 & \omega_3' & -\omega_2' \\ -\omega_3' & 0 & \omega_1' \\ \omega_2' & -\omega_1' & 0 \end{pmatrix}, \tag{10.7}$$

with ω_1', ω_2' and ω_3', the components of the angular velocity in the primed basis i.e.,

$$\omega = \omega_1' \mathbf{i}' + \omega_2' \mathbf{j}' + \omega_3' \mathbf{k}'. \tag{10.8}$$

As shown in (F.9), ω can be expressed in terms of Euler's angles and their time variation.

Having introduced the previous concepts, we can now relate the velocity of a particle in both reference frames

$$\frac{d\mathbf{r}_\alpha}{dt} = \frac{d\mathbf{c}}{dt} + \frac{d\mathbf{r}_\alpha'}{dt}, \tag{10.9}$$

which explicitly (in matrix form) reads

$$(\mathbf{i}, \mathbf{j}, \mathbf{k}) \begin{pmatrix} dx_\alpha/dt \\ dy_\alpha/dt \\ dz_\alpha/dt \end{pmatrix} = (\mathbf{i}, \mathbf{j}, \mathbf{k}) \begin{pmatrix} dc_x/dt \\ dc_y/dt \\ dc_z/dt \end{pmatrix} + (\mathbf{i}, \mathbf{j}, \mathbf{k}) R^T \begin{pmatrix} dx_\alpha'/dt \\ dy_\alpha'/dt \\ dz_\alpha'/dt \end{pmatrix}$$

$$+ (\mathbf{i}, \mathbf{j}, \mathbf{k}) \frac{dR^T}{dt} \begin{pmatrix} x_\alpha' \\ y_\alpha' \\ z_\alpha' \end{pmatrix}. \tag{10.10}$$

As mentioned previously, the $\mathcal{O}'$ reference frame is at rest with the particle system, and so

$$v_{\alpha,x}' = \frac{dx_\alpha'}{dt} = v_{\alpha,y}' = \frac{dy_\alpha'}{dt} = v_{\alpha,z}' = \frac{dz_\alpha'}{dt} = 0. \tag{10.11}$$

Hence, we obtain

$$(\mathbf{i}, \mathbf{j}, \mathbf{k}) \begin{pmatrix} v_{\alpha,x} \\ v_{\alpha,y} \\ v_{\alpha,z} \end{pmatrix} = (\mathbf{i}, \mathbf{j}, \mathbf{k}) \begin{pmatrix} V_x \\ V_y \\ V_z \end{pmatrix} + (\mathbf{i}', \mathbf{j}', \mathbf{k}') \Omega^T \begin{pmatrix} x_\alpha' \\ y_\alpha' \\ z_\alpha' \end{pmatrix}, \tag{10.12}$$

where $V_i = dc_i/dt$ with $i = x, y, z$. Let us now analyse the last term of the previous expression. Again, as deduced in Appendix F,

$$(\mathbf{i}', \mathbf{j}', \mathbf{k}') \Omega^T \begin{pmatrix} x_\alpha' \\ y_\alpha' \\ z_\alpha' \end{pmatrix} = \omega \times \mathbf{r}_\alpha'. \tag{10.13}$$

We are left with the following simple expression

$$\mathbf{v}_\alpha = \mathbf{V} + \omega \times \mathbf{r}'_\alpha. \tag{10.14}$$

We can now introduce the total kinetic energy of the system and relate it to ω

$$
\begin{aligned}
K &= \frac{1}{2} \sum_\alpha m_\alpha \mathbf{v}_\alpha^2 \\
&= \frac{1}{2} \sum_\alpha m_\alpha \left(\mathbf{V} + \omega \times \mathbf{r}'_\alpha\right)^2 \\
&= \frac{1}{2} \sum_\alpha m_\alpha \mathbf{V}^2 + \sum_\alpha m_\alpha \mathbf{V} \cdot (\omega \times \mathbf{r}'_\alpha) + \frac{1}{2} \sum_\alpha m_\alpha (\omega \times \mathbf{r}'_\alpha)^2 \\
&= \frac{1}{2} M \mathbf{V}^2 + M \mathbf{V} \cdot (\omega \times \mathcal{R}') + \frac{1}{2} \sum_\alpha m_\alpha (\omega \times \mathbf{r}'_\alpha)^2, \tag{10.15}
\end{aligned}
$$

where $\mathcal{R}'$ are the coordinates of the CM with respect to the primed basis and M the total mass of the system. As the first two terms from the last line of the previous expression present no complications, we shall focus our analysis on the third term, that we shall call *rotational kinetic energy*

$$K_r = \frac{1}{2} \sum_\alpha m_\alpha (\omega \times \mathbf{r}'_\alpha)^2. \tag{10.16}$$

Using the following vector identity

$$(\mathbf{a} \times \mathbf{b})^2 = \mathbf{a}^2 \mathbf{b}^2 - (\mathbf{a} \cdot \mathbf{b})^2, \tag{10.17}$$

we obtain

$$K_r = \frac{1}{2} \sum_\alpha m_\alpha \left(\omega^2 \mathbf{r}'^2_\alpha - (\omega \cdot \mathbf{r}'_\alpha)^2\right). \tag{10.18}$$

Let us now introduce the following notation

$$x'_{1,\alpha} \equiv x'_\alpha, \qquad x'_{2,\alpha} \equiv y'_\alpha, \qquad x'_{3,\alpha} \equiv z'_\alpha. \tag{10.19}$$

We obtain

$$
\begin{aligned}
K_r = \frac{1}{2} \sum_\alpha m_\alpha \Big(&(\omega'^2_1 + \omega'^2_2 + \omega'^2_3)(x'^2_{1,\alpha} + x'^2_{2,\alpha} + x'^2_{3,\alpha}) \\
&- (\omega'_1 x'_{1,\alpha} + \omega'_2 x'_{2,\alpha} + \omega'_3 x'_{3,\alpha})^2 \Big). \tag{10.20}
\end{aligned}
$$

The reader is invited to check that the previous expression can be written in the compact form

$$K_r = \frac{1}{2} \sum_{i,j} \left[\sum_\alpha m_\alpha \left(\delta_{ij} \sum_k x_{k,\alpha}'^2 - x_{i,\alpha}' x_{j,\alpha}' \right) \right] \omega_i' \omega_j' , \tag{10.21}$$

where, the inertia tensor of the system, with respect to the primed basis, is defined as

$$I_{ij}' \equiv \sum_\alpha m_\alpha \left(\delta_{ij} \sum_k x_{k,\alpha}'^2 - x_{i,\alpha}' x_{j,\alpha}' \right) . \tag{10.22}$$

Therefore the rotational kinetic energy can be expressed as the matrix product

$$K_r = \frac{1}{2} \sum_{i,j} I_{ij}' \omega_i' \omega_j' \equiv \frac{1}{2} \omega^T I \omega . \tag{10.23}$$

We shall try to find a similar expression for the angular momentum. The total angular momentum of the system is given by

$$\begin{aligned}
\mathbf{L} &= \sum_\alpha m_\alpha (\mathbf{r}_\alpha \times \mathbf{v}_\alpha) \\
&= \sum_\alpha m_\alpha \left((\mathbf{c} + \mathbf{r}_\alpha') \times (\mathbf{V} + \omega \times \mathbf{r}_\alpha') \right) \\
&= M \left(\mathbf{c} \times \mathbf{V} + \mathbf{c} \times (\omega \times \mathcal{R}') + \mathcal{R}' \times \mathbf{V} \right) + \mathbf{L}_r , \tag{10.24}
\end{aligned}$$

where we have introduced the rotational angular momentum $\mathbf{L}_r$ as

$$\mathbf{L}_r = \sum_\alpha \mathbf{r}_\alpha' \times (\omega \times \mathbf{r}_\alpha') . \tag{10.25}$$

It is left for the reader as a simple exercise to show that, the rotational angular momentum can be written as

$$L_{r,i} = \sum_j I_{ij}' \omega_j' \quad \Rightarrow \quad \mathbf{L}_r = I \omega . \tag{10.26}$$

From the previous results we conclude that we can also write K_r in terms of the angular momentum i.e.,

$$K_r = \frac{1}{2} \mathbf{L}_r \cdot \omega . \tag{10.27}$$

In order to obtain the equations of motion of a rotating rigid solid, we have to calculate the torque

$$\boldsymbol{\tau}_r = \frac{d\mathbf{L}_r}{dt} \, . \tag{10.28}$$

If we write $\mathbf{L}_r$ in the primed basis we obtain

$$\boldsymbol{\tau}_r = (\mathbf{i}', \mathbf{j}', \mathbf{k}') \begin{pmatrix} dL'_{r,1}/dt \\ dL'_{r,2}/dt \\ dL'_{r,3}/dt \end{pmatrix} + \boldsymbol{\omega} \times \mathbf{L}_r \, . \tag{10.29}$$

Using (10.26) and performing the vector product, the expression we find for $\boldsymbol{\tau}_r$ is

$$\boldsymbol{\tau}_r = \tau'_{r,1} \mathbf{i}' + \tau'_{r,2} \mathbf{j}' + \tau'_{r,3} \mathbf{k}' \, , \tag{10.30}$$

with

$$\tau'_{r,1} = \sum_j I'_{1j} \dot{\omega}'_j + (L'_{r,3} \omega'_2 - L'_{r,2} \omega'_3) \, ,$$

$$\tau'_{r,2} = \sum_j I'_{2j} \dot{\omega}'_j + (L'_{r,1} \omega'_3 - L'_{r,3} \omega'_1) \, , \tag{10.31}$$

$$\tau'_{r,3} = \sum_j I'_{3j} \dot{\omega}'_j + (L'_{r,2} \omega'_1 - L'_{r,1} \omega'_2) \, .$$

These last three expressions represent Euler's equations of motion for a rigid body, in the primed basis.

10.1.1   Principal Axes

Note that the inertia tensor is a generic 3×3 matrix that is, in general, non-diagonal. Let us consider that there is a reference system in which the inertia tensor is diagonal. We shall denote the orthonormal reference system associated to this basis as $\{\mathcal{O}', \mathbf{i}'_P, \mathbf{j}'_P, \mathbf{k}'_P\}$ (where P stands for *principal* axes). If we consider that the two bases are related through an orthogonal matrix B (with $B^T = B^{-1}$) then

$$\begin{pmatrix} \mathbf{i}'_P \\ \mathbf{j}'_P \\ \mathbf{k}'_P \end{pmatrix} = B \begin{pmatrix} \mathbf{i}' \\ \mathbf{j}' \\ \mathbf{k}' \end{pmatrix} , \tag{10.32}$$

and

$$\begin{pmatrix} \omega'_{P,1} \\ \omega'_{P,2} \\ \omega'_{P,3} \end{pmatrix} = B \begin{pmatrix} \omega'_1 \\ \omega'_2 \\ \omega'_3 \end{pmatrix}. \tag{10.33}$$

Both the principal axes (thus the matrix B) and the diagonal terms of the inertia tensor (with respect to the principal axes) can be obtained by diagonalizing I'_{ij}. We will denote the eigenvalues of I'_{ij} as $I'_{P,1}$, $I'_{P,2}$, $I'_{P,3}$, which will be the components of the diagonal inertia tensor

$$I_P = \begin{pmatrix} I'_{P,1} & 0 & 0 \\ 0 & I'_{P,2} & 0 \\ 0 & 0 & I'_{P,3} \end{pmatrix}. \tag{10.34}$$

The eigenvectors of I'_{ij}, except normalization factors, represent the the principal axes.[2] Using the basis $\{\mathcal{O}', \mathbf{i}'_P, \mathbf{j}'_P, \mathbf{k}'_P\}$ Euler's equations of motion simplify to

$$\begin{aligned} \tau'_{r,1} &= I'_{P,1} \dot{\omega}'_{P,1} + \omega'_{P,2} \omega'_{P,3}(I'_{P,3} - I'_{P,2}), \\ \tau'_{r,2} &= I'_{P,2} \dot{\omega}'_{P,2} + \omega'_{P,1} \omega'_{P,3}(I'_{P,1} - I'_{P,3}), \\ \tau'_{r,3} &= I'_{P,3} \dot{\omega}'_{P,3} + \omega'_{P,1} \omega'_{P,2}(I'_{P,2} - I'_{P,1}). \end{aligned} \tag{10.35}$$

If all three eigenvalues are different then the object is said to be asymmetric. If we have one doubly degenerate eigenvalue then we say that the body presents cylindrical symmetry. If all three eigenvalues coincide then we say the object has spherical symmetry. It is left for the reader as an exercise to prove that a sphere and a cube (of uniform constant mass densities) are spherically symmetric with respect to their center of mass.

Some further comments are required. For the case of one doubly-degenerate eigenvalue, the eigenvectors that correspond to the degenerate eigenvalue are not, in general, mutually orthogonal. In this case one can choose as two principal axes, any pair of two mutual orthogonal axes that belong to plane formed by the eigenvectors that correspond to the degenerate eigenvalue. The third axis will be of course given by the eigenvector corresponding to the non-degenerate eigenvalue. If the inertia tensor has one triply-degenerate eigenvalue, then one can chose any three arbitrary orthogonal axes as the principal axes.

10.1.2 Huygens–Steiner Theorem

This theorem relates the inertia tensor with respect to some reference system $\{\mathcal{O}, \mathbf{i}, \mathbf{j}, \mathbf{k}\}$ with the tensor of inertia with respect to the center of mass (CM) and

---

[2]As the inertia tensor is a symmetric hermitian matrix $I = I^T$, one can prove that the eigenvectors that correspond to non degenerate eigenvalues are orthogonal.

Fig. 10.2 An arbitrary
frame $\mathcal{O}$ and the center of
mass frame with same axes,
obtained through a
translation

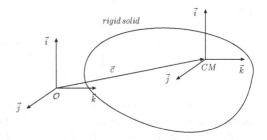

same axes i.e., $\{CM, \mathbf{i}, \mathbf{j}, \mathbf{k}\}$, as shown in the Fig. 10.2. The theorem states

$$I_{ij}^{\mathcal{O}} = I_{ij}^{CM} + M \left(\mathbf{c}^2 \, \delta_{ij} - c_i \, c_j\right), \tag{10.36}$$

where M is the total mass of the body. The proof of this theorem is left for the reader
as an exercise.

The previous expression is valid both in the discrete and the continuum case. In
the following we shall discuss the continuum generalization. As we shall see, it will
turn out to be rather trivial and most of the previous results will remain unmodified.

10.2   Continuum Generalization

Taking a careful look at all the previous equations, one should realize that in the
continuum limit only the calculation of the coordinates of the center of mass and the
inertia tensor are affected. For the continuum case we have the following expressions

$$\mathcal{R} = \frac{1}{\int dm} \int \mathbf{r} \, dm, \tag{10.37}$$

for the CM coordinates with respect to the non-primed basis (and a similar expression
for $\mathcal{R}'$) and

$$I_{ij}' = \int dm \left(\delta_{ij} \sum_k x_k'^2 - x_i' x_j'\right), \tag{10.38}$$

for the inertia tensor. The quantity dm can be expressed as $dm = \rho \, dV$ for a volume
mass distribution of density ρ, $dm = \sigma \, ds$ for a two-dimensional superficial mass
distribution of density σ, or $dm = \mu \, dr$ for a one-dimensional mass distribution of
density μ.

So far we have distinguished between the rotational angular momentum and
kinetic energy, and the total angular momentum and the total kinetic energy i.e.,
as in (10.15) and (10.24). However, in what is left of this analysis, we shall pay close
attention to the case $\mathbf{L} = \mathbf{L}_r$ and $K = K_r$. Thus:

> In the remaining part of this chapter, if not stated otherwise we shall consider that
> - $\mathcal{O} = \mathcal{O}'$, thus $\mathbf{c} = \mathbf{0}$ and,
> - $\mathbf{V} = d\mathbf{c}/dt = \mathbf{0}$,
>   therefore $K = K_r$ and $\mathbf{L} = \mathbf{L}_r$.

As a reminder, it should be obvious that, as $\mathcal{O} = \mathcal{O}'$ and there is no relative linear velocity $\mathbf{V}$ between the two reference systems, the quantities $\mathbf{L}$, $\boldsymbol{\tau}$, $\mathbf{r}$, $\mathbf{v}$, $\dot{\mathbf{v}}$ and $\boldsymbol{\omega}$ behave as vectors, when transforming their components from the primed to the non-primed reference frame and vice-versa, at each time instant t.

10.3 Stable Solutions for the Torque-Free Motion

Except for a few particular cases, Euler's equations of motion have no generic analytical solution. In the remaining sections we shall present the solutions for the equations of motion for the torque-free motion and constant $\boldsymbol{\omega}$, and for the force-free and heavy spinning top problem (with one pint fixed), which are three representative examples in the study of rigid body motion.

In this section we shall perform our analysis in the principal basis thus, for simplicity **we shall drop the primed notation and the** P **lower index**.

Let us now focus our attention on the first case. Consider the Euler equations of motion for the torque-free motion of a rigid body written in the principal basis. The equations (10.35) for $\boldsymbol{\tau} = \boldsymbol{\tau}_r = \mathbf{0}$ are given by

$$
\begin{aligned}
I_1\,\dot{\omega}_1 + \omega_2\,\omega_3(I_3 - I_2) &= 0\,, \\
I_2\,\dot{\omega}_2 + \omega_1\,\omega_3(I_1 - I_3) &= 0\,, \\
I_3\,\dot{\omega}_3 + \omega_1\,\omega_2(I_2 - I_1) &= 0\,.
\end{aligned}
\tag{10.39}
$$

Here, we are interested in stable solutions with constant $\boldsymbol{\omega}$ thus

$$
\begin{aligned}
\omega_2\,\omega_3(I_3 - I_2) &= 0\,, \\
\omega_1\,\omega_3(I_1 - I_3) &= 0\,, \\
\omega_1\,\omega_2(I_2 - I_1) &= 0\,.
\end{aligned}
\tag{10.40}
$$

From the previous expressions we conclude that:

1. if $I_1 \neq I_2 \neq I_3$ then we have three possible solutions (with C an arbitrary real constant)

   - $\omega_1 = C$ and $\omega_2 = \omega_3 = 0$
   - $\omega_2 = C$ and $\omega_1 = \omega_3 = 0$

- $\omega_3 = C$ and $\omega_1 = \omega_2 = 0$.

2. if $I_1 = I_2 = I_3$, any constant ω is a solution.
3. if we have one doubly degenerate principal eigenvalue, say $I_1 = I_2 \neq I_3$ then there are two possible configurations: either $\omega_1 = \omega_2 = 0$ and $\omega_3 = C_3$ or $\omega_3 = 0$ and $\omega_1 = C_1$, $\omega_2 = C_2$ (with $C_{1,2,3}$ arbitrary real constants).

In the following, we shall analyse the stability of the previous solutions by considering small perturbations about the equilibrium state. Let us start with the first case $I_1 \neq I_2 \neq I_3$. Without loss of generality we can consider $I_1 < I_2 < I_3$. The first solution is given by $\omega_1 = C$ and $\omega_2 = \omega_3 = 0$. In order to calculate if this solution is stable we shall set $\omega_2 = \epsilon_2$ and $\omega_3 = \epsilon_3$ with $|\epsilon_{2,3}| \ll |\omega_1| \neq 0$ and use (10.39) to calculate ω_1. We obtain

$$
\begin{aligned}
I_1 \dot{\omega}_1 + \epsilon_2 \epsilon_3 (I_3 - I_2) &= 0, \\
I_2 \dot{\epsilon}_2 + \omega_1 \epsilon_3 (I_1 - I_3) &= 0, \\
I_3 \dot{\epsilon}_3 + \omega_1 \epsilon_2 (I_2 - I_1) &= 0.
\end{aligned}
\tag{10.41}
$$

From the first equation we get

$$
\omega_1 = C + \mathcal{O}(\epsilon^2) \simeq C,
\tag{10.42}
$$

with C an arbitrary real constant. We now have to check that the quantities $\epsilon_{2,3}$ remain small as $t \to \infty$. Taking the time derivative of the last two equations of the previous system, we get to the following system of differential equations

$$
\begin{aligned}
I_2 \ddot{\epsilon}_2 + \omega_1 \dot{\epsilon}_3 (I_1 - I_3) &= 0, \\
I_3 \ddot{\epsilon}_3 + \omega_1 \dot{\epsilon}_2 (I_2 - I_1) &= 0.
\end{aligned}
\tag{10.43}
$$

Introducing the expressions of $\dot{\epsilon}_2$ and $\dot{\epsilon}_3$ from (10.41) and simplifying, the previous system becomes

$$
\ddot{\epsilon}_i + \beta \epsilon_i = 0,
\tag{10.44}
$$

with $i = 2, 3$ and

$$
\beta = \omega_1^2 \frac{(I_1 - I_2)(I_1 - I_3)}{I_2 I_3} > 0.
\tag{10.45}
$$

The solution of the differential equations (10.44) read

$$
\epsilon_i = A_i \cos\left(\sqrt{\beta} \, t\right) + B_i \sin\left(\sqrt{\beta} \, t\right),
\tag{10.46}
$$

with A_i and B_i two real integration constants ($i = 2, 3$). Hence, $\epsilon_{2,3}$ remain small as $t \to \infty$ and, the system oscillates (with small oscillations) about the equilibrium solution. It is left for the reader as an exercise to analyse the remaining two cases.

Let us now move on to the spherically symmetrical case $I_1 = I_2 = I_3$. In this case any three mutually orthogonal axes are principal axes. Therefore we can always find a set of axes for which ω_1 is an arbitrary constant and $\omega_2 = \omega_3 = 0$. For the stability analysis we set $\omega_{2,3} = \epsilon_{2,3}$ with $|\epsilon_{2,3}| \ll |\omega_1| \neq 0$. We thus need to study the behaviour of

$$I_1 \dot{\omega}_1 = I_2 \dot{\epsilon}_2 = I_3 \dot{\epsilon}_3 = 0.$$

Note that the previous equations are exact (not only valid up to $\mathcal{O}(\epsilon^2)$). We thus obtain

$$\omega_1 = C_1, \qquad \omega_2 = \epsilon_2 = C_2, \qquad \omega_3 = \epsilon_3 = C_3, \qquad (10.47)$$

with $C_{1,2,3}$ real constants. Therefore, the rotation axis slightly changes its orientation but, the motion remains stable about the new axis. Note that in this case, the previous conclusions are valid even for large perturbations.

Let us thus finally consider the cylindrically symmetric case $I_1 = I_2 \neq I_3$. The first solution is given by $\omega_3 = C \neq 0$, with C an arbitrary real constant and $\omega_1 = \omega_2 = 0$. Introducing small perturbations $\omega_{1,2} = \epsilon_{1,2}$ with $|\epsilon_{1,2}| \ll |\omega_3|$, we need to study the behaviour of

$$
\begin{aligned}
I_1 \dot{\epsilon}_1 + \epsilon_2 \omega_3 (I_3 - I_2) &= 0, \\
I_2 \dot{\epsilon}_2 + \epsilon_1 \omega_3 (I_1 - I_3) &= 0, \qquad (10.48) \\
I_3 \dot{\omega}_3 &= 0.
\end{aligned}
$$

From the last equation we obtain $\omega_3 = C$, with C an arbitrary constant and, from the first two equations we obtain

$$\epsilon_i = A_i \cos\left(\sqrt{\beta}\,t\right) + B_i \sin\left(\sqrt{\beta}\,t\right), \qquad (10.49)$$

with A_i and B_i integration constants ($i = 2, 3$) and with β given by

$$\beta = \omega_3^2 \frac{(I_1 - I_3)^2}{I_1^2} > 0. \qquad (10.50)$$

The motion is therefore stable. Finally, let us consider $\omega_3 = 0$ and $\omega_{2,3} = C_{2,3}$, with $C_{2,3} \neq 0$, two real constants. As any two orthogonal axes of the (ω_1, ω_2) plane are principal axes, we can always find a set of axes for which $\omega_1 = C_1 \neq 0$ and $\omega_2 = 0$. By introducing small perturbations we obtain

Fig. 10.3 Schematic representation of the precession of the Earth's angular velocity ω described in the Earth's reference system

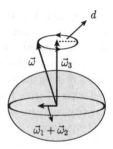

$$I_1 \dot{\omega}_1 + \epsilon_2 \epsilon_3 (I_3 - I_2) = 0,$$
$$I_2 \dot{\epsilon}_2 + \omega_1 \epsilon_3 (I_1 - I_3) = 0,$$
$$I_3 \dot{\epsilon}_3 = 0. \tag{10.51}$$

The solution corresponds to $\epsilon_3 = C_3$ with C_3 a (small) constant and also, $\omega_1 = C_1$ (constant) up to $\mathcal{O}(\epsilon^2)$. As for ϵ_2 we have

$$\epsilon_2(t) = C_2 + \frac{\omega_1 \epsilon_3 (I_1 - I_3)}{I_2} t, \tag{10.52}$$

and so ϵ_2 does not remain small as $t \to \infty$. This configuration is therefore unstable.

10.3.1   Earth's Precession

As one already knows, the Earth is not perfectly spherical. It is in fact better approximated by a rigid body with cylindrical symmetry. It rotates about its principle $\mathbf{k}'_p$ axis (considering $I_1 = I_2 \neq I_3$) with angular velocity ω_3, and due to small ω_1 and ω_2 components, the total angular velocity ω precesses about the $\mathbf{k}'_p$ axis. This is schematically shown in Fig. 10.3. Therefore the solution for the equations of motion is simply given by (10.49), which is valid both for small and large values of $\omega_{1,2}$.

Considering the boundary conditions, we can chose the integration constants of (10.49) so that the solution can be written in the form

$$\omega_1 = d \cos\left(\sqrt{\beta} t + \phi\right),$$
$$\omega_2 = d \sin\left(\sqrt{\beta} t + \phi\right). \tag{10.53}$$

This is the precession of the angular velocity ω, as seen from the Earth's reference system, where the rotation axes are the principal axes. Note that, in order to obtain this result we have not used any external *fixed* reference system and so, it was not necessary to make use of the Euler angles. Thus, one **should not confuse the pre-**

cession of ω (Chandler wobble) described in the Earth's reference system, with the Equinoctial precession, which is the precession of the $\mathbf{k}'_P$ axis about a *fixed*[3] $\mathbf{k}$ axis. This phenomenon will be studied in the following subsection.

Back to our case, for the Earth we approximately have

$$\frac{I_3 - I_1}{I_1} \approx \frac{1}{300}, \tag{10.54}$$

and so the precession frequency is given by

$$\sqrt{\beta} = \frac{\omega_3}{300} \simeq \frac{\omega}{300} \simeq \frac{2\pi/(1\,\text{day})}{300} = \frac{2\pi}{300\,\text{days}}. \tag{10.55}$$

Hence, the precession period is $T \simeq 300$ days. There are experimental evidences that this period is actually $T \approx 400$ days. This is due to different factors. First of all, the Earth is not a perfect rigid body. While getting closer to the core, the Earth becomes elastic. Due to internal friction forces, there is also a non-negligible damping factor. However, as experimental data indicates, there must also exist some additional *excitation* forces that compensate for the damping so that the period is maintained. This phenomenon however, is not entirely understood.

In the next section we shall analyse the force-free symmetric spinning top, which is another representative example of stable motion about a fixed point.

10.4 Free Symmetric Spinning Top

In the following we will analyse *precession*, *nutation* and *spin* for the case of a force-free symmetric ($I'_1 = I'_2 \neq I'_3$) spinning top with one point fixed, in terms of Euler's angles. We shall choose the primed axes to be the principal axes,[4] as shown in Fig. 10.4. As explained in Appendix F, the spin (rotation) is given by the ψ angle (about the $\mathbf{k}'$ axis) and so the spinning rate is given by $\dot{\psi}$. The precession is given by the angle ϕ (about the $\mathbf{k}$ axis) therefore the precession rate is given by $\dot{\phi}$. Finally, the nutation is given by the θ angle (about the $\mathbf{i}_1$ axis) and so the nutation rate is given by $\dot{\theta}$.

When expressing the components of the angular velocity in the same orthonormal reference frame i.e., in the rigid solid's reference frame (whose axes in this case are the primed principal axes) we obtain mixed contributions

---

[3] With respect to the *fixed* far away stars.

[4] Thus we will drop the P subindex.

Fig. 10.4 The force-free symmetric spinning top with one point fixed, in terms of Euler's angles

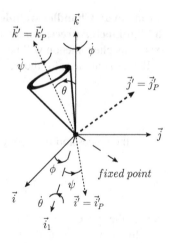

$$\omega_1' = \dot{\phi}\sin\theta\sin\psi + \dot{\theta}\cos\psi\,,$$
$$\omega_2' = \dot{\phi}\sin\theta\cos\psi - \dot{\theta}\sin\psi\,, \qquad (10.56)$$
$$\omega_3' = \dot{\phi}\cos\theta + \dot{\psi}\,.$$

Remember that in the previous subsection we obtained that, for a rigid solid with zero-torque, the motion was nutation-free (because amplitude of ω_1 and ω_2 was constant). Therefore we conclude that θ is constant and the components of the angular velocity simplify to

$$\omega_1' = \dot{\phi}\sin\theta\sin\psi\,,$$
$$\omega_2' = \dot{\phi}\sin\theta\cos\psi\,, \qquad (10.57)$$
$$\omega_3' = \dot{\phi}\cos\theta + \dot{\psi}\,.$$

From the third equation we obtain

$$\psi = (\omega_3' - \dot{\phi}\cos\theta)\,t + \psi_0\,, \qquad (10.58)$$

and so

$$\omega_1' = \dot{\phi}\sin\theta\sin(\omega_3'\,t - \dot{\phi}\cos\theta\,t + \psi_0)\,,$$
$$\omega_2' = \dot{\phi}\sin\theta\cos(\omega_3'\,t - \dot{\phi}\cos\theta\,t + \psi_0)\,, \qquad (10.59)$$

which is consistent with the results obtained in the previous section for constant $\dot{\phi}$ and constant ω_3'.

Let us now perform this analysis by expressing $\boldsymbol{\omega}$ in the non-primed reference frame. Remember that $\boldsymbol{\omega}$ is given by

Fig. 10.5 Precession of ω about the **k** axis for a free spinning top with one fixed point

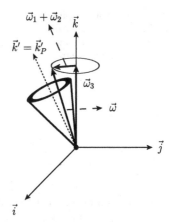

$$\omega = \dot{\phi}\mathbf{k} + \dot{\theta}\mathbf{i}_1 + \dot{\psi}\mathbf{k}'. \tag{10.60}$$

Rotating all the components into the non-primed basis we obtain

$$\omega = \omega_1\mathbf{i} + \omega_2\mathbf{j} + \omega_3\mathbf{k}, \tag{10.61}$$

with

$$\begin{aligned}
\omega_1 &= \dot{\theta}\cos\phi + \dot{\psi}\sin\theta\sin\phi, \\
\omega_2 &= \dot{\theta}\sin\phi - \dot{\psi}\sin\theta\cos\phi, \\
\omega_3 &= \dot{\phi} + \dot{\psi}\cos\theta.
\end{aligned} \tag{10.62}$$

For constant θ, constant $\dot{\phi}$ and $\dot{\psi}$ (thus constant ω_1 and constant $|\omega_2 + \omega_2|$) we obtain

$$\begin{aligned}
\omega_1 &= \dot{\psi}\sin\theta\sin\phi, \\
\omega_2 &= -\dot{\psi}\sin\theta\cos\phi, \\
\omega_3 &= \dot{\phi} + \dot{\psi}\cos\theta,
\end{aligned} \tag{10.63}$$

which means that ω (thus also **L** and $\mathbf{k}'_p$, as we shall prove in the following) precesses about the **k** axis, as shown schematically in Fig. 10.5.

Let us now prove that **L**, ω and $\mathbf{k}'_p$ belong to the same plane. The angular momentum in terms of the principal axes for a symmetric top with $I'_1 = I'_2 \neq I'_3$ (where the components of the primed basis are the principal axes) can be written as

$$\begin{aligned}
\mathbf{L} &= I'_1(\omega'_1\mathbf{i}' + \omega'_2\mathbf{j}') + I'_3\omega'_3\mathbf{k}' \\
&= I'_1\omega + (I'_3 - I'_1)\omega'_3\mathbf{k}',
\end{aligned} \tag{10.64}$$

Fig. 10.6 Rotation (spin), precession and nutation of the Earth described in the non-primed reference frame

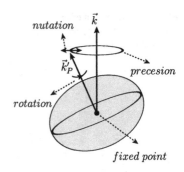

which indeed proves that $\mathbf{L}$, $\boldsymbol{\omega}$ and $\mathbf{k}'_p$ belong to the same plane. Hence, $\mathbf{L}$, $\boldsymbol{\omega}$ and $\mathbf{k}'_p$ precess about the $\mathbf{k}$ axis at the same rate.

If the motion is not torque-free, then $\dot{\theta} \neq 0$ and the equations of motion must be solved numerically. In the case of the Earth, there are small torque components which are of gravitational origin, which introduce nutation. This is schematically shown in Fig. 10.6. This precession (of the angular velocity, angular momentum and the $\mathbf{k}'_p$ axis, at the same rate, about the $\mathbf{k}$ axis) is called **Equinoctial precession**.

In the case of the Earth, the θ angle (between $\mathbf{k}$ and $\mathbf{k}'_p$) is approx 23.5°. The precession has a period of approximately 26,000 years. As for the nutation, its principal origin is the motion of the Moon. It has a period of approximately 18.6 years and the amplitude (without getting into further details) is of the order of arcseconds.

The final section of this chapter will be dedicated to a brief analysis of the heavy symmetric top with one point fixed.

10.5 Heavy Symmetric Spinning Top

Before ending this chapter with the Proposed Exercises section, we shall briefly analyse, the symmetric spinning top ($I'_1 = I'_2 \neq I'_3$) with one point fixed with non-zero torque components (that are of gravitational origin), as shown in Fig. 10.7. The net effect of the gravitational field over the top is a force

$$\mathbf{F}_g = -Mg\,\mathbf{k}, \tag{10.65}$$

that acts on the center of mass of the body (where M is the total mass).

One might result confused, but the net force calculated previously, that acts on the CM does not affect ω'_1. If not intuitively understood, this result is analytically deduced, without substantial effort, in Chap. 13, when introducing the Lagrangian formulation of mechanics (more precisely in Exercise 13.4). Thus $\tau'_3 = 0$, which is a curious result as $\tau_3 = 0$ also (in the non-primed reference frame), as we shall see in the following. Introducing the continuum generalization for the torque in the non-primed reference frame

Fig. 10.7 Heavy spinning top with one point fixed where the primed axes are the principal axes

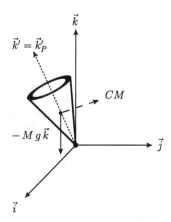

$$\boldsymbol{\tau} = \iiint_V \rho_V(\mathbf{r})\, dV\, (\mathbf{r} \times \mathbf{a})\,, \tag{10.66}$$

one can easily check that

$$\tau_3 = \iiint_V \rho_V(\mathbf{r})\, dV\, (x\, a_{g,y} - y\, a_{g,x}) = 0\,, \tag{10.67}$$

as the components of the gravitational acceleration $a_{g,y} = a_{g,x} = 0$. One should not be confused about this result either. One intuitively expects that if, as in this case, the third component of a vector in a certain basis is zero i.e., τ_3, after performing a proper orthogonal rotation, the third component in the new basis i.e., τ_3', should not be zero any more. This intuitive interpretation however turns out to be false. In some cases, as in the one we are dealing with here, we can simultaneously have $\tau_3 = \tau_3' = 0$. By looking closely at the corresponding transformation

$$\begin{pmatrix} \tau_1' \\ \tau_2' \\ 0 \end{pmatrix} = \begin{pmatrix} c\phi\,c\psi - s\phi\,c\theta\,s\psi & s\phi\,c\psi + c\phi\,c\theta\,s\psi & s\theta\,s\psi \\ -c\phi\,s\psi - s\phi\,c\theta\,c\psi & -s\phi\,s\psi + c\phi\,c\theta\,c\psi & s\theta\,c\psi \\ s\phi\,s\theta & -c\phi\,s\theta & c\theta \end{pmatrix} \begin{pmatrix} \tau_1 \\ \tau_2 \\ 0 \end{pmatrix}, \tag{10.68}$$

we can observe that the transformation is indeed consistent if

$$s\phi\,s\theta\,\tau_1 - c\phi\,s\theta\,\tau_2 = 0\,. \tag{10.69}$$

Geometrically, $\boldsymbol{\tau}$ lays in the intersection of the plane (x, y) with the plane (x', y').

Having made the previous clarifications, as in the primed reference frame

$$I_3'\,\dot{\omega}_3{}' = 0\,, \tag{10.70}$$

we conclude that

$$\omega_3' = \dot{\psi} + \dot{\phi}\cos\theta, \tag{10.71}$$

is a constant of motion. A second constant of motion can be found straightforwardly. As we have just mentioned $\tau_3 = 0$ in the fixed (non-primed) reference frame, and so the component L_3 of the angular momentum is also constant. We can obtain the explicit expression of L_3 by performing the rotation

$$\begin{pmatrix} L_1 \\ L_2 \\ L_3 \end{pmatrix} = R_{\psi\theta\phi}^{-1} \begin{pmatrix} L_1' \\ L_2' \\ L_3' \end{pmatrix} = R_{\psi\theta\phi}^{T} \begin{pmatrix} I_1'\,\omega_1' \\ I_2'\,\omega_2' \\ I_3'\,\omega_3' \end{pmatrix}. \tag{10.72}$$

We obtain the following expression for L_3

$$\begin{aligned} L_3 &= \sin\theta\,(I_1'\,\omega_1'\sin\psi + I_2'\,\omega_2'\cos\psi) + I_3'\,\omega_3'\,\cos\theta \\ &= I_1'\sin\theta\,(\dot{\phi}\sin\theta\sin^2\psi + \dot{\theta}\cos\psi\sin\psi + \dot{\phi}\sin\theta\cos^2\psi - \dot{\theta}\sin\psi\cos\psi) \\ &\quad + I_3'\,\omega_3'\cos\theta \\ &= I_1'\sin^2\theta\,\dot{\phi} + I_3'\,\omega_3'\cos\theta, \end{aligned} \tag{10.73}$$

which is the expression we were looking for.

As mentioned previously, as we shall see in Chap. 13 (Exercise 13.4), where we will introduce the Lagrangian formulation of classical mechanics, these constants of motion arise naturally in a few lines without having to investigate the angular momentum or the torque in the two reference frames.

The last constant of motion associated to the system is obviously the total energy. As there are no friction forces acting upon the system, the total energy is conserved. The gravitational potential energy of a N-particle system is given by

$$U = \sum_{\alpha} m_\alpha\,g\,h_\alpha, \tag{10.74}$$

where h_α is the height of the particle α. Thus, the final result is

$$U = Mg\,h_{CM}, \tag{10.75}$$

where h_{CM} is the height of the center of mass. The result for a continuous mass distribution is obviously identical. We can thus write U as

$$U = Mgl\cos\theta, \tag{10.76}$$

where l is the distance from the origin to the CM. The kinetic energy is given by

$$K = \frac{1}{2}I_1' \left(\omega_1'^2 + \omega_2'^2\right) + \frac{1}{2}I_3'\omega_3'^2$$
$$= \frac{1}{2}I_1'(\dot{\phi}^2 \sin^2\theta + \dot{\theta}^2) + \frac{1}{2}I_3'(\dot{\psi} + \dot{\phi}\cos\theta)^2 .\tag{10.77}$$

We can now introduce in this expression the previously deduced constants of motion. Following the notation from standard textbooks we define the following constants

$$a \equiv \omega_3' \frac{I_3'}{I_1'} = (\dot{\psi} + \dot{\phi}\cos\theta)\frac{I_3'}{I_1'} ,\tag{10.78}$$

and

$$b \equiv \frac{L_3}{I_1'} = \dot{\phi}\sin^2\theta + \frac{I_3'}{I_1'}\omega_3'\cos\theta = \dot{\phi}\sin^2\theta + a\cos\theta .\tag{10.79}$$

The kinetic energy finally simplifies to

$$K = \frac{1}{2}I_1' \left(\frac{b - a\cos\theta}{\sin\theta}\right)^2 + \frac{1}{2}I_1'\dot{\theta}^2 + \frac{1}{2}I_3'\omega_3'^2 .\tag{10.80}$$

Note that, for the previous expression, by using the constants of motion we have managed to eliminate the dependency on the ϕ, ψ and their time derivatives. Finally, the total energy takes the form

$$E = K + U ,\tag{10.81}$$

which is constant. We can further re-arrange the terms from the previous expression into

$$E' = \frac{1}{2}I_1'\dot{\theta}^2 + U_{eff}(\theta) ,\tag{10.82}$$

with

$$E' = E - \frac{1}{2}I_3'\omega_3'^2 ,$$
$$U_{eff}(\theta) = \frac{1}{2}I_1' \left(\frac{b - a\cos\theta}{\sin\theta}\right)^2 + Mgl\cos\theta ,\tag{10.83}$$

and so our problem finally reduces to a one-dimensional problem.

Several exercises will be proposed, to help the reader further investigate on the solutions of the previous equation and thus, better understand the allowed motion of the heavy spinning top with one point fixed. We shall also analyse the *fast* spinning top, or the gyroscope.

10.6 Proposed Exercises

10.1 Prove Steiner's theorem.

Solution: Even if it might seem difficult at first sight, this proof is rather straight-forward. Let us consider some arbitrary reference frame $\{\mathcal{O}, \mathbf{i}, \mathbf{j}, \mathbf{k}\}$. With respect to this coordinate system we have the following expression for the inertia tensor (for an N-particle system, $\alpha = 1, \ldots, N$)

$$I_{ij} = \sum_{\alpha} m_\alpha \left(\delta_{ij} \sum_k x^2_{k,\alpha} - x_{i,\alpha} x_{j,\alpha} \right) , \tag{10.84}$$

with $i, j = 1, 2, 3$. We thus have to check that under the translation of the origin (with no rotation of the axes) given by $\mathbf{c}$ i.e.,

$$\mathbf{r}_\alpha = \mathbf{c} + \mathbf{r}'_\alpha , \tag{10.85}$$

where $\mathbf{c}$ are the coordinates of the CM with respect to $\{\mathcal{O}, \mathbf{i}, \mathbf{j}, \mathbf{k}\}$ and $\mathbf{r}'_\alpha$, the coordinates of the particle α with respect to $\{CM, \mathbf{i}, \mathbf{j}, \mathbf{k}\}$, the inertia tensor can be expressed as

$$I_{ij} = I_{ij}^{CM} + M \left(\mathbf{c}^2 \delta_{ij} - c_i c_j \right) . \tag{10.86}$$

The coordinate transformation (10.85) in terms of $x_{i,\alpha}$ reads

$$x_{i,\alpha} = c_i + x'_{i,\alpha} . \tag{10.87}$$

Inserting this expression into (10.84) we obtain

$$
\begin{aligned}
I_{ij} &= \sum_\alpha m_\alpha \left(\delta_{ij} \sum_k (c_k + x'_{k,\alpha})^2 - (c_i + x'_{i,\alpha})(c_j + x'_{j,\alpha}) \right) \\
&= \sum_\alpha m_\alpha \left(\delta_{ij} \sum_k (c_k^2 + x'^2_{k,\alpha} + 2c_k x'_{k,\alpha}) - c_i c_j - x'_{i,\alpha} x'_{j,\alpha} - c_i x'_{j,\alpha} - c_j x'_{i,\alpha} \right) \\
&= \sum_\alpha m_\alpha \left(\delta_{ij} \sum_k x'^2_{k,\alpha} - x'_{i,\alpha} x'_{j,\alpha} \right) + \sum_\alpha m_\alpha \left(\delta_{ij} \sum_k c_k^2 - c_i c_j \right) \\
&\quad + \sum_\alpha m_\alpha \left(\delta_{ij} \sum_k 2c_k x'_{k,\alpha} - c_i x'_{j,\alpha} - c_j x'_{i,\alpha} \right) \\
&= I_{ij}^{CM} + M \left(\mathbf{c}^2 \delta_{ij} - c_i c_j \right) + \Delta_{ij} . \tag{10.88}
\end{aligned}
$$

The only thing left now is to prove that the last term $\Delta_{ij} = 0$. Summing over α we obtain

$$\Delta_{ij} = M(2\delta_{ij}\mathbf{c}\cdot\mathbf{c}' - c_i c_j' - c_j c_i'),\qquad(10.89)$$

where $\mathbf{c}'$ are the coordinates of the CM with respect to the center of mass, which are zero (as proven in Chap. 5, where the coordinates of the CM with respect to the CM were denoted by $\mathcal{R}'$). Hence $\Delta_{ij} = 0$. The continuum case is equally straightforward.

10.2 Calculate the inertia tensor with respect to the coordinate axes $\{\mathbf{i}, \mathbf{j}, \mathbf{k}\}$ for a cube of uniform mass density and side a, that is placed as shown in Fig. 10.8.

Solution: Let us explicitly calculate, step by step the first component

$$I_{11} = \rho\int_V dV\left(\delta_{11}(x^2+y^2+z^2)-x^2\right) = \rho\int_V dV(y^2+z^2).\qquad(10.90)$$

Introducing the integration limits

$$\begin{aligned}
I_{11} &= \rho\int_0^a\int_0^a\int_0^a dx\,dy\,dz\,(y^2+z^2)\\
&= \rho a\int_0^a\int_0^a dy\,dz\,(y^2+z^2)\\
&= \rho a\left(a\frac{1}{3}a^3 + a\frac{1}{3}a^3\right)\\
&= \frac{2}{3}\rho a^5,
\end{aligned}\qquad(10.91)$$

or in terms of the mass

$$I_{11} = \frac{2}{3}Ma^2.\qquad(10.92)$$

Fig. 10.8 An uniform cube placed with one corner at the origin of coordinates and whose sides are parallel to one of the axes

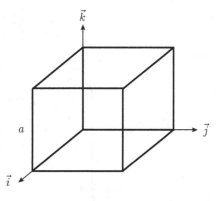

One can straightforwardly check that $I_{11} = I_{22} = I_{33}$. One can also observe from the definition of the inertia tensor that it is, in general, symmetric i.e., $I_{ij} = I_{ji}$. Let us now calculate the remaining components. We obtain

$$I_{12} = I_{21} = -\rho \int_0^a \int_0^a \int_0^a dx \, dy \, dz \, x \, y = -\frac{1}{4} \rho a^5 = -\frac{1}{4} M a^2 . \quad (10.93)$$

The remaining components are equal to the previously calculated one. Hence, we can write the inertia tensor in matrix form as

$$I = M a^2 \begin{pmatrix} \dfrac{2}{3} & -\dfrac{1}{4} & -\dfrac{1}{4} \\ -\dfrac{1}{4} & \dfrac{2}{3} & -\dfrac{1}{4} \\ -\dfrac{1}{4} & -\dfrac{1}{4} & \dfrac{2}{3} \end{pmatrix} . \quad (10.94)$$

In the following we shall make use of Steiner's theorem in order to directly obtain the inertia tensor with respect to the CM and find out some interesting results with respect to the cube's symmetry axes.

10.3 Using Steiner's theorem, calculate the inertia tensor with respect to the CM and obtain the principal axes of inertia with respect to the CM.

Solution: As seen previously, Steiner's theorem states:

$$I_{ij}^{CM} = I_{ij} - M \left(\mathbf{c}^2 \delta_{ij} - c_i c_j \right) . \quad (10.95)$$

As the cube has uniform mass density, the cube's CM is the center of the cube. Its coordinates (with respect to $\{\mathcal{O}, \mathbf{i}, \mathbf{j}, \mathbf{k}\}$) are simply

$$\mathbf{c} = \frac{a}{2} \mathbf{i} + \frac{a}{2} \mathbf{j} + \frac{a}{2} \mathbf{k} . \quad (10.96)$$

In matrix form (10.95) reads

$$I^{CM} = M a^2 \begin{pmatrix} \dfrac{2}{3} & -\dfrac{1}{4} & -\dfrac{1}{4} \\ -\dfrac{1}{4} & \dfrac{2}{3} & -\dfrac{1}{4} \\ -\dfrac{1}{4} & -\dfrac{1}{4} & \dfrac{2}{3} \end{pmatrix} - M a^2 \begin{pmatrix} \dfrac{1}{2} & -\dfrac{1}{4} & -\dfrac{1}{4} \\ -\dfrac{1}{4} & \dfrac{1}{2} & -\dfrac{1}{4} \\ -\dfrac{1}{4} & -\dfrac{1}{4} & \dfrac{1}{2} \end{pmatrix}$$

$$= M a^2 \frac{1}{6} I_3 , \quad (10.97)$$

where I_3 is the 3×3 identity matrix. We can observe two things. First, that the inertia tensor is diagonal, and second, that its three eigenvalues are equal. We can therefore conclude that the the cube is spherically symmetric with respect to its center of mass, and so, any three mutually orthogonal axes with origin in the CM will be principal axes of inertia.

10.4 Calculate the inertia tensor of a sphere with respect to its CM.

Solution: In spherical coordinates we have

$$x = x_1 = r \sin \theta \cos \phi,$$
$$y = x_2 = r \sin \theta \sin \phi, \tag{10.98}$$
$$z = x_3 = r \cos \theta, \tag{10.99}$$

and so, the expression for the inertia tensor reduces to

$$I_{ij}^{CM} = \rho \int_V dV (\delta_{ij} r^2 - x_i x_j). \tag{10.100}$$

One can check, by direct integration (over ϕ) that the non-diagonal terms

$$-\rho \int_V dV x_i x_j = 0, \tag{10.101}$$

for $i \neq j$ (as expected). For the first term we obtain

$$I_{11}^{CM} = \rho \int_V dV r^2 (1 - \sin^2 \theta \cos^2 \phi)$$
$$= \rho \int_0^R r^2 \int_0^\pi \sin \theta d\theta \int_0^{2\pi} d\phi \, r^2 (1 - \sin^2 \theta \cos^2 \phi)$$
$$= \rho \frac{R^5}{5} \int_0^\pi \sin \theta d\theta \, (2\pi - \pi \sin^2 \theta)$$
$$= \rho \frac{R^5}{5} \left(4\pi - \frac{4\pi}{3} \right)$$
$$= \frac{8\pi \rho R^5}{15}$$
$$= \frac{2}{5} M R^2, \tag{10.102}$$

where R is the radius of the sphere. One can also check that $I_{11}^{CM} = I_{22}^{CM} = I_{33}^{CM}$, hence the sphere is also (obviously) spherically symmetrical and, any mutually

Fig. 10.9 Discrete rigid
solid disposed on a square of
length a, formed by four
masses held together by rods
of negligible mass

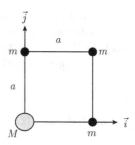

orthogonal axes with origin at the center of the sphere will be principal axes of
inertia.

10.5 Calculate the tensor of inertia of the discrete rigid solid shown in
Fig. 10.9, neglecting the mass of the rods that hold together the four masses.
Obtain the principal axes of inertia.

Solution: As the masses are disposed within the (x, y) plane, $z = 0$ and so, we can
straightforwardly conclude that

$$I_{13} = I_{31} = I_{23} = I_{32} = 0. \tag{10.103}$$

As for I_{11} we have

$$\begin{aligned} I_{11} &= M(0 - 0) + m(a^2 - a^2) + m(a^2 - 0) + m(2a^2 - a^2) \\ &= 2\,m\,a^2, \end{aligned} \tag{10.104}$$

where, inside the parentheses we have written the two contributing terms for each
mass m_α (the one corresponding to $\delta_{ij} \sum_k x_{k,\alpha}^2$ and the one corresponding to $x_{i,\alpha} x_{j,\alpha}$).
Similarly one obtains

$$I_{11} = I_{22}, \qquad I_{33} = 4\,m\,a^2, \qquad I_{12} = I_{21} = -m\,a^2. \tag{10.105}$$

In matrix form

$$I = m\,a^2 \begin{pmatrix} 2 & -1 & 0 \\ -1 & 2 & 0 \\ 0 & 0 & 4 \end{pmatrix} \tag{10.106}$$

In order to obtain the principal axes of inertia we have to diagonalize the 2×2
sub-matrix from the upper left corner. We thus have to solve

$$\begin{vmatrix} 2 - \lambda & -1 \\ -1 & 2 - \lambda \end{vmatrix} = 0 = (2 - \lambda)^2 - 1. \tag{10.107}$$

The corresponding eigenvalues are given by $\lambda_1 = 1$ and $\lambda_2 = 3$. For obtaining the eigenvectors we have to solve

$$\begin{pmatrix} 2 & -1 \\ -1 & 2 \end{pmatrix} \begin{pmatrix} v_{x,i} \\ v_{y,i} \end{pmatrix} = \lambda_i \begin{pmatrix} v_{x,i} \\ v_{y,i} \end{pmatrix},$$ (10.108)

with $i = 1, 2$. The eigenvectors that we find are

$$\begin{pmatrix} 1 \\ 1 \end{pmatrix}, \quad \begin{pmatrix} -1 \\ 1 \end{pmatrix}.$$ (10.109)

We thus conclude that the (normalized) principal axes of inertia are

$$\mathbf{i}' = \frac{1}{\sqrt{2}}(\mathbf{i} + \mathbf{j}), \quad \mathbf{j}' = \frac{1}{\sqrt{2}}(-\mathbf{i} + \mathbf{j}),$$ (10.110)

and $\mathbf{k} = \mathbf{k}'$ (it remains unchanged). This corresponds to a positive (passive) rotation about the $\mathbf{k}$ axis of $45°$. The inertia tensor with respect to the principal axes will be simply given by the matrix

$$I' = m\,a^2 \begin{pmatrix} 1 & 0 & 0 \\ 0 & 3 & 0 \\ 0 & 0 & 4 \end{pmatrix}.$$ (10.111)

One should note that the eigenvectors are determined up to a global sign. This sign should be chosen, as we did here, in order to have the new axes forming a right handed reference frame. The way one labels the new axes also counts. In our case we could have labelled

$$\frac{1}{\sqrt{2}}(\mathbf{i} + \mathbf{j})$$ (10.112)

as $\mathbf{j}'$. However, this choice would not correspond to a right handed coordinate system, as one can easily check.

10.6 Analyse the stability of the solutions for the torque free motion for a rigid body with $I_1 < I_2 < I_3$ (for the remaining two cases)

- $\omega_2 = C$ and $\omega_1 = \omega_3 = 0$,
- $\omega_3 = C$ and $\omega_1 = \omega_2 = 0$,

with $C \neq 0$ a positive real constant, by considering small perturbations about the equilibrium state.

Solution: For the first case we set $\omega_1 = \epsilon_1$ and $\omega_3 = \epsilon_3$ with $|\epsilon_{1,3}| \ll |\omega_2| \neq 0$, and use (10.39) to calculate ω_2. We obtain

$$
\begin{aligned}
I_1 \dot{\epsilon}_1 + \omega_2 \epsilon_3 (I_3 - I_2) &= 0, \\
I_2 \dot{\omega}_2 + \epsilon_1 \epsilon_3 (I_1 - I_3) &= 0, \\
I_3 \dot{\epsilon}_3 + \epsilon_1 \omega_2 (I_2 - I_1) &= 0.
\end{aligned}
\tag{10.113}
$$

From the second equation we get

$$
\omega_2 = C + \mathcal{O}(\epsilon^2) \simeq C,
\tag{10.114}
$$

with $C \neq 0$, an arbitrary real constant. Let us check if the quantities $\epsilon_{1,3}$ remain small as $t \to \infty$. Taking the time derivative of the first and third equation, we obtain the following system of differential equations

$$
\begin{aligned}
I_1 \ddot{\epsilon}_1 + \omega_2 \dot{\epsilon}_3 (I_3 - I_2) &= 0, \\
I_3 \ddot{\epsilon}_3 + \dot{\epsilon}_1 \omega_2 (I_2 - I_1) &= 0.
\end{aligned}
\tag{10.115}
$$

Introducing the expressions of $\dot{\epsilon}_1$ and $\dot{\epsilon}_3$ and simplifying, the previous system becomes

$$
\ddot{\epsilon}_i + \beta \epsilon_i = 0,
\tag{10.116}
$$

with $i = 1, 3$ and

$$
\beta = \omega_2^2 \frac{(I_1 - I_2)(I_3 - I_2)}{I_1 I_3} < 0.
\tag{10.117}
$$

As analysed previously, the solutions are of the type

$$
\epsilon_i = A_i \cos\left(\sqrt{\beta}\, t\right) + B_i \sin\left(\sqrt{\beta}\, t\right),
\tag{10.118}
$$

with A_i and B_i two real integration constants ($i = 2, 3$). If $\beta < 0$ the previous expressions can be re-written as

$$
\begin{aligned}
\epsilon_i &= A_i \cos\left(i\sqrt{|\beta|}\, t\right) + B_i \sin\left(i\sqrt{|\beta|}\, t\right) \\
&= A_i \cosh\left(\sqrt{|\beta|}\, t\right) + B_i \sinh\left(\sqrt{|\beta|}\, t\right),
\end{aligned}
\tag{10.119}
$$

and so, in comparison with (10.46) where $\beta > 0$, the current solutions are no longer oscillatory but of the exponential type. Hence, we conclude that this solution is unstable i.e., $\epsilon_{1,3} \to \infty$ as $t \to \infty$.

For the last case $\omega_3 = C$ and $\omega_1 = \omega_2 = 0$, the reader should be able to straightforwardly deduce in a similar way, that the motion is stable.

Fig. 10.10 Effective
potential for the heavy
symmetric spinning top with
one point fixed

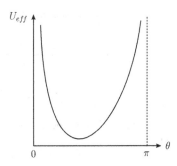

The following exercises will be dedicated to the analysis of the motion of the heavy
symmetric spinning top ($I_1' = I_2' \neq I_3'$) with one point fixed. With a little guidance
we shall be able to draw a few interesting conclusions about its motion.

10.7 Consider the one-dimensional energy equation for the heavy spinning
top with one point fixed, given by the expression

$$E' = \frac{1}{2}I_1'\,\dot{\theta}^2 + U_{eff}(\theta),$$ (10.120)

with E' a constant, and where the effective potential reads

$$U_{eff}(\theta) = \frac{1}{2}I_1'\left(\frac{b - a\cos\theta}{\sin\theta}\right)^2 + Mgl\cos\theta.$$ (10.121)

Obtain the shape of the effective potential (without calculating the local
extrema). Find the analytical expression for the local extrema as a function
of $c \equiv b - a\cos\theta$. Comment upon the solutions.

Solution: One should be able to immediately check that for $\theta \to \{0, \pi\}$ we have

$$\lim_{\theta \to 0} U_{eff} = \lim_{\theta \to \pi} U_{eff} = \infty.$$ (10.122)

Without having to calculate anything else, from the expression of U_{eff}, we can
conclude that the approximate shape of the effective potential is similar to the one
shown in Fig. 10.10. This means that U_{eff} has a minimum at $\theta = \theta_0$ and two turning
angles θ_1^R and θ_2^R, given by the usual condition

$$U_{eff}(\theta_i^R) = E'.$$ (10.123)

Let us now calculate the expression for the minimum θ_0. We have

$$
\begin{aligned}
\frac{\partial U_{eff}}{\partial \theta} &= -\frac{I_1' \cos \theta (b - a \cos \theta)^2}{\sin^3 \theta} + \frac{a I_1'(b - a \cos \theta)}{\sin \theta} - M g l \sin \theta \\
&= -c^2 \frac{I_1' \cos \theta}{\sin^3 \theta} + c \frac{a I_1'}{\sin \theta} - M g l \sin \theta \,,
\end{aligned}
\tag{10.124}
$$

where we have introduced $c \equiv b - a \cos \theta$. The minimum θ_0 will therefore have to satisfy

$$
c^2 \left(\frac{-I_1' \cos \theta_0}{\sin^3 \theta_0} \right) + c \left(\frac{a I_1'}{\sin \theta_0} \right) - M g l \sin \theta_0 = 0 \,.
\tag{10.125}
$$

Rearranging terms we obtain

$$
c^2 - c \left(\frac{a \sin^2 \theta_0}{\cos \theta_0} \right) + \frac{M g l \sin^4 \theta_0}{I_1' \cos \theta_0} = 0 \,,
\tag{10.126}
$$

whose solution is

$$
c = \frac{a \sin^2 \theta_0}{2 \cos \theta_0} \left(1 \pm \sqrt{1 - \frac{4 M g l \cos \theta_0}{a^2 I_1'}} \right) .
\tag{10.127}
$$

We thus observe that we have two solutions. As the effective potential presents only one minimum, we can safely conclude that only one of the previous solutions will be the physical one, depending on the configuration of the remaining angles.

10.8 Study the previous solutions for c, in terms of $\dot{\phi}$ and $\dot{\psi}$, for $\theta = \theta_0 < \pi/2$. Study also the limit $a^2 \gg M g l \cos \theta_0 / I_1'$ (with θ_0 the minimum of U_{eff}) by making a Taylor expansion.

Solution: The first thing we have to notice is that, as $\cos \theta_0 > 0$ ($\theta_0 < \pi/2$), and the physical solution must be real, we have the following constraint

$$
\frac{4 M g l \cos \theta_0}{a^2 I_1'} \leq 1 \,,
\tag{10.128}
$$

which can be re-written as

$$
|\omega_3'| \geq \frac{2}{I_3'} \sqrt{M g l I_1' \cos \theta_0} \,,
\tag{10.129}
$$

where we have used the explicit expression for a i.e., $a = I_3' \omega_3'/I_1'$. Let us also remember the expression of b and ω_3'. For $\theta = \theta_0$ we have

$$b = \dot{\phi} \sin^2 \theta_0 + a \cos \theta_0 , \qquad \omega_3' = \dot{\psi} + \dot{\phi} \cos \theta_0 . \tag{10.130}$$

As θ_0, a, b and ω_3' are constants, we conclude that $\dot{\theta} = 0$ ($\theta = \theta_0$ constant), $\dot{\phi} = \dot{\phi}_0$ and $\dot{\psi} = \dot{\psi}_0$ are also constants. Hence, the motion is nutation free, with steady (constant) rotation and precession. From the expression of b we can isolate $\dot{\phi}$. We obtain

$$\dot{\phi} = \dot{\phi}_0 = \frac{b - a \cos \theta_0}{\sin^2 \theta_0} = \frac{c}{\sin^2 \theta_0} . \tag{10.131}$$

Hence, there are two possible values for $\dot{\phi}_0$ corresponding to the $+$ or $-$ sign of the two solutions of c

$$\dot{\phi}_0^{\pm} = \frac{c^{\pm}}{\sin^2 \theta_0} . \tag{10.132}$$

If $\omega_3' > 0$, $\dot{\phi}_0^{+}$ is called fast precession and $\dot{\phi}_0^{-}$ is called slow precession, and vice-versa for $\omega_3' < 0$. The spin $\dot{\psi}$ is highly related. From the expression of ω_3' we obtain

$$\dot{\psi} = \dot{\psi}_0 = \omega_3' - \dot{\phi}_0 \cos \theta_0 , \tag{10.133}$$

which also depends on the solutions $c^{\pm}$. Depending on the sign of $\dot{\phi}_0^{\pm}$ the spin can also be fast or slow.

Let us now analyse the last case corresponding to $a^2 \gg 4 M g l \cos \theta_0/I_1'$, which corresponds to

$$|\omega_3'| \gg \frac{2}{I_3'} \sqrt{M g l I_1' \cos \theta_0} . \tag{10.134}$$

In this case we have

$$c \simeq \frac{a \sin^2 \theta_0}{2 \cos \theta} \left(1 \pm 1 \mp \frac{2 M g l \cos \theta_0}{a^2 I_1'} \right) . \tag{10.135}$$

For the slow precession we simply have

$$c^{-} \simeq \frac{M g l \sin^2 \theta_0}{a I_1'} = \frac{M g l \sin^2 \theta_0}{I_3' \omega_3'} , \tag{10.136}$$

and so

$$\dot{\phi}_o^{-} \simeq \frac{M g l}{I_3' \omega_3'} . \tag{10.137}$$

As for the fast precession (we can safely ignore the $M \, g \, l$ term compared to 2) and so

$$\dot{\phi}_o^+ \simeq \frac{I_3' \, \omega_3'}{I_1' \, \cos \theta_0} \, . \qquad (10.138)$$

10.9 Repeat the previous exercise for $\theta_0 > \pi/2$.

Solution: The first thing one has to note is that, as $\pi \geq \theta > \pi/2$, $\cos \theta_0$ is negative, and so the squared root corresponding to the solution of c is always positive. Therefore there is no restriction on ω_3'. One should be able, with no further complication, to repeat the previous calculations for this case. One other thing is worth commenting upon. As one is used to imagine a spinning top on a platform, the case corresponding to $\theta > \pi/2$ could result counter-intuitive. This case is obviously possible, however, with an *upside-down top* with a point fixed *underneath some platform* or, for a gyroscope-like device.

We are going to end this chapter by analysing the gyroscopic limit of the heavy symmetric top with one point fixed.

10.10 Study the gyroscopic limit $\dot{\psi} \gg \dot{\phi}$ of the heavy spinning top for $\theta = \theta_0$. Obtain the relation between $\dot{\psi}$ and $\dot{\phi}$ (hint: use directly the relation

$$c^2 - c \left(\frac{a \sin^2 \theta_0}{\cos \theta_0} \right) + \frac{M \, g \, l \, \sin^4 \theta_0}{I_1' \cos \theta_0} = 0 \, , \qquad (10.139)$$

and not its roots).

Solution: The gyroscopic limit is obviously called so because a gyroscope's spin is much faster than its precession. In this limit we simply obtain

$$\omega_3' = \dot{\psi} + \dot{\phi} \cos \theta_0 \simeq \dot{\psi} \, , \qquad (10.140)$$

and

$$a = \frac{I_3'}{I_1'} \omega_3' \simeq \frac{I_3'}{I_1'} \dot{\psi} \, ,$$

$$b = \dot{\phi} \sin^2 \theta_0 + a \cos \theta_0 \simeq \frac{I_3' \cos \theta_0}{I_1'} \dot{\psi} \, . \qquad (10.141)$$

In terms of $\dot{\phi}$, with no approximation, the expression for c reads

$$c = b - a \cos \theta_0 = \dot{\phi} \sin^2 \theta_0 + a \cos \theta_0 - a \cos \theta_0 = \dot{\phi} \sin^2 \theta_0. \qquad (10.142)$$

Hence, the expression (10.139) in terms of $\dot{\phi}$ and $\dot{\psi}$ can be written as

$$\dot{\phi}^2 \sin^4 \theta_0 - \dot{\phi} \dot{\psi} \frac{I_3'}{I_1'} \frac{\sin^4 \theta_0}{\cos \theta_0} + \frac{M g l}{I_1'} \frac{\sin^4 \theta_0}{\cos \theta_0} = 0. \qquad (10.143)$$

Neglecting the $\dot{\phi}^2$ term and simplifying we obtain

$$\dot{\phi} = \frac{M g l}{I_3' \dot{\psi}}. \qquad (10.144)$$

So far we have only studied the cases corresponding to the minimum of the effective potential $\theta = \theta_0$. For small perturbations, as θ_0 is a minimum, the previously obtained solutions are stable, and can be considered as good approximations. Additionally in this case we would have nutation, corresponding to the small variation of theta in between the two turning angles. For the generic case, i.e., for large theta variations, there is no generic solution and the equations can normally be modelled numerically. However, as the inter-dependence of the angles is non-linear, in many cases we can even have chaotic motion, changes of the sense of the precession (of the sign of $\dot{\phi}$) etc. Thus, as the generic case is highly complex we will not go further with this analysis and end this chapter here.

Further Reading

1. J.V. José, E.J. Saletan, *Classical Dynamics: A Contemporary Approach*. Cambridge University Press
2. S.T. Thornton, J.B. Marion, *Classical Dynamics of Particles and Systems*
3. H. Goldstein, C. Poole, J. Safko, *Classical Mechanics*, 3rd edn. Addison Wesley
4. J.R. Taylor, *Classical Mechanics*
5. D.T. Greenwood, *Classical Dynamics*. Prentice-Hall Inc.
6. D. Kleppner, R. Kolenkow, *An Introduction to Mechanics*
7. C. Lanczos, *The Variational Principles of Mechanics*. Dover Publications Inc.
8. W. Greiner, *Classical Mechanics: Systems of Particles and Hamiltonian Dynamics*. Springer
9. H.C. Corben, P. Stehle, *Classical Mechanics*, 2nd edn. Dover Publications Inc.
10. T.W.B. Kibble, F.H. Berkshire, *Classical Mechanics*. Imperial College Press
11. M.G. Calkin, *Lagrangian and Hamiltonian Mechanics*
12. A.J. French, M.G. Ebison, *Introduction to Classical Mechanics*

Chapter 11
Special Theory of Relativity

Abstract After some brief historical comments, we start this chapter by introducing the Lorentz transformations from basic principles and present some relevant examples for a better conceptual understanding of the underlying phenomena. We reformulate the Newtonian equations of motion in order to obtain their Relativistic Lorentz-invariant formulation. We analyse and comment upon the Newtonian limit $c \to \infty$ in all cases. In parallel we give a tensor formulation of all equations. This chapter is therefore intended to introduce all readers to the realm of Special Relativity, step by step, but also to give a more complete vision to a more advanced reader that is already familiar with the subject.

11.1 Introduction

Maxwell's equations (presented in 1864) predicted that light is an electromagnetic wave that moves at a constant speed c. However, a peculiar aspect of these equations is that they are not invariant under Galilean transformations. This has severe implications. It means that if they hold in a certain reference frame $\mathcal{O}$, they will not hold in another one $\mathcal{O}'$, related to the first one by a Galilean transformation. As we have discussed previously, equations that only hold in some privileged reference frames, that do not remain invariant under some set of transformations, can difficultly be regarded as physical laws. However, Maxwell's equations successfully unified electricity with magnetism, explained the nature of light and described all electromagnetic phenomena with high precision. This was indeed an impasse.

In order to solve this apparent conflict, the existence of the *ether* was postulated. A substance that fills the entire Universe and that constitutes the medium in which the electromagnetic waves are propagating. Therefore Maxwell's equations would remain invariant only under a subset of Galilean transformations i.e., only in the reference frames that are at rest with respect to the ether. The Michelson-Morley experiment (1887) proved otherwise. The limits on the ether $drag$ (friction with the ether) of the electromagnetic waves, obtained by the experiment, were much smaller then the theory predicted. Perhaps there was something wrong with the Galilean transformations.

© Springer Nature Switzerland AG 2020

V. Ilisie, *Lectures in Classical Mechanics*, Undergraduate Lecture Notes in Physics, https://doi.org/10.1007/978-3-030-38585-9_11

With the findings of Lorentz, Fitzgerald, Larmor and Poincaré among others, a set of new transformations were derived, the Lorentz transformations, under which Maxwell's equations remained invariant. These equations implied, roughly speaking, that moving objects were contracted and moving clocks ran slow.

In 1905, it was Einstein who proposed the Principle of Special Relativity. The two postulates of Special Relativity state:

- Equations of motion must remain invariant in all inertial reference frames.
- The speed of light in the vacuum c, is constant in all reference frames (inertial and non-inertial).

With these two postulates the Lorentz transformations arise naturally. With this new set of rules we are now in possession of a more *generic* set of transformations that leave Maxwell's equations invariant and describe the coordinate transformations between any two inertial reference frames. According to these new principles, inertial frames are no longer associated to Galilean transformations or Newton's equations, but to a deeper concept. The concept of the frames in which *equations of motion hold in their simplest form*, as we previously mentioned in Chap. 4. However, practically speaking, we can still consider as inertial reference frames, the ones at rest, or moving with constant velocity with respect to the apparently fixed, far away stars.

As the new transformations between inertial reference frames are no longer Galilean, Newton's equations will not remain invariant under these transformations. This will not be an issue though, as we will be able to, almost trivially, reformulate them in order to obtain Lorentz invariant equations. As we shall deduce later on, Galilean transformations are recovered for reference systems or particles, moving with relative velocities much smaller that c (i.e., in the limit $c \to \infty$). A question that might naturally arise regarding the previous considerations is, why are Maxwell's equations Lorentz invariant but not invariant under Galilei's transformations? The answer will be given in the next section.

In the following we shall introduce Lorentz transformations, or more generally, Poincaré transformations.

11.2 Lorentz–Poincaré Transformations

As we will observe, Lorentz transformations not only imply transformations among spatial coordinates but also involve time. Consider an inertial reference frame $\mathcal{O}$. In this reference system we send a light pulse from a point A to B, as shown in Fig. 11.1 (left). If the squared travelled distance from A to B is $\Delta l^2 = \Delta x^2 + \Delta y^2 + \Delta z^2$ and the squared time interval for the pulse to reach from A to B is given by Δt^2, then the measure Δs^2 defined below, must be obviously zero:

$$\Delta s^2 = c^2 \Delta t^2 - \Delta x^2 - \Delta y^2 - \Delta z^2 = 0, \qquad (11.1)$$

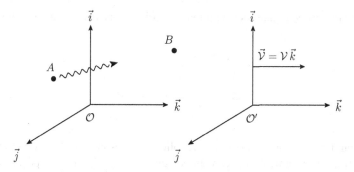

Fig. 11.1 A light pulse travelling from A to B and two reference systems used to describe the same physical phenomena: $\mathcal{O}$ and $\mathcal{O}'$ (that moves with constant relative velocity $\mathcal{V} = \mathcal{V}\mathbf{k}$ with respect to $\mathcal{O}$)

where c is the speed of light in the vacuum. Consider another observer, at rest in another reference frame $\mathcal{O}'$ (with same axes as $\mathcal{O}$), that moves with constant velocity $\mathcal{V} = \mathcal{V}\mathbf{k}$ with respect to $\mathcal{O}$, as shown in Fig. 11.1 (right). In this reference frame, based on the two Special Relativity postulates, the measure $\Delta s'$, performed in the $\mathcal{O}'$ frame must equally be zero:

$$\Delta s'^2 = c^2\Delta t'^2 - \Delta x'^2 - \Delta y'^2 - \Delta z'^2 = 0. \tag{11.2}$$

The transformations that the space and time (space-time) intervals must obey in order to satisfy

$$\begin{aligned}\Delta s^2 &= c^2\Delta t^2 - \Delta x^2 - \Delta y^2 - \Delta z^2 \\ &= c^2\Delta t'^2 - \Delta x'^2 - \Delta y'^2 - \Delta z'^2 = \Delta s'^2, \end{aligned} \tag{11.3}$$

are the following Lorentz transformations which we shall call a *Lorentz boost* (along the $\mathbf{k}$ axis)

$$\begin{aligned} \Delta x' &= \Delta x, \\ \Delta y' &= \Delta y, \\ \Delta z' &= \gamma\Delta z - \gamma\beta c\,\Delta t, \\ c\,\Delta t' &= \gamma c\,\Delta t - \gamma\beta\,\Delta z, \end{aligned} \tag{11.4}$$

where we have introduced

$$\beta = \mathcal{V}/c, \qquad \gamma = \frac{1}{\sqrt{1 - \mathcal{V}^2/c^2}} = (1 - \beta^2)^{-1}. \tag{11.5}$$

Note that the Lorentz boost described previously is a passive transformation (see Appendix C for details). The inverse transformations will be given by simply interchanging the primed quantities with the non-primed ones and switching the sign of V (of β)

$$\Delta x = \Delta x',$$
$$\Delta y = \Delta y',$$
$$\Delta z = \gamma \Delta z' + \gamma \beta c \Delta t',$$
$$c \Delta t = \gamma c \Delta t' + \gamma \beta \Delta z'. \qquad (11.6)$$

This should come as no surprise for the following reason. For an observer at rest in the frame $\mathcal{O}'$, the frame $\mathcal{O}$ moves with velocity $V = -V\mathbf{k}$ with respect to $\mathcal{O}'$, and therefore if the transformations (11.4) hold, then the previous ones must necessarily hold also. In this aspect, they are similar to Galileo's transformations. However, it will be a nice exercise to check that (11.6) are consistent with (11.4) i.e., that when substituting the expressions for $\Delta z'$ and $\Delta t'$ into the expressions of Δz and Δt we obtain $\Delta z = \Delta z$ and $\Delta t = \Delta t$. Several other things should be noted.

First of all, we have skipped a rather tedious and unnecessary deduction of the previous transformations. Instead, as a proposed exercise one will have to check that for the previously given set of rules, we obtain $\Delta s^2 = \Delta s'^2$.

Second, the measure Δs^2 was defined for a light-pulse (travelling from A to B) and was found to be zero. For other type of events A and B, occurring at a squared distance $\Delta l^2 = \Delta x^2 + \Delta y^2 + \Delta z^2$ and at some squared time interval Δt^2 (in some reference frame $\mathcal{O}$), we will define the same measure Δs^2. We thus have three categories of events[1]

- light-like events: those for which $\Delta s^2 = 0$.
- causal events (or time-like): those for which $\Delta s^2 > 0$.
- non-causal events (or space-like): those for which $\Delta s^2 < 0$.

We call non-causal events, events for which $\Delta s^2 < 0$ which means that $c \Delta t^2 < \Delta l^2$. This implies that a light pulse (which is the fastest information carrier), travelling at c during Δt, cannot connect events A and B and therefore, these events can't possibly be related in any way by *cause and effect*. Events with $\Delta s^2 > 0$ are obviously called causal because they can be *connected* by a signals travelling at $v < c$. Depending on the author, light-like events (that are connected by light-pulses travelling at c) can be considered (or not) causal. Here we shall not be concerned with this rather philosophical digression and leave this issue open.

A natural question that might now arise is the following. Will $\Delta s^2 \neq 0$ also remain invariant under Lorentz transformations? The answer is yes, and it is just a straightforward consequence of the fact that $\Delta s = \Delta s'$, under Lorentz transformations (without making any assumptions on the value of Δs).

Third issue is related to the fact that we have considered a reference frame $\mathcal{O}'$ that moves with constant relative velocity along the $\mathbf{k}$ axis with respect to $\mathcal{O}$ system. This is not the most general case, as we could also perform a rotation a translation of the

---

[1]Some authors use a different metric i.e., a different definition of Δs^2, which is $\Delta s^2 = -c \Delta t^2 + \Delta l^2$ and therefore, the following considerations in terms of the sign of Δs^2 will be the opposite.

reference system $\mathcal{O}'$, and also consider an arbitrary direction and sense of $\mathcal{V}$. Such cases are unnecessarily complicated and there is nothing new to learn from them (about Special Relativity). We shall thus only consider simple cases. More generic configurations can be easily found in the literature.

Finally, the fourth and last comment before moving on, is again related to Δs^2. There are several other transformations that leave the interval Δs^2 invariant. These transformations are: rotations, parity, time inversion, spatial translations and time translations (space-time translations). Invariance under rotations and space-time translations should not come as a surprise. Nature's laws cannot possibly have a preferred direction, position or cannot depend at which time instant we perform the experiments. This is also valid for Newton's laws, i.e., the corresponding equations are invariant under rotations and/or space-time translations. As for time inversion and parity, these concepts are more related to quantum mechanics and particle physics and we shall no be concerned about them here. Lorentz boosts together with rotations, parity, and time inversion are generically called Lorentz transformations and, in more advanced text books (on particle physics usually) we talk about the Lorentz group of transformations. If we additionally consider space-time translations then we refer to the Poincaré group.

Before skipping to the next section, we still need to clarify one crucial aspect of Lorentz boosts. Consider the (Galilean) limit $\mathcal{V} \ll c$ or equivalently $c \to \infty$. In this limit we have

$$\beta \to 0, \qquad \beta c \to \mathcal{V}, \qquad \gamma \to 1, \tag{11.7}$$

and therefore the Lorentz transformations become

$$
\begin{aligned}
\Delta x' &= \Delta x, \\
\Delta y' &= \Delta y, \\
\Delta z' &= \Delta z - \mathcal{V} \Delta t, \\
c \, \Delta t' &= c \, \Delta t,
\end{aligned}
\tag{11.8}
$$

which are nothing but Galileo's transformations rules. However, if this is so, if the Galilean transformations are recovered in this limit, why aren't Maxwell's equations invariant under these transformations. The answer to this question has to do with the velocity addition rules in Special Relativity, which is something we have not yet treated. In Galilean relativity, if an object moves with velocity $\mathbf{v}$ with respect to a reference frame $\mathcal{O}$, the velocity of this object measured in a reference frame $\mathcal{O}'$ that moves with $\mathcal{V}$ with respect to $\mathcal{O}$, will be given by $\mathbf{v}' = \mathbf{v} - \mathcal{V}$. For an electromagnetic field this obviously does not hold, as it moves at c in both reference frames according to the postulates of Special Relativity. Even if $\mathcal{V} \ll c$, the previous Galilean transformations would imply that in one of the reference frames the electromagnetic field would not move at c.

11.3 Velocity Addition Rules

It will be more useful in what follows to consider the Lorentz boost (11.4) in differential form

$$
\begin{aligned}
dx' &= dx, \\
dy' &= dy, \\
dz' &= \gamma dz - \gamma \beta c \, dt, \\
c \, dt' &= \gamma c \, dt - \gamma \beta \, dz.
\end{aligned}
\tag{11.9}
$$

These transformations were defined for two events separated by a squared distance given by $dx^2 + dy^2 + dz^2$ that occur at a squared time interval dt^2 in some reference frame $\mathcal{O}$, as related to the same two events as seen from another reference frame $\mathcal{O}'$. Consider now that the events consist in an object moving from $x \rightarrow x + dx$, $y \rightarrow y + dy$ and $z \rightarrow z + dx$ in a time interval $t \rightarrow t + dt$ i.e., the first event corresponds to an object having coordinates (ct, x, y, z) and the second corresponds to the same object reaching $(ct + cdt, x + dx, y + dy, z + dz)$. The velocity of this moving object in the reference frame $\mathcal{O}$ will be thus given by

$$
\mathbf{v} = \frac{dx}{dt}\mathbf{i} + \frac{dy}{dt}\mathbf{j} + \frac{dz}{dt}\mathbf{k} = v_x\,\mathbf{i} + v_y\,\mathbf{j} + v_z\,\mathbf{k}.
\tag{11.10}
$$

The velocity of the body in the primed reference frame $\mathcal{O}'$, that moves with velocity $\mathcal{V} = V\mathbf{k}$ with respect to $\mathcal{O}$, will be similarly given by

$$
\mathbf{v}' = \frac{dx'}{dt'}\mathbf{i} + \frac{dy'}{dt'}\mathbf{j} + \frac{dz'}{dt'}\mathbf{k} = v_x'\,\mathbf{i} + v_y'\,\mathbf{j} + v_z'\,\mathbf{k}.
\tag{11.11}
$$

In order to relate these two quantities we can use the previous transformations (11.9). Consider the quotient of dx' and $c\,dt'$. We obtain

$$
\frac{dx'}{c\,dt'} = \frac{1}{c}v_x' = \frac{dx}{\gamma c\,dt - \gamma \beta\,dz}.
\tag{11.12}
$$

Simplifying we get

$$
v_{x,y}' = \frac{v_{x,y}}{\gamma(1 - V v_z/c^2)}.
\tag{11.13}
$$

Similarly, for v_z' we obtain

$$
v_z' = \frac{v_z - V}{1 - V v_z/c^2}.
\tag{11.14}
$$

When calculating the γ function related to a Lorentz boost, one should not confuse $\mathcal{V}$ with $\mathbf{v}$ or $\mathbf{v}'$. As defined previously, the γ function corresponding to a boost, is a function of $\mathcal{V}$, $\gamma = \gamma(\mathcal{V})$, and not a function of v or v'!

We shall check with an explicit exercise that these transformations become much more transparent when introducing the notion of four-vectors in tensor notation. By construction, the previous expressions maintain the speed of light invariant i.e., if the magnitude of the velocity $v = (v_x^2 + v_y^2 + v_z^2)^{1/2} = c$ then, it remains invariant under these transformations i.e., $v' = (v_x'^2 + v_y'^2 + v_z'^2)^{1/2} = c$ (we shall check as an exercise that this holds true). It will also hold true that, if $v < c$ then $v' < c$. This is simply due to the fact that ds^2 is an invariant quantity. If in a given inertial reference frame $v < c$, then $ds^2 > 0$, and this will be the case in all inertial reference frames, as ds^2 will always have the same value.

11.4 Minkowski Space-Time and Four-Vectors

In the first chapter, we have insisted upon the notion of intrinsic properties of a vector. We have defined a vector as an entity with a basis-invariant magnitude. However, this did not hold under Galilean transformations (i.e., the magnitude of the velocity vector varies). In Special Relativity this is going to be slightly different. We will be able to define four dimensional space-time vectors (four-vectors) with invariant magnitude under space-time translations, rotations and Lorentz boosts.

The space-time coordinates in a given reference frame will be given by x^μ, that can be written in matrix form as

$$\begin{pmatrix} x^0 \\ x^1 \\ x^2 \\ x^3 \end{pmatrix} \equiv \begin{pmatrix} ct \\ x \\ y \\ z \end{pmatrix}. \tag{11.15}$$

Even though this is the correct way of writing coordinates or vector components (as column matrices, as explained in Chap. 1 and in Appendix A), we shall oftenly make an **abuse of notation** and write

$$x^\mu = (x^0, x^1, x^2, x^3) \equiv (ct, x^i) \equiv (ct, \mathbf{x}). \tag{11.16}$$

At this stage, the interested reader is invited to read Appendix A for a brief introduction to tensors in Cartesian coordinates. However, the concepts introduced in this chapter are intended to be self-consistent, and therefore the lecture of this appendix is **optional**.

Note that previously, we have introduced an upper index notation. The reason for this will be understood shortly. Also note that we use Greek letters $\mu = 0, \ldots, 4$ for space-time coordinates and Roman letters $i = 1, \ldots, 3$ only for the spatial coordinates. A space-time differential interval, will given by dx^μ.

We define the Minkowski space, as the four-dimensional space $\mathbb{R}^4$ with the following metric

$$
g_{\mu\nu} = \begin{pmatrix} 1 & 0 & 0 & 0 \\ 0 & -1 & 0 & 0 \\ 0 & 0 & -1 & 0 \\ 0 & 0 & 0 & -1 \end{pmatrix}. \tag{11.17}
$$

Thus, the invariant measure can be written using Einstein's summation convention (an implicit summation is performed over repeated indices) as

$$
ds^2 = g_{\mu\nu} dx^\mu dx^\nu. \tag{11.18}
$$

Note that the previous indices are mute indices used for the summation. Therefore one can freely change these indices without altering the summation i.e., $g_{\mu\nu} dx^\mu dx^\nu = g_{\alpha\beta} dx^\alpha dx^\beta$.

Introducing explicitly the non-zero components of the metric, for the previous expression (11.18) we obtain

$$
\begin{aligned}
ds^2 &= g_{00}(dx^0)^2 + g_{11}(dx^1)^2 + g_{22}(dx^2)^2 + g_{33}(dx^3)^2 \\
&= (dx^0)^2 - (dx^1)^2 - (dx^2)^2 - (dx^3)^2 \\
&= c^2 dt^2 - d\mathbf{x}^2 \\
&= c^2 dt^2 - dx^2 - dy^2 - dz^2.
\end{aligned} \tag{11.19}
$$

In matrix form we simply have the following

$$
ds^2 = (dx^0, dx^1, dx^2, dx^3) \begin{pmatrix} 1 & 0 & 0 & 0 \\ 0 & -1 & 0 & 0 \\ 0 & 0 & -1 & 0 \\ 0 & 0 & 0 & -1 \end{pmatrix} \begin{pmatrix} dx^0 \\ dx^1 \\ dx^2 \\ dx^3 \end{pmatrix}. \tag{11.20}
$$

As ds^2 is a basis-invariant quantity, and we want to define vectors in Minkowski space as entities with basis invariant magnitudes under Lorentz (or more generically Poincaré) transformations, we can naturally identify **the quantities dx^μ as the components of a vector in Minkowski space**. The quantity ds^2 will be then, the squared magnitude of the vector. Using this formalism the scalar product of any two vectors u and v with components u^μ and v^μ (in some inertial reference frame) in Minkowski space is given by

$$
u \cdot v = g_{\mu\nu} u^\mu v^\nu. \tag{11.21}
$$

This product will be basis invariant (note that we have suppressed the bold vector notation for u and v in order to distinguish them from three-dimensional vectors). In three-dimensional space, in Cartesian coordinates, as explained in Appendix A, the scalar product of two vectors can be expressed in terms of the Euclidean metric

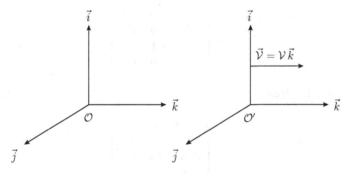

Fig. 11.2 Two inertial reference systems $\mathcal{O}$ and $\mathcal{O}'$, where $\mathcal{O}'$ moves with constant $\mathcal{V} = V\mathbf{k}$ with respect to $\mathcal{O}$

$$g_{ij} = \begin{pmatrix} 1 & 0 & 0 \\ 0 & 1 & 0 \\ 0 & 0 & 1 \end{pmatrix} \tag{11.22}$$

and so

$$\mathbf{u} \cdot \mathbf{v} = g_{ij} u^i v^j . \tag{11.23}$$

We can therefore re-interpret the allowed transformations in Special Relativity, as the ones that leave the Minkowski metric invariant i.e.,

$$ds = g_{\mu\nu} dx^\mu dx^\nu = g'_{\mu\nu} dx'^\mu dx'^\nu = ds'^2 , \tag{11.24}$$

with $g'_{\mu\nu} = g_{\mu\nu}$ and therefore

$$\begin{aligned} ds^2 &= c^2 dt^2 - dx^2 - dy^2 - dz^2 \\ &= c^2 dt'^2 - dx'^2 - dy'^2 - dz'^2 = ds'^2 . \end{aligned} \tag{11.25}$$

As mentioned earlier, these transformations can be a Lorentz transformation and/or a space-time translation (more generically, a Poincaré transformation) i.e.,

$$x'^\mu = \Lambda^\mu_\nu x^\nu + a^\mu , \tag{11.26}$$

where a^μ is a constant space-time translation. The matrix Λ^μ_ν can be a Lorentz boost or a proper orthogonal rotation.[2] A Lorentz boost along the $\mathbf{k}$ axis (Fig. 11.2) will be given by

---

[2] As we have also mentioned, other type of transformations such as parity or time-reversal that fall into the Lorentz group category, shall not be treated here.

$$\Lambda^\mu_\nu = \begin{pmatrix} \gamma & 0 & 0 & -\gamma\beta \\ 0 & 1 & 0 & 0 \\ 0 & 0 & 1 & 0 \\ -\gamma\beta & 0 & 0 & \gamma \end{pmatrix}.$$ (11.27)

A rotation will be given by

$$\Lambda^\mu_\nu = \begin{pmatrix} 1 & 0 & 0 & 0 \\ 0 & R_{11} & R_{12} & R_{13} \\ 0 & R_{21} & R_{22} & R_{23} \\ 0 & R_{31} & R_{32} & R_{33} \end{pmatrix},$$ (11.28)

where R_{ij} are the components of a three-dimensional proper rotation matrix. Using matrix notation, the transformation (11.26) will read

$$\begin{pmatrix} x'^0 \\ x'^1 \\ x'^2 \\ x'^3 \end{pmatrix} = \begin{pmatrix} \Lambda^0_0 & \Lambda^0_1 & \Lambda^0_2 & \Lambda^0_3 \\ \Lambda^1_0 & \Lambda^1_1 & \Lambda^1_2 & \Lambda^1_3 \\ \Lambda^2_0 & \Lambda^2_1 & \Lambda^2_2 & \Lambda^2_3 \\ \Lambda^3_0 & \Lambda^3_1 & \Lambda^3_2 & \Lambda^3_3 \end{pmatrix} \begin{pmatrix} x^0 \\ x^1 \\ x^2 \\ x^3 \end{pmatrix} + \begin{pmatrix} a^0 \\ a^1 \\ a^2 \\ a^3 \end{pmatrix}.$$ (11.29)

11.4.1 Covariant and Contravariant Transformations

Some readers might find this subsection rather difficult. If this is so, one can skip the following discussion go directly to the next subsection, **Summary**, which contains the relevant results that are going to be needed in order to proceed with this topic.

Differentiating on both sides of (11.26), and due to the fact that $d\Lambda^\mu_\nu = 0$ (the matrix components are constants, they only depend on c and on $\mathcal{V}$ which is the magnitude of the relative velocity of the two inertial frames, which is constant) we obtain

$$dx'^\mu = \Lambda^\mu_\nu dx^\nu,$$ (11.30)

which are transformations that we are already familiar with i.e, (11.9) if Λ^μ_ν corresponds to a Lorentz boost. The previous transformation represents the transformation law for the components of a vector. More generically, any four-vector u^μ, under a change of coordinates (11.26) will transform as

$$u'^\mu = \Lambda^\mu_\nu u^\nu.$$ (11.31)

This is called the contravariant law of transformation, and is the transformation law that the components of a vector obey under a change of reference frame. As previously seen in Chaps. 1 and 2, we must insist that, for the same reasons as presented in the mentioned two chapters, **we must not consider coordinates as components of a vector**.

Let us now define the following quantity. Given the vector components u^μ we define

$$u_\nu = g_{\mu\nu} u^\mu . \tag{11.32}$$

Following this notation, the scalar product can be expressed as

$$u \cdot v = g_{\mu\nu} u^\mu v^\nu = u_\nu v^\nu . \tag{11.33}$$

Knowing that the previously defined quantity is basis invariant (under Poincaré transformations) and knowing the transformation law of v^ν, we can obtain the transformation law for u_ν as follows. Consider that under a Poincaré transformation $x^\mu \to x'^\mu$, the components of u transform as $u'_\mu = \bar\Lambda_\mu^\beta u_\beta$, where $\bar\Lambda$ is some unknown matrix. Therefore

$$u'_\mu v'^\mu = (\bar\Lambda_\mu^\beta u_\beta)(\Lambda_\alpha^\mu v^\alpha) = \bar\Lambda_\mu^\beta \Lambda_\alpha^\mu u_\beta v^\alpha = u_\beta v^\beta . \tag{11.34}$$

This is only true if and only if

$$\bar\Lambda_\mu^\beta \Lambda_\alpha^\mu = \delta_\alpha^\beta , \tag{11.35}$$

where δ_α^β is the usual Kronecker delta written in tensor form. Thus we conclude that $\bar\Lambda = \Lambda^{-1}$ or equivalently

$$\bar\Lambda_\nu^\mu = (\Lambda^{-1})_\nu^\mu . \tag{11.36}$$

The transformation

$$u'_\nu = (\Lambda^{-1})_\nu^\mu u_\nu , \tag{11.37}$$

is called the **covariant law** of transformation and where u_μ are the components of a **covector**. This is the reason why we use upper indices and lower indices. In order to distinguish vectors from covectors or, contravariant quantities from covariant quantities, that transform differently.

11.4.2 Summary

Under a Poincaré transformation, the coordinates change as

$$\boxed{x'^\mu = \Lambda_\nu^\mu x^\nu + a^\mu} . \tag{11.38}$$

The quantities dx^μ are the components of vectors in Minkowski space. dx^μ or any other vector with components u^μ, under the previous transformation will transform as

$$\boxed{u'^\mu = \Lambda^\mu_\nu u^\nu}\,,\tag{11.39}$$

which is called the contravariant law of transformation. The quantity defined as

$$\boxed{u_\nu = g_{\mu\nu}u^\mu}\,,\tag{11.40}$$

under a Poincaré transformation will transform as

$$\boxed{u'_\nu = (\Lambda^{-1})^\mu_\nu u_\mu}\,,\tag{11.41}$$

which is called the covariant law of transformation. We reserve the upper indices for the coordinates (even if they are not components of a vector) and for the components of vectors (which are contravariant quantities). The lower indices will be used for the components of covectors (which are covariant quantities).

Following these transformation laws, the scalar product of two vectors u and v defined as

$$\boxed{u \cdot v = g_{\mu\nu}u^\mu v^\nu = u_\mu v^\mu}\,,\tag{11.42}$$

it is a Poincaré invariant quantity. In particular the magnitude of a vector, or equivalently the squared magnitude defined as

$$u^2 = g_{\mu\nu}u^\mu u^\nu = u^\mu u_\mu\,,\tag{11.43}$$

is a Poincaré invariant quantity.

Following our (abusive) short-hand notation, if we write the components of a vector as

$$u^\mu = (u^0, u^1, u^2, u^3) = (u^0, \mathbf{u}) = (u^0, u^i)\,,\tag{11.44}$$

then the components of a covector will be given by

$$
\begin{aligned}
u_\mu &= (u_0, u_1, u_2, u_3)\\
&= (u^0, -u^1, -u^2, -u^3)\\
&= (u^0, -\mathbf{u})\\
&= (u^0, -u^i)\,,
\end{aligned}\tag{11.45}
$$

and so the scalar product $u^\mu v_\mu$ can be written as

$$\boxed{u^\mu v_\mu = u^0 v^0 - \mathbf{u} \cdot \mathbf{v}}. \tag{11.46}$$

Before moving on, a few more comments are required. The first one has to do, again, with the implicit summation rules. As we have mentioned, one can replace mute indices (repeated indices) at will. For example

$$u^\mu v_\mu = u^\beta v_\beta = u^\nu v_\nu = \cdots \tag{11.47}$$

Also, note that the metric is symmetric i.e., $g_{\mu\nu} = g_{\nu\mu}$ and therefore

$$\begin{aligned}
g_{\mu\nu} u^\mu v^\nu &= (g_{\mu\nu} u^\mu) v^\nu = u_\nu v^\nu \\
&= (g_{\mu\nu} v^\nu) u^\mu = u^\mu v_\mu = \cdots
\end{aligned} \tag{11.48}$$

The second comment has to do with the coordinate basis. In the first chapters, we normally studied in parallel, the transformation of the basis i.e., of $\{\mathbf{i}, \mathbf{j}, \mathbf{k}\}$. We could also define in our case a basis e_μ and therefore, a vector would be written as

$$v = v^\mu e_\mu, \tag{11.49}$$

where $e_\mu = \{\mathbf{e}_0, \mathbf{i}, \mathbf{j}, \mathbf{k}\}$, with $\mathbf{e}_0$ the temporal axis. Similar considerations can be made for covectors. This would be rather tedious and not at all necessary for the topics treated here, therefore, unless explicitly needed, we shall not deal with coordinate axes.

As a last comment within this section, note that all the concepts we are developing are in Cartesian coordinates. The generalization to curvilinear coordinates would suppose some additional complications i.e., $d\Lambda^\mu_\nu \neq 0$ in general, and we would have to modify all equations accordingly. However, this would not suppose any difference for the fundamental concepts presented herein. Thus, we shall not treat this topic here.

11.5 Four-Velocity, Acceleration and Force

For the mechanics of Special Relativity, we are going to define four-vectors corresponding to the velocity (or equivalently, the momentum), acceleration and force, that will have Lorentz (or more generically Poincaré) invariant magnitudes. This constitutes a fundamental difference with Newtonian mechanics in the following aspects. Consider the comparison between Galilean and Lorentz boosts:

- In Newtonian mechanics, velocity and momentum were not Galilean invariant quantities, therefore their magnitudes were not invariant either. The acceleration and the force on the other hand, were the same in all inertial reference frames, therefore their magnitude also.

- In Special relativity, all the components of any four-vector will transform (according to the contravariant law) under Lorentz boosts (in all cases: velocity, momentum, acceleration, force) but their magnitudes will remain invariant. Furthermore we will also respect the Special Relativity postulate that states that all equations of motion will be written in the same form in any inertial reference frame. This is also slightly different from Newtonian mechanics, where all equations of motion were written in the same form because the force was the same in all (inertial) reference frames.
- If in our comparison, we limit ourselves to rotations and/or translations, then both mechanics behave the same way. Under a translation, all equations remain the same because all physical magnitudes are not modified. Under a rotation the components of the vectors change in both cases (Newtonian mechanics and Special Relativity) but the equations of motions are written in the same form. The corresponding magnitudes of the vectors remain invariant in both cases.

The previous explanations will become more clear in this section, when constructing step by step, the equations of motion in Special Relativity. In order to define the four velocity we first need to introduce the proper time. Consider again the expression for the invariant measure ds^2 in some reference frame

$$ds^2 = c^2dt^2 - dx^2 - dy^2 - dz^2. \qquad (11.50)$$

If, again, we consider that the previous interval corresponds to a moving body (that moves from (ct, x, y, z) to $(ct + cdt, x + dx, y + dy, z + dz)$) we can re-write the previous expression in terms of the velocity of the body, as follows

$$ds^2 = c^2dt^2 - \left(\frac{dx^2}{dt^2} + \frac{dy^2}{dt^2} + \frac{dz^2}{dt^2}\right)dt^2 = c^2dt^2 - v^2dt^2. \qquad (11.51)$$

We define the **proper time** interval, as the time interval measured in a reference frame where the object is at rest. There is a unique reference frame[3] where this holds, therefore the proper time is and invariant magnitude! Thus, the proper time interval can be obtained from the previous expression by setting $v = 0$. We denote this time interval by $d\tau$ in order to distinguish it from any other time interval measured in other reference frame. Thus

$$\boxed{ds = c\, d\tau}. \qquad (11.52)$$

Let us now remember that we have identified our Lorentz (Poincaré) invariant four-vectors with dx^μ. These vectors have invariant magnitudes, therefore in order to maintain this property when defining the four-velocity, we have to differentiate with respect to an invariant magnitude. The natural election is of course $d\tau$. We therefore define the four-velocity vector as

---

[3]Up to constant space-time translations (which leave v and dt invariant).

$$u^\mu = \frac{dx^\mu}{d\tau}.$$ (11.53)

We shall now further investigate the properties of $d\tau$ and how we can relate it to the measured time interval in some arbitrary inertial reference frame. As $d\tau^2 = ds^2/c$,

$$
\begin{aligned}
d\tau^2 &= dt^2 - \frac{1}{c^2}(dx^2 + dy^2 + dx^2) \\
&= dt^2 - \frac{v^2}{c^2}dt^2 \\
&= dt^2\left(1 - \frac{v^2}{c^2}\right) \\
&= \frac{dt^2}{\gamma^2(v)}.
\end{aligned}
$$ (11.54)

In some other inertial reference frame $\mathcal{O}'$ we will obtain $d\tau^2 = dt'^2/\gamma^2(v')$. We can therefore conclude that in other inertial reference frames that move with arbitrary velocities (with magnitudes given by v, v', etc.) with respect to the reference frame in which the object is at rest, the relation among time intervals will be given by

$$d\tau = \frac{dt}{\gamma(v)} = \frac{dt'}{\gamma(v')}.$$ (11.55)

This phenomenon is called time dilation. It means that for a moving observer time intervals are larger ($dt = \gamma(v)\,d\tau$) than time intervals measured in the frame where *the clock is at rest* (corresponding to $d\tau$).

Having found this relation we can express the four velocity u^μ in terms of local variables. In a reference frame in which an object moves with velocity

$$\mathbf{v} = \frac{dx}{dt}\mathbf{i} + \frac{dy}{dt}\mathbf{j} + \frac{dz}{dt}\mathbf{k} = v_x\,\mathbf{i} + v_y\,\mathbf{j} + v_z\,\mathbf{k},$$ (11.56)

the four velocity will be given by

$$u^\mu = \frac{dx^\mu}{d\tau} = \gamma\frac{dx^\mu}{dt} = \gamma\,(c, v_x, v_y, v_z) = (\gamma c, \gamma \mathbf{v}),$$ (11.57)

where we have introduced the short-hand notation $\gamma = \gamma(v)$. In some other inertial reference frame the four-velocity will be given by

$$u'^\mu = \frac{dx'^\mu}{d\tau} = \gamma'\frac{dx'^\mu}{dt'} = (\gamma' c, \gamma' \mathbf{v}'),$$ (11.58)

where we have used the notation $\gamma' = \gamma(v')$. As $dx'^\mu = \Lambda^\mu_\nu dx^\nu$ and because $d\tau$ is an invariant quantity, both velocities will be related by the Lorentz transformation

$$u'^{\mu} = \Lambda^{\mu}_{\nu} u^{\nu}. \tag{11.59}$$

We are going to check as an exercise the previous results with an explicit example. One should also check that the previous transformations are equivalent to (11.13) and (11.14).

Note that in the previous expressions, there are three γ functions involved: $\gamma(v)$, $\gamma(v')$ and, $\gamma(\mathcal{V})$ which is the gamma function associated to the Lorentz boost, and the reader should make a clear distinction among them.

If u^{μ} are the components of the four-velocity vector

$$u^{\mu} = (u^0, u^1, u^2, u^3) = (\gamma c, \gamma v_x, \gamma v_y, \gamma v_z) = (\gamma c, \gamma \mathbf{v}), \tag{11.60}$$

then, the quantity u_{μ} will be given by

$$\begin{aligned} u_{\mu} &= g_{\mu\nu} u^{\nu} \\ &= (u_0, u_1, u_2, u_3) \\ &= (u^0, -u^1, -u^2, -u^3) \\ &= (\gamma c, -\gamma v_x, -\gamma v_y, -\gamma v_z) \\ &= (\gamma c, -\gamma \mathbf{v}). \end{aligned} \tag{11.61}$$

Its magnitude should be a Lorentz invariant quantity. Indeed

$$u^2 = u^{\mu} u_{\mu} = \gamma^2 c^2 - \gamma^2 v^2 = \frac{c^2 - v^2}{1 - v^2/c^2} = c^2. \tag{11.62}$$

Let us now introduce the four-momentum. Similar to the three-dimensional case it will by given by the product of the mass and the four velocity

$$p^{\mu} = m u^{\mu} = (\gamma mc, \gamma m \mathbf{v}). \tag{11.63}$$

The previous mass is the *rest mass* of the particle (the mass measured by an observer at rest with the particle).[4] Several things should be noted. First, the magnitude of p^{μ} is another Lorentz invariant quantity

$$p^2 = p^{\mu} p_{\mu} = m^2 c^2. \tag{11.64}$$

Second, as we have argued previously, the four-velocity will transform as (11.59). As m is an invariant quantity, the four-momentum will follow the same rule i.e.,

$$p'^{\mu} = \Lambda^{\mu}_{\nu} p^{\nu}, \tag{11.65}$$

---

[4]Some authors define the rest mass as m_0 and $m \equiv \gamma m_0$, which is somehow a dynamical mass. This is not going to be our case.

and therefore p^μ is also a well defined, well behaved vector. If we denote the non-relativistic three-momentum as $\mathbf{p}_{NR} = m\mathbf{v}$, then

$$\mathbf{p} = \gamma m\mathbf{v} = \gamma \mathbf{p}_{NR}, \tag{11.66}$$

is our relativistic three-momentum.

Finally, we must note that the element $p^0 = \gamma mc$ has units of E/c where E is some energy of yet unknown signification, that we are going to analyse in the following. In Newtonian mechanics, there is a well known relation between the (non-relativistic) kinetic energy and the (non-relativistic) momentum

$$K_{NR} = \frac{1}{2}m\mathbf{v}^2 = \frac{\mathbf{p}_{NR}^2}{2m}. \tag{11.67}$$

Let us thus therefore, further investigate the relation between the energy

$$E = p^0 c = \gamma mc^2, \tag{11.68}$$

and the (relativistic) three-momentum $\mathbf{p}$. Squaring the previous expression we obtain

$$E^2 = \gamma^2 m^2 c^4 = m^2 c^2 \frac{c^2}{1 - v^2/c^2}$$
$$= m^2 c^2 \frac{c^4}{c^2 - v^2}. \tag{11.69}$$

Adding and subtracting $(m^2 c^2)c^2 v^2/(c^2 - v^2)$ we obtain

$$E^2 = m^2 c^2 \frac{c^4 + c^2 v^2 - c^2 v^2}{c^2 - v^2}$$
$$= m^2 c^2 \frac{c^2 v^2 + c^2(c^2 - v^2)}{c^2 - v^2}$$
$$= m^2 c^4 + \frac{m^2 c^4 v^2}{c^2 - v^2}$$
$$= m^2 c^4 + \gamma^2 m^2 v^2 c^2$$
$$= m^2 c^4 + \mathbf{p}^2 c^2. \tag{11.70}$$

Therefore we finally obtain the following relation

$$E = \gamma mc^2 = \sqrt{m^2 c^4 + \mathbf{p}^2 c^2}. \tag{11.71}$$

For an object at rest ($\mathbf{p} = 0$) we have

$$E = mc^2, \tag{11.72}$$

which is Einstein's famous equation of the energy for a particle at rest. Thus we can define a relativistic kinetic energy by subtracting the rest energy from the total energy. We obtain

$$K = \gamma mc^2 - mc^2 = (\gamma - 1)mc^2 = \sqrt{m^2c^4 + \mathbf{p}^2c^2} - mc^2 . \tag{11.73}$$

Let us now calculate the limit $c \to \infty$ of the previous expression. We obtain

$$
\begin{aligned}
K &= (\gamma - 1)mc^2 \\
&= mc^2 \left(\frac{1}{\sqrt{1 - v^2/c^2}} - 1 \right) \\
&= mc^2 \left(1 + \frac{1}{2}\frac{v^2}{c^2} + \cdots - 1 \right) \\
&\approx \frac{1}{2}mv^2 ,
\end{aligned}
\tag{11.74}
$$

which is exactly the expression of the non-relativistic kinetic energy. Therefore, our previous relativistic construction of the four momentum is consistent with the studied Newtonian dynamics in the non-relativistic limit.

The remaining task is to define the acceleration and the force. As we have used $d\tau$ for defining the four velocity, it is natural to also define the four-acceleration as

$$a^\mu = \frac{d^2x^\mu}{d\tau^2} = \frac{du^\mu}{d\tau} . \tag{11.75}$$

We define the Minkowski four-force as

$$f^\mu = \frac{dp^\mu}{d\tau} . \tag{11.76}$$

We have seen that the four-velocity and momentum transform following the contravariant law i.e.,

$$u'^\mu = \Lambda^\mu_\nu u^\nu , \qquad p'^\mu = \Lambda^\mu_\nu p^\nu . \tag{11.77}$$

By taking the time derivative $d/d\tau$ on both sides of the previous expression, again keeping in mind that $d\tau$ is an invariant quantity and that $d\Lambda^\mu_\nu = 0$, we obtain the transformation laws for the acceleration and force

$$a'^\mu = \Lambda^\mu_\nu a^\nu , \qquad f'^\mu = \Lambda^\mu_\nu f^\nu . \tag{11.78}$$

In local coordinates the expression of a^μ takes the form

$$a^\mu = \gamma \frac{d}{dt}(\gamma c, \gamma \mathbf{v}) = \gamma \frac{d\gamma}{dt}(c, \mathbf{v}) + \gamma^2(0, \mathbf{a}) , \tag{11.79}$$

where $\mathbf{a}$ is the usual expression for the acceleration $\mathbf{a} = d\mathbf{v}/dt$. We observe that the previous expression contains a term proportional to $d\gamma/dt$ (one should not get confused with the gamma function from the Lorentz transformations $\gamma = \gamma(V)$ that depends on the relative velocity of two inertial frames, which is constant). We shall leave the explicit calculation of the previous expression as a proposed exercise.

11.5.1 Massless Particles

Up to this point, we have developed the previous expressions and calculations for a massive particle with four momentum $p^\mu = mu^\mu$. We have finally obtained the following simple expressions

$$p^\mu = (E/c, \mathbf{p}), \qquad E = \sqrt{m^2 c^4 + \mathbf{p}^2 c^2}. \qquad (11.80)$$

Without going into too many details, in turns out that the previous two expressions (written in this form) are general. They are valid for both massive and massless particles. For the case of massive particles, we have $E = \gamma m c^2$, $\mathbf{p} = \gamma m \mathbf{v}$ and $p^\mu p_\mu = m^2 c^2$. For massless particles i.e., a photon, we know from quantum mechanics that the expression for the energy is given by

$$E = \frac{hc}{\lambda} = h\nu = \hbar\omega, \qquad (11.81)$$

where h is Plank's constant, $\hbar = h/2\pi$ is the reduced Plank constant, λ is the wavelength of the photon, ν is the frequency of the wave and $\omega = 2\pi\nu$ is the so-called angular frequency. The momentum of a photon is given by

$$\mathbf{p} = \hbar\mathbf{k}, \qquad (11.82)$$

where $\mathbf{k}$ is the wave-vector (do not confuse it with the $\mathbf{k}$ axis) with its magnitude given by

$$k = |\mathbf{k}| = \frac{2\pi}{\lambda} = \frac{2\pi\nu}{c} = \frac{\omega}{c}. \qquad (11.83)$$

Therefore we obtain

$$\mathbf{p} = \frac{\hbar\omega}{c}\mathbf{u}_k, \qquad (11.84)$$

where we have defined $\mathbf{u}_k = \mathbf{k}/k$, the unitary vector corresponding to the direction and sense of the photon propagation. We can conclude that a photon also satisfies the expressions given in (11.80) in the limit $m \to 0$. This is also valid for the magnitude of the four-momentum i.e.,

$$p^2 = p^\mu p_\mu = 0. \tag{11.85}$$

Note that here we cannot distinguish between relativistic and non-relativistic momentum or energy because a photon is intrinsically a relativistic particle. It always moves at c and there is no classical limit that we can apply. We shall study with an explicit exercise, how the components of the four momentum of a photon

$$p^\mu = \frac{\hbar\omega}{c}(1, \mathbf{u}_k), \tag{11.86}$$

transform under a Lorentz boost. In the next chapter we shall also study collisions between photons and massive particles, such as Compton scattering.

One should also note that we have introduced directly the four-momentum of a photon and not its four velocity. As $d\tau = 0$ in the case of a light pulse, we cannot use $dx^\mu/d\tau$ as the definition of the four-velocity. We would have to introduce some other parameter in order to obtain u^μ. This is however, unnecessary. For practically all calculations we only need the four-momentum.

So far we have developed the needed formalism to describe the dynamics of a single particle. In the following chapter we are going to study systems formed by more than one particle, the corresponding conservation laws, and other phenomena such as particle creation and destruction in collisions or decays, the mass-energy correspondence and others, that are unique characteristics, or better said predictions, of Special Relativity.

Next, some exercises are proposed in order to better establish some of the previously introduced concepts, and to also introduce some new ones that will arise naturally. Some of them are the proper length and simultaneity. This last concept is highly affected by the mechanics of Special Relativity. We shall see that simultaneity is no longer an absolute notion, but depends on the observer. Limitations of Special Relativity applied to Rigid Solid mechanics are also commented upon and one representative example is given. Finally, after the Proposed Exercises section some final comments and reflections are made.

11.6 Proposed Exercises

11.1 Obtain the Lorentz boost relating the space-time coordinates of an inertial reference frame $\mathcal{O}$ with the space-time coordinates of another inertial reference frame $\mathcal{O}'$, that moves with $\mathcal{V} = \mathcal{V}\mathbf{k}$, $(\mathcal{V} < 0)$ with respect to $\mathcal{O}$.

Solution: Note that the equations (11.4) and (11.6) refer to a frame $\mathcal{O}'$ that moves with $\mathcal{V} = \mathcal{V}\mathbf{k}$ with respect to $\mathcal{O}$. However, when deducing these equations we haven't specified if $\mathcal{V}$ was positive or negative. Therefore these expressions are valid in both cases i.e., if $\mathcal{V} > 0$ then

$$\beta = \frac{\mathcal{V}}{c} > 0, \tag{11.87}$$

and vice-versa, if (as in this case) $\mathcal{V} < 0$ then $\beta < 0$. If one however, wishes to keep track of all the signs and express the equations as sums and subtractions of positive terms, then we can express $\beta = -|\beta|$ and so the Lorentz transformations for $\beta < 0$ read

$$\Delta z' = \gamma \Delta z + \gamma |\beta| c \Delta t,$$
$$c \Delta t' = \gamma c \Delta t + \gamma |\beta| \Delta z, \tag{11.88}$$

and the inverse

$$\Delta z = \gamma \Delta z' - \gamma |\beta| c \Delta t',$$
$$c \Delta t = \gamma c \Delta t' - \gamma |\beta| \Delta z'. \tag{11.89}$$

In case one has any doubt about the correct sign of β, one can simply calculate the Galilean limit as in (11.8) where the sign of $\mathcal{V}$ becomes obvious.

11.2 Consider a light source and a light detector device placed over a platform as shown in Fig. 11.3 (left). The trajectory of a light pulse for an observer at rest with the platform is also shown in Fig. 11.3 (left). For a second observer for which the platform is moving with $\mathcal{V} = \mathcal{V}\mathbf{k}$ ($\mathcal{V} > 0$), the trajectory is shown Fig. 11.3 (right). If the speed of light is the same in both reference frames, calculate using Pythagoras' theorem the relation between the measured time intervals (given by the emission and detection of the photon) in both reference frames. Check that the same result is obtained by applying a Lorentz boost.

Solution: Let us denote the time interval between the emission and detection in the reference frame at rest with the platform as Δt. If d is the distance between the emission and detection points, then

$$d = c\Delta t. \tag{11.90}$$

From the point of view of an observer for which the platform is moving with $\mathcal{V} = \mathcal{V}\mathbf{k}$, the travelled distance by the light pulse will be given by

$$d' = c\Delta t', \tag{11.91}$$

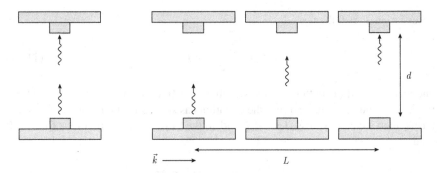

Fig. 11.3 Emission and detection of a light pulse for an observer at rest with the platform (left) and for an observer with respect to which, the platform is moving with $\mathcal{V} = V\mathbf{k}$ (right)

where d' is given by Pythagoras' theorem

$$d'^2 = L^2 + d^2 . \tag{11.92}$$

The quantity L is simply the travelled distance during Δt, thus $L = V \Delta t'$ and therefore we obtain

$$d'^2 = c^2 \Delta t'^2 = V^2 \Delta t'^2 + c^2 \Delta t^2 . \tag{11.93}$$

From the previous expression we obtain

$$c^2 \Delta t^2 = c^2 \Delta t'^2 \left(1 - \frac{V^2}{c^2} \right) = \frac{c^2 \Delta t'^2}{\gamma^2} . \tag{11.94}$$

This simplifies to

$$\Delta t' = \gamma \Delta t , \tag{11.95}$$

which means that the time interval in between the two events, for an observer for which the platform is moving, is larger than the time interval measured by an observer at rest with the platform. This phenomenon is known as time dilation, and it was already discussed in (11.55).

Let us now check that we obtain the same result by using Lorentz transformations. In our case we have measured Δt in the reference frame $\mathcal{O}$ and we need to obtain $\Delta t'$ in a reference frame $\mathcal{O}'$, with respect to which $\mathcal{O}$ moves with $\mathcal{V} = V\mathbf{k}$, $V > 0$ (this means that $\mathcal{O}'$ moves with $\mathcal{V} = -V\mathbf{k}$, $V > 0$ with respect to $\mathcal{O}$). As discussed in the previous exercise this corresponds to the transformations given in (11.4) and (11.6) with $\beta = -|\mathcal{V}| < 0$.

Going back to our problem, for the observer at rest with the platform there is no motion of the photon along the $\mathbf{k}$ axis. Therefore $\Delta z = 0$ and so, by using the set of equations (11.6) we obtain ($\beta < 0$)

$$\gamma \Delta z' = -\gamma \beta c \, \Delta t' . \tag{11.96}$$

Inserting this into the expression of $c\Delta t$ from (11.6)

$$c\Delta t = c\Delta t'(\gamma - \gamma \beta^2) = c\Delta t'\gamma(1 - \mathcal{V}^2/c^2) = \frac{c\Delta t'}{\gamma} , \tag{11.97}$$

which is the result we were looking for. Note that we could have also used the last expression of (11.4)

$$c \, \Delta t' = \gamma c \, \Delta t - \gamma \beta \, \Delta z . \tag{11.98}$$

For $\Delta z = 0$ we directly obtain the needed expression.

Note that the final result does not depend on the sign of β, thus, the same result corresponds to $\beta > 0$.

This time dilation, among other numerous experiments, has been verified with the mean life-time of muons. High energy gamma rays from outer-space interact with the atmosphere producing many types of particles. Among these particles we can find muons, whose life-time is well known. According to the measured life-time (in the laboratory) most of the muons should decay before reaching the sea-level. However, this is not the case experimentally. A much greater number is detected at ground (or sea) level, then expected by their life-time. The explanation of this phenomenon is compatible with the time-dilation factor predicted by Special Relativity. Given that they travel at approximately $0.99c$ the gamma factor is roughly 7 and all the experiments are compatible with this Special Relativity prediction.

> **11.3** The proper length of an object is defined as the length of the object measured in a reference system at rest with it. Consider that Δz is the proper length (along the **k** axis in the frame $\mathcal{O}$). Using Lorentz transformations, relate it to the length measured by another observer in $\mathcal{O}'$ that moves with $\mathcal{V} = \mathcal{V}\mathbf{k}$ ($\mathcal{V} > 0$) with respect to $\mathcal{O}$. Note: as the object with respect to the reference frame $\mathcal{O}'$ is moving, an observer in this reference frame must perform an instantaneous measurement of the length (of the object along the **k** axis).

Solution: This problem can be easily addressed the following way. We know that the proper length is Δz and this measurement does not involve time, as the observer performing the measurement is at rest with the object. On the other hand, the observer in $\mathcal{O}'$ needs to perform an instantaneous length measurement of the object. Thus $\Delta t' = 0$, and $\Delta z'$ is the length of the object instantaneously measured in $\mathcal{O}'$. The Lorentz transformations in our case correspond to (11.4) and (11.6) with $\beta > 0$. The expression that directly relates Δz with $\Delta z'$ and $\Delta t'$ is

$$\Delta z = \gamma \Delta z' + \gamma \beta c \, \Delta t'. \tag{11.99}$$

For $\Delta t' = 0$ we simply obtain

$$\Delta z' = \frac{\Delta z}{\gamma}, \tag{11.100}$$

which means that the length along the **k** axis, measured by a moving observer is contracted with respect to the proper length. Again, note that the final result does not depend on the sign of β and so, the same result corresponds to $\beta < 0$.

Alternatively one can use any other expressions from (11.4) and (11.6) that indirectly relate Δz with $\Delta z'$. For example one can choose

$$\begin{aligned} c \, \Delta t &= \gamma c \, \Delta t' + \gamma \beta \, \Delta z' \\ \Delta z' &= \gamma \Delta z - \gamma \beta c \, \Delta t. \end{aligned} \tag{11.101}$$

By imposing $\Delta t' = 0$ the first expression simplifies to

$$c \, \Delta t = \gamma \beta \, \Delta z'. \tag{11.102}$$

Inserting this expression into the second one we obtain

$$\Delta z' = \gamma \Delta z - \gamma^2 \beta^2 \, \Delta z'. \tag{11.103}$$

Manipulating, the previous expression reads

$$\begin{aligned} \Delta z &= \frac{\Delta z'}{\gamma}(1 + \gamma^2 \beta^2) \\ &= \frac{\Delta z'}{\gamma}\left(1 + \frac{\beta^2}{1 - \beta^2}\right) \\ &= \frac{\Delta z'}{\gamma}\left(\frac{1 - \beta^2 + \beta^2}{1 - \beta^2}\right) \\ &= \frac{\Delta z'}{\gamma}\gamma^2, \end{aligned} \tag{11.104}$$

and therefore $\Delta z' = \Delta z / \gamma$.

This result can also be related to the muon experiment. Note that we have defined the proper length in the same reference frame in which we have defined the proper time interval. We have seen that for an observer on the surface of the Earth, the lifetime of the muon is larger than its rest life-time (by a factor γ). An observer at rest with the muon, would measure the rest life-time, which is shorter. Thus, according to this observer, if *nothing else happens* the muon should not reach the ground. The fact is that indeed, *something else happens*. For the observer at rest with the muon, the distance between the atmosphere and the surface of the Earth (or the sea level) is shorter (contracted by a factor γ). Thus, there will be no contradiction in between an observer on the ground and one travelling with the muon. For both of them, the muon reaches the ground.

11.4 Consider an isosceles right-angled triangle, placed as in Fig. 11.4, in some reference frame $\mathcal{O}$ at rest with the triangle. How will another observer in some inertial frame $\mathcal{O}'$ (moving with $\mathcal{V} = \mathcal{V}\mathbf{k}$ with respect to $\mathcal{O}$) describe the same triangle? Consider $h = 1$ (m) and $\mathcal{V} = 0.9c$.

Solution: The solution of this exercise only requires a straightforward application of the previously studied length contraction. The triangle, in the $\mathcal{O}'$ frame will have its hypotenuse contracted by a factor gamma i.e., $b' = b/\gamma$. As there is no motion along the other axes, the height h of the triangle will remain unchanged. Thus, one can safely affirm that the triangle is not right-angled in the second reference frame. We can easily calculate the length of the other two sides in $\mathcal{O}'$. We have

$$a' = \sqrt{h^2 + \left(\frac{b'}{2}\right)^2} = \sqrt{h^2 + \left(\frac{b}{2\gamma}\right)^2}. \tag{11.105}$$

As for $\cos\theta'$, it will be given by

$$\cos\theta' = \frac{b'/2}{a'} = \frac{b}{2\,a'\gamma}. \tag{11.106}$$

If we call the right angle ϕ in the reference frame $\mathcal{O}$, then in $\mathcal{O}'$ we have $\phi' = \pi - 2\arccos\theta'$. Let us now numerically perform the calculation. In the reference frame $\mathcal{O}$

$$a\cos(\pi/4) = h \quad \Rightarrow \quad a = \frac{h}{\cos(\pi/4)} = \sqrt{2}\,(m), \tag{11.107}$$

therefore $b/2 = 1$ (m) and so $b = 2$ (m). The gamma factor for $\mathcal{V} = 0.9c$ is $\gamma \simeq 2.29$ and so $b' \simeq 0.87$ (m) and $a' \simeq 1.1$ (m). Finally $\theta' \simeq 66.5°$ angle and so $\phi' \simeq 47°$ which is almost half of the initial 90° angle. As an exercise the reader can find $\mathcal{V}$ for which the triangle as seen from $\mathcal{O}'$ is equilateral.

In the next exercise we shall deal with another concept that is affected by Lorentz boosts, and that is simultaneity. As we shall next, events that are simultaneous on some reference frame, they will not be simultaneous in other reference frames.

11.5 Consider a train moving with velocity $\mathcal{V} = \mathcal{V}\mathbf{k}$ ($\mathcal{V} > 0$) along its railway tracks. An observer (at rest with the train, whose reference frame we shall

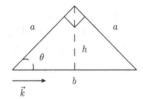

Fig. 11.4 An isosceles right-angled triangle in a reference frame at rest with the triangle, with its hypotenuse placed along the **k** axis

Fig. 11.5 Emission of two light pulses by an observer at rest with the train, that moves with $\mathcal{V} = \mathcal{V}\,\mathbf{k}\ (\mathcal{V} > 0)$ along its tracks with respect to some external observer

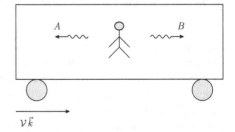

denote by $\mathcal{O}$) is standing in the middle of one of the wagons with a light source. At some time instant t_0 the observer emits two pulses (along the **k** axis), one towards each end of the wagon (labelled as A and B), as shown in Fig. 11.5. For the observer from the train, both pulses (A and B) will reach both ends of the wagon simultaneously, at some time instant $t_1 = t_1^A = t_1^B$. What will an observer outside the train have to say about the order in which the pulses reach the end of the wagon? How about if the train moved in the opposite sense? Use Lorentz transformations.

Solution: If we call the reference frame of the observer outside the train $\mathcal{O}'$, then $\mathcal{O}'$ is moving with $-\mathcal{V}\,\mathbf{k}$ (with $\mathcal{V} > 0$) with respect to $\mathcal{O}$. The corresponding Lorentz transformations, are given by (11.4) with $\beta = -|\mathcal{V}| < 0$. If we denote the corresponding time difference of the detection times t_1^A and t_1^B in the $\mathcal{O}$ reference frame as Δt, then $\Delta t = 0$. The time interval in $\mathcal{O}'$ will thus be given by

$$c\,\Delta t' = -\gamma\,\beta\,\Delta z \neq 0, \qquad (11.108)$$

and so the two events are no longer simultaneous.

If, on the other hand, the train moved in the opposite sense, the same expression is valid, but with $\beta > 0$. We can observe thus, that the time sequence of the two events is the opposite when compared to the first case.

It is left for a reader as an exercise, to argue why $t_1^A > t_1^B$ in the first case.

11.6 Check that the Lorentz transformations (11.6) are the inverse transformations of (11.4).

Solution: We have to check that by inserting

$$\Delta z = \gamma \Delta z' + \gamma \beta c \, \Delta t',$$
$$c \, \Delta t = \gamma c \, \Delta t' + \gamma \beta \, \Delta z', \tag{11.109}$$

into

$$\Delta z' = \gamma \Delta z - \gamma \beta c \, \Delta t,$$
$$c \, \Delta t' = \gamma c \, \Delta t - \gamma \beta \, \Delta z, \tag{11.110}$$

we obtain $\Delta z' = \Delta z'$ and $c \Delta t' = c \Delta t'$. For the first expression we have

$$\begin{aligned}
\Delta z' &= \gamma(\gamma \Delta z' + \gamma \beta c \, \Delta t') - \gamma \beta (\gamma c \, \Delta t' + \gamma \beta \, \Delta z') \\
&= \gamma^2 \Delta z' - \gamma^2 \beta^2 \, \Delta z' \\
&= \gamma^2 \Delta z'(1 - \beta^2) \\
&= \Delta z'. \tag{11.111}
\end{aligned}$$

The second expression is equally straightforward and it is left for the reader.

11.7 Check that the quantity Δs^2 is invariant under the Lorentz transformations (11.6).

Solution: The expression for Δs^2 as a function of Δx, Δy, Δz and Δt is given by

$$\Delta s^2 = c^2 \Delta t^2 - \Delta x^2 - \Delta y^2 - \Delta z^2. \tag{11.112}$$

Inserting the expressions from (11.6) we obtain

$$\begin{aligned}
\Delta s^2 &= (\gamma c \, \Delta t' + \gamma \beta \, \Delta z')^2 - \Delta x'^2 - \Delta y'^2 - (\gamma \Delta z' + \gamma \beta c \, \Delta t')^2 \\
&= \gamma^2 c^2 \Delta t'^2 + \gamma^2 \beta^2 \Delta z'^2 + 2\gamma^2 \beta c \, \Delta t' \, \Delta z' - \Delta x'^2 - \Delta y'^2 \\
&\quad - \gamma^2 \Delta z'^2 - \gamma^2 \beta^2 c^2 \, \Delta t'^2 - 2\gamma^2 \beta c \, \Delta t' \, \Delta z' \\
&= \gamma^2 (1 - \beta^2) c^2 \Delta t'^2 - \Delta x'^2 - \Delta y'^2 + \gamma^2 (\beta^2 - 1) \Delta z'^2 \\
&= c^2 \Delta t'^2 - \Delta x'^2 - \Delta y'^2 - \Delta z'^2 \\
&= \Delta s'^2. \tag{11.113}
\end{aligned}$$

An equivalent approach for obtaining the same result is by writing the expression for $\Delta s'^2$ and inserting the expressions (11.4).

11.8 Check that for a light-pulse moving in an arbitrary direction with $\mathbf{v} = v_x\,\mathbf{i} + v_y\,\mathbf{j} + v_z\,\mathbf{k}$ where $v = (v_x^2 + v_y^2 + v_z^2)^{1/2} = c$, the velocity addition formulae (11.13) and (11.14) leave the magnitude of the photon velocity invariant i.e., $v' = (v_x'^2 + v_y'^2 + v_z'^2)^{1/2} = c$. How about if $v < c$? For this last case consider $v = c - \Delta$.

Solution: Note that this is just another way of stating the previous exercise. Squaring and summing the expressions for v_x', v_y' and v_z' we obtain

$$
\begin{aligned}
v_x'^2 + v_y'^2 + v_z'^2 &= \frac{v_x^2 + v_y^2}{\gamma^2(1 - \mathcal{V}v_z/c^2)^2} + \frac{(v_z - \mathcal{V})^2}{(1 - \mathcal{V}v_z/c^2)^2} \\
&= \frac{v_x^2 + v_y^2 + \gamma^2(v_z - \mathcal{V})^2}{\gamma^2(1 - \mathcal{V}v_z/c^2)^2} \\
&= \frac{c^2 - v_z^2 + \gamma^2(v_z - \mathcal{V})^2}{\gamma^2(1 - \mathcal{V}v_z/c^2)^2} \\
&= c^2\frac{1 - v_z^2/c^2 + \gamma^2 v_z^2/c^2 + \gamma^2\mathcal{V}^2/c^2 - 2\gamma^2\mathcal{V}v_z/c^2}{\gamma^2 + \gamma^2\mathcal{V}^2 v_z^2/c^4 - 2\gamma^2\mathcal{V}v_z/c^2} \\
&\equiv c^2\frac{A}{B}\,.
\end{aligned}
\tag{11.114}
$$

Expressing the gamma function in terms of $\mathcal{V}$ and simplifying we obtain the numerator

$$
A = \frac{(c^2 - \mathcal{V}v_z)^2}{c^4 - c^2\mathcal{V}^2}\,.
\tag{11.115}
$$

As for the denominator, we obtain

$$
B = \frac{(c^2 - \mathcal{V}v_z)^2}{c^4 - c^2\mathcal{V}^2} = A\,,
\tag{11.116}
$$

and therefore $v' = c$, as expected. Similarly we can consider $v < c$ i.e., $c = v + \Delta$, with $\Delta > 0$ and demonstrate that $v' < c$.

11.9 Check that for a massive particle the transformations $u'^\mu = \Lambda^\mu_\nu u^\nu$, with Λ^μ_ν a Lorentz boost, are equivalent to the velocity addition rules (11.13) and (11.14).

Solution: The transformation

$$u'^{\mu} = \Lambda^{\mu}_{\nu} u^{\nu} , \tag{11.117}$$

written in matrix form reads

$$\begin{pmatrix} \gamma' c \\ \gamma' v'_x \\ \gamma' v'_y \\ \gamma' v'_z \end{pmatrix} = \begin{pmatrix} \gamma(\mathcal{V}) & 0 & 0 & -\gamma(\mathcal{V})\beta(\mathcal{V}) \\ 0 & 1 & 0 & 0 \\ 0 & 0 & 1 & 0 \\ -\gamma(\mathcal{V})\beta(\mathcal{V}) & 0 & 0 & \gamma(\mathcal{V}) \end{pmatrix} \begin{pmatrix} \gamma c \\ \gamma v_x \\ \gamma v_y \\ \gamma v_z \end{pmatrix} , \tag{11.118}$$

where $\gamma = \gamma(v)$, $\gamma' = \gamma(v')$ with v and v' the magnitudes of $\mathbf{v}$ and $\mathbf{v}'$, and where we have explicitly written down the dependence of the gamma and the beta functions (on $\mathcal{V}$) for the Lorentz boost, in order to make the appropriate distinctions. We obtain

$$\begin{aligned} \gamma' c &= \gamma(\mathcal{V})\gamma c - \gamma(\mathcal{V})\beta(\mathcal{V})\gamma v_z , \\ \gamma' v'_x &= \gamma v_x , \\ \gamma' v'_y &= \gamma v_y , \\ \gamma' v'_z &= -\gamma(\mathcal{V})\beta(\mathcal{V})\gamma c + \gamma(\mathcal{V})\gamma v_z . \end{aligned} \tag{11.119}$$

Dividing the second and the third expression by the first one we obtain

$$\frac{v'_{x,y}}{c} = \frac{v_{x,y}}{\gamma(\mathcal{V}) c (1 - \mathcal{V} v_z/c^2)} . \tag{11.120}$$

Dividing the last expression by the first one

$$\frac{v'_z}{c} = \frac{v_z - \mathcal{V}}{c (1 - \mathcal{V} v_z/c^2)} . \tag{11.121}$$

These last two expressions correspond to (11.13) and (11.14).

11.10 A photon is emitted with angular frequency ω in the $\mathcal{O}$ reference frame, with the propagation direction and sense given by $\mathbf{u}_k = \mathbf{k}$ (where $\mathbf{k}$ in this particular case is the $\mathbf{k}$ axis). Calculate four-momentum transformation $p'^{\mu} = \Lambda^{\mu}_{\nu} p^{\nu}$ of the photon under a Lorentz boost (along the $\mathbf{k}$ axis).

Solution: In the initial reference frame, the four-momentum of the photon is given by

$$p^{\mu} = \frac{\hbar \omega}{c}(1, 0, 0, 1) , \tag{11.122}$$

as the photon is propagating along the **k** axis. Applying a Lorentz boost along the **z** axis (11.27) we obtain

$$p'^{\mu} = \Lambda^{\mu}_{\nu} p^{\nu} = \frac{\hbar\omega}{c}(\gamma - \gamma\beta)(1, 0, 0, 1) = \frac{\hbar\omega'}{c}(1, 0, 0, 1). \qquad (11.123)$$

We thus obtain the following relation between the two frequencies

$$\omega' = \omega\gamma(1 - \beta) = \omega\frac{\sqrt{(1 - \beta)^2}}{\sqrt{1 - \beta^2}} = \omega\sqrt{\frac{(1 - \beta)^2}{(1 + \beta)(1 - \beta)}}. \qquad (11.124)$$

Simplifying, the previous expression reads

$$\omega' = \omega\sqrt{\frac{1 - \beta}{1 + \beta}} = \omega\sqrt{\frac{c - \mathcal{V}}{c + \mathcal{V}}}. \qquad (11.125)$$

This relation is called the Relativistic Doppler effect. It accounts for the variation of the frequency of a light pulse (or photon) depending on the motion of the observer. If $\beta > 0$, the observer is moving away from the photon and so $\omega' < \omega$. Thus the wavelength $\lambda' > \lambda$. This phenomenon is called red-shift (the displacement of the λ towards larger wavelengths). If, on the contrary $\beta < 0$, the observer is moving towards the photon and so $\omega' > \omega$ and therefore $\lambda' < \lambda$. This phenomenon is called blue-shift for obvious reasons. Both cases are schematically shown in Fig. 11.6.

The Doppler effect is fundamental for determining cosmological scale distances, such as distances between galaxies. It was also used by Hubble to determine that the Universe is expanding.

Fig. 11.6 A photon of frequency ω emitted in $\mathcal{O}$ along the positive **k** axis. An observer from $\mathcal{O}'$ that moves along the positive **k** axis (away from the photon) will observe a red-shift. An observer moving along the negative **k** axis (towards the photon) will observe a blue-shift

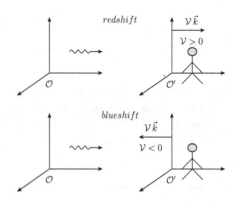

11.11 Obtain the explicit expression for the four-acceleration a^μ, and for the Minkowski force f^μ for a particle with constant mass. What can we say about the product $u \cdot a = u^\mu a_\mu$? Calculate the $v \ll c$ limit for a^μ, or equivalently for f^μ.

Solution: The acceleration four-vector was given in (11.79)

$$a^\mu = \gamma \frac{d\gamma}{dt} (c, \mathbf{v}) + \gamma^2 (0, \mathbf{a}). \tag{11.126}$$

The expression for $d\gamma/dt$ reads

$$\frac{d\gamma(v)}{dt} = \frac{d}{dt} \left(1 - \frac{v^2}{c^2}\right)^{-1/2} = -\frac{1}{2} \left(1 - \frac{v^2}{c^2}\right)^{-3/2} (-1) \frac{2v}{c^2} \frac{dv}{dt}. \tag{11.127}$$

Remember from Chap. 3, that the variation of the magnitude of the velocity is the tangential velocity i.e., $dv/dt = a_\tau$. Therefore, the previous expression simplifies to

$$\frac{d\gamma(v)}{dt} = \frac{\gamma^3}{c^2} v \, a_\tau. \tag{11.128}$$

Remember also that the velocity can be written as

$$\mathbf{v} = v \boldsymbol{\tau}, \tag{11.129}$$

and the acceleration

$$\mathbf{a} = a_\tau \boldsymbol{\tau} + a_n \mathbf{n}, \tag{11.130}$$

where $\boldsymbol{\tau}$ is the tangential vector, $\mathbf{n}$ is the normal vector and $\boldsymbol{\tau} \cdot \mathbf{n} = 0$. Therefore $v \, a_\tau$ can be expressed as the scalar product

$$v \, a_\tau = \mathbf{v} \cdot \mathbf{a}. \tag{11.131}$$

We finally obtain

$$\frac{d\gamma(v)}{dt} = \frac{\gamma^3 \mathbf{v} \cdot \mathbf{a}}{c^2}. \tag{11.132}$$

Defining $\boldsymbol{\beta} = \mathbf{v}/c$ we get to the following expression for the four-acceleration

$$a^\mu = (\gamma^4 \boldsymbol{\beta} \cdot \mathbf{a}, \gamma^4 (\boldsymbol{\beta} \cdot \mathbf{a})\boldsymbol{\beta} + \gamma^2 \mathbf{a}). \tag{11.133}$$

As for the Minkowski force for a particle with constant mass, we simply have

$$f^\mu = ma^\mu = (f^0, \mathbf{f}) = (\gamma^4 \boldsymbol{\beta} \cdot \mathbf{F}, \gamma^4 (\boldsymbol{\beta} \cdot \mathbf{F})\boldsymbol{\beta} + \gamma^2 \mathbf{F}), \tag{11.134}$$

where we have introduced the Newtonian force $\mathbf{F} = m\mathbf{a}$. Note that, following the previous notation $\mathbf{f} \neq \mathbf{F}$. Let us now analyse the product $u \cdot a = u^\mu a_\mu$. We obtain

$$u \cdot a = c\gamma^5 \boldsymbol{\beta} \cdot \mathbf{a} - \gamma^5 (\boldsymbol{\beta} \cdot \mathbf{a})(\boldsymbol{\beta} \cdot \mathbf{v}) - \gamma^3 \mathbf{v} \cdot \mathbf{a} . \tag{11.135}$$

Writing everything in terms of $\mathbf{v}$ (and not $\boldsymbol{\beta}$) we have

$$u \cdot a = \gamma^5 \mathbf{v} \cdot \mathbf{a} - \gamma^5 (\mathbf{v} \cdot \mathbf{a})\frac{v^2}{c^2} - \gamma^3 \mathbf{v} \cdot \mathbf{a} . \tag{11.136}$$

Grouping the first two terms and simplifying

$$u \cdot a = \gamma^5 (\mathbf{v} \cdot \mathbf{a}) \left(1 - v^2/c^2\right) - \gamma^3 \mathbf{v} \cdot \mathbf{a}$$
$$= \frac{\gamma^5}{\gamma^2}(\mathbf{v} \cdot \mathbf{a}) - \gamma^3 \mathbf{v} \cdot \mathbf{a} = 0 . \tag{11.137}$$

Therefore we obtain that the four-velocity and the four-acceleration (or equivalently the four-force) are orthogonal

$$u^\mu a_\mu = u^\mu f_\mu = 0 . \tag{11.138}$$

As the previous scalar-product is invariant, the previous relation holds in any inertial reference frame.

Let us now consider the limit $v \ll c$. In this limit $\beta = (v/c) \to 0$ and $\gamma \to 1$ thus, from (11.133) we simply obtain

$$\lim_{c \to \infty} a^\mu = (0, \mathbf{a}) , \tag{11.139}$$

and so

$$\lim_{c \to \infty} f^\mu = (0, \mathbf{F}) , \tag{11.140}$$

where $\mathbf{F} = m\,\mathbf{a}$. Thus, in this limit one obtains Newton's law.

11.12 Consider a very large rigid solid rod that has one extreme resting on a motion sensor and another extreme placed on a device that, under the action of an observer, it pushes the rod towards the motion sensor. Consider also that, next to the rod, there is laser-emitting device which emits light-pulses that propagate parallel to the rod (again, under the action of an observer). If next to the motion sensor there is a light detector (as in Fig. 11.7) and the observer decides to simultaneously push the rod and send a light signal, will the motion sensor be activated before the light pulse is detected?

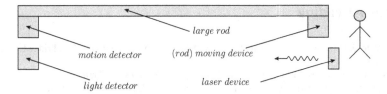

Fig. 11.7 An observer that simultaneously sends a light signal and activates the moving device for the rod

Solution: This is a tricky question that sheds light on the limitations, not of Special Relativity, but on the notion of rigid body. For a rigid body, by definition, the distance between any two of its points is constant. Therefore, if it starts moving due to a force acting on one extreme, the other extreme also starts moving instantaneously. This *instantaneous* action at a distance is no longer compatible with Special Relativity. In many cases we can neglect these effects, however in the present case it explicitly violates the postulates of Special Relativity. As there is nothing that can travel faster than light, the light-pulse will reach the detector before the rod activates the motion sensor. One can calculate the delay between the two events by modelling more precisely the motion of a rigid body (i.e., as the propagation of some perturbation induced by the application of an external force). There are some other examples slightly more complex, in which the notion of *rigidity* does not hold (see for example Born rigidity), however we shall not get into further details here.

11.7 Final Comments

Most results we just derived in Special Relativity might seem rather counterintuitive, or might even seem to belong to the realm of science fiction. This is because they do not correspond to day-to-day experience. A similar impression one might get from quantum mechanics. When unifying quantum mechanics with Special Relativity a new theory arises, the Quantum Filed Theory. Based on this new paradigm, the Standard Model of particle physics was build (in the last decades). It unifies the electromagnetic force with the weak force (responsible among other things for nuclear decays) and also incorporates the strong interaction (which is responsible, roughly speaking for holding together the nuclei). All Standard Model predictions have been precisely verified at large particle colliders such as Tevatron or the LHC. All the corresponding kinematics are relativistic and no deviation has been ever observed or reported. We can therefore trust that, at least for the experimentally probed energy ranges, Special Relativity is not science fiction.

Further Reading

1. J.V. José, E.J. Saletan, *Classical Dynamics: A Contemporary Approach*. Cambridge University Press
2. S.T. Thornton, J.B. Marion, *Classical Dynamics of Particles and Systems*
3. H. Goldstein, C. Poole, J. Safko, *Classical Mechanics*, 3rd edn. Addison Wesley
4. J.R. Taylor, *Classical Mechanics*
5. D.T. Greenwood, *Classical Dynamics*. Prentice-Hall Inc.
6. D. Kleppner, R. Kolenkow, *An Introduction to Mechanics*
7. C. Lanczos, *The Variational Principles of Mechanics*. Dover Publications Inc.
8. W. Greiner, *Classical Mechanics: Systems of Particles and Hamiltonian Dynamics*. Springer
9. H.C. Corben, P. Stehle, *Classical Mechanics*, 2nd edn. Dover Publications Inc.
10. T.W.B. Kibble, F.H. Berkshire, *Classical Mechanics*. Imperial College Press
11. M.G. Calkin, *Lagrangian and Hamiltonian Mechanics*
12. A.J. French, M.G. Ebison, *Introduction to Classical Mechanics*

Chapter 12
Relativistic Collisions and Decays

Abstract As the title suggests, this chapter will be dedicated to the study of the kinematics corresponding to relativistic collisions and decays. The concept of collisions in Special Relativity, far from being a straightforward generalization of non-relativistic processes, introduces a totally new phenomenon, conversion of matter into energy and vice-versa. In the realm of subatomic particles, we can collide two particles and obtain two (or more) other different particles. Some other processes such as, nuclear decays, have no classical correspondence either. Kinematically, they are purely relativistic processes. Here we shall study a great deal of such processes, introduce the needed Lorentz-invariant kinematic variables and study the relation between the angles of the scattered particles in the Laboratory and Center of Mass reference frames.

12.1 Conservation Laws and Kinematic Invariants

In the previous chapter we have generically talked about invariance under Poincaré transformations, which were composed by a Lorentz transformation (a rotation or a boost) and/or a space-time translation. As kinematics is not affected by constant space-time translations or rotations, in this chapter we shall speak of Lorentz invariance as referred to Lorentz boost invariance, in order to follow more closely the terminology used in standard literature.

It might seem, at first sight, that relativistic kinematics is a much more complicated subject than the previously studied *classical* kinematics. This is very far from the truth. Conceptually, relativistic processes are simpler. They do not distinguish between elastic and non-elastic scattering or particle decays. In all cases the only conservation law we have to apply is four-momentum conservation.

Consider that we initially have N colliding particles with four-momenta (in some arbitrary reference frame) given by

$$p_{i,ini}^{\mu},$$ (12.1)

© Springer Nature Switzerland AG 2020
V. Ilisie, *Lectures in Classical Mechanics*, Undergraduate Lecture
Notes in Physics, https://doi.org/10.1007/978-3-030-38585-9_12

where *ini* stands for *initially* and $i = 1, \ldots, N$. Consider that the final state is given by M outgoing particles with momenta (in the same reference frame) given by

$$p^{\mu}_{j,fin} \, ,\tag{12.2}$$

where *fin* stands for *final state* and $j = 1, \ldots, M$. We can thus define an initial and final (total) four-momentum of the system as

$$\mathcal{P}^{\mu}_{ini} = \sum_{i}^{N} p^{\mu}_{i,ini} \, , \qquad \mathcal{P}^{\mu}_{fin} = \sum_{j}^{M} p^{\mu}_{j,fin} \, .\tag{12.3}$$

The four-momentum conservation theorem then simply states: *if no external force acts upon the system, the four-momentum is conserved.* This means

$$\mathcal{P}^{\mu}_{ini} = \mathcal{P}^{\mu}_{fin} \, .\tag{12.4}$$

Roughly speaking, this is all that it is needed, in order to describe Lorentz invariant relativistic kinematics. Next, we shall apply this theorem to different cases and, together with the introduction of Lorentz-invariant quantities, we will be able to derive many useful results.

Before moving on, we have to define the center of mass (CM) reference system. Similar to the non-relativistic case, is given by the frame where the total four-momentum of the system is

$$\mathcal{P}^{\mu} = (\mathcal{P}^{0}, \mathbf{0}) \, .\tag{12.5}$$

As $\mathcal{P}^{\mu}$ is conserved, the previous expression will be valid both before, and after the collision (in the CM).

12.2   Decays

Consider a particle with mass m_a. In the CM reference system this particle is at rest. Its four-momentum will be given by

$$p^{\mu}_a = (m_a c, \mathbf{0}) \, .\tag{12.6}$$

Due to the weak interaction, it decays into two particles with masses m_1 and m_2. In the same reference frame, after the decay, the four-momenta of the daughter particles are

$$p^{\mu}_1 = (E_1/c, \mathbf{p}) \, ,$$
$$p^{\mu}_2 = (E_2/c, -\mathbf{p}) \, ,\tag{12.7}$$

Fig. 12.1 A particle of mass m_a that decays into two particles, as seen from the CM reference frame

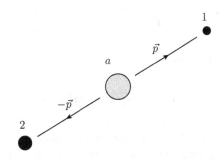

as schematically shown in Fig. 12.1, with

$$m_a c = (E_1 + E_2)/c \,, \tag{12.8}$$

due to four-momentum conservation, and where

$$\mathbf{p} = \gamma_1 m_1 \mathbf{v}_1 = -\gamma_2 m_2 \mathbf{v}_2 \,, \quad E_1/c = \gamma_1 m_1 c \,, \quad E_2/c = \gamma_2 m_2 c \,. \tag{12.9}$$

Let us now try to deduce simplified expressions for E_1, E_2 and $|\mathbf{p}|$ in terms of the masses. The squared magnitude of p_1^μ is given by

$$p_1^2 = p_1^\mu \, p_{\mu,1} = m_1^2 c^2 \,. \tag{12.10}$$

On the other hand $p_1^\mu = p_a^\mu - p_2^\mu$, therefore, we obtain

$$\begin{aligned}
(p_a - p_2)^2 &\equiv (p_a^\mu - p_2^\mu)(p_{\mu,a} - p_{\mu,2}) \\
&= p_a^\mu \, p_{\mu,a} + p_2^\mu \, p_{\mu,2} - 2 \, p_a^\mu \, p_{\mu,2} \\
&= p_a^2 + p_2^2 - 2 \, p_a \cdot p_2 \\
&= m_a^2 c^2 + m_2^2 c^2 - 2 \, E_2 m_a \,.
\end{aligned} \tag{12.11}$$

Thus we obtain the following simple expression for E_2

$$E_2 = \frac{1}{2m_a}(m_a^2 + m_2^2 - m_1^2)c^2 \,. \tag{12.12}$$

Similarly, we find

$$E_1 = \frac{1}{2m_a}(m_a^2 + m_1^2 - m_2^2)c^2 \,. \tag{12.13}$$

In order to obtain $|\mathbf{p}|$ we need to remember the relation between the total energy and the three-momentum: $E^2 = m^2 c^4 + \mathbf{p}^2 c^2$. In our case, we can use either E_1 or E_2. Using E_1 we obtain

$$\mathbf{p}^2 = \frac{E_1^2}{c^2} - m_1^2 c^2 = \frac{1}{4m_a^2}(m_a^2 + m_1^2 - m_2^2)^2 c^2 - m_1^2 c^2 . \tag{12.14}$$

Expanding

$$\mathbf{p}^2 = \frac{1}{4m_a^2}(m_1^4 + m_2^4 + m_a^4 - 2m_1^2 m_2^2 - 2m_1^2 m_a^2 - 2m_2^2 m_a^2)c^2 . \tag{12.15}$$

In order to maintain the notation short, and because the previous expression will appear repeatedly, we introduce the Kallen lambda function

$$\lambda(x, y, z) \equiv x^2 + y^2 + z^2 - 2xy - 2xz - 2yz . \tag{12.16}$$

Therefore $|\mathbf{p}|$ can be simply expressed as

$$|\mathbf{p}| = \frac{c}{2m_a}\lambda^{1/2}(m_a^2, m_1^2, m_2^2) . \tag{12.17}$$

It will be left for the reader as an exercise, to calculate the previous kinematic variables for a particle that decays in flight i.e., in a reference frame $\mathcal{O}'$ with respect to which, the particle moves with $\mathbf{v} = v\,\mathbf{k}$.

Three-body decays (and more generically N-body decays) can also occur, however, as they are slightly more complicated cases, they are usually analysed in more advanced particle physics courses and thus, we shall not further advance on this subject. For the interested reader we present the relevant formulae for three-body decays in Appendix G.

One should note that, from the fact that $\mathbf{p} = -\mathbf{p}$ in the CM frame, the process takes place along a straight path. It is also worth mentioning that there is a mass threshold relation in order for the process to be kinematically allowed. This relation is $m_a \geq m_1 + m_2$. If $m_a = m_1 + m_2$, in the CM frame, this corresponds to the mother particle decaying into two daughter particles at rest.

All sorts of particle decays occur in Nature, in the laboratory and in particle colliders and their kinematics obey the predictions of Special Relativity. For a list of (possibly) all experimentally measured decays one can consult the Particle Data Group (http://pdg.lbl.gov/) database.

Next, we will analyse the $2 \to 2$ frontal collision process in the CM and Laboratory (Lab) reference frames. The relation between the two scattering angles (corresponding to the two reference frames) will be left for the reader as an exercise.

12.3 $2 \to 2$ Frontal Collisions

We shall start our analysis in the CM reference frame. This collision is represented in Fig. 12.2. For this process we consider the most generic configuration where all masses are different. The four-momenta of the particles before the collision are given by

Fig. 12.2 The $1 + 2 \rightarrow 3 + 4$ process in the center of mass reference frame

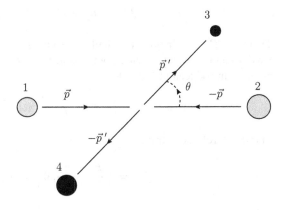

$$p_1^\mu = (E_1/c, \mathbf{p}) \,,$$
$$p_2^\mu = (E_2/c, -\mathbf{p}) \,, \qquad (12.18)$$

and after the collision

$$p_3^\mu = \left(E_3/c, \mathbf{p}'\right) \,,$$
$$p_4^\mu = \left(E_4/c, -\mathbf{p}'\right) \,, \qquad (12.19)$$

where

$$E_1 + E_2 = E_3 + E_4 \,. \qquad (12.20)$$

As in the non-relativistic case, this reaction takes places in the same plane. For this process, there are three useful Lorentz invariant quantities that we can define (independently of the reference frame, obviously) in terms of the four-momenta. These quantities are called Mandelstam variables and they are given by

$$s \equiv (p_1 + p_2)^2 = (p_3 + p_4)^2 \,,$$
$$t \equiv (p_1 - p_3)^2 = (p_2 - p_4)^2 \,,$$
$$u \equiv (p_1 - p_4)^2 = (p_2 - p_3)^2 \,. \qquad (12.21)$$

It is left for the reader as an exercise to show that

$$s + t + u = \sum_i m_i^2 c^2 \,. \qquad (12.22)$$

Let us now calculate the corresponding energies and relativistic three-momenta as functions of the Mandelstam variables and the involved masses. The total four-momentum of the system is

$$\mathcal{P}^\mu = p_1^\mu + p_2^\mu = p_3^\mu + p_4^\mu. \tag{12.23}$$

Therefore $\mathcal{P}^2 = s$. In order to calculate the energies we simply proceed as previously. On one hand $p_2^2 = m_2^2 c^2$. On the other hand $p_2^\mu = \mathcal{P}^\mu - p_1^\mu$ and so (using the compact notation introduced in (12.11))

$$(\mathcal{P} - p_1)^2 = \mathcal{P}^2 + p_1^2 - 2\mathcal{P} \cdot p_1 = s + m_1^2 c^2 - 2\mathcal{P} \cdot p_1. \tag{12.24}$$

In the CM reference frame

$$\mathcal{P}^\mu = \left(\frac{E_1 + E_2}{c}, \mathbf{0}\right) = (\sqrt{s}, \mathbf{0}), \tag{12.25}$$

and therefore

$$\mathcal{P} \cdot p_1 = \sqrt{s}\,\frac{E_1}{c}. \tag{12.26}$$

We thus obtain

$$m_2^2 c^2 = s + m_1^2 c^2 - 2\sqrt{s}\,\frac{E_1}{c}, \tag{12.27}$$

or equivalently

$$E_1 = \frac{c}{2\sqrt{s}}(s + m_1^2 c^2 - m_2^2 c^2). \tag{12.28}$$

Similarly one can obtain the expressions for the three remaining energies

$$E_2 = \frac{c}{2\sqrt{s}}(s + m_2^2 c^2 - m_1^2 c^2),$$
$$E_3 = \frac{c}{2\sqrt{s}}(s + m_3^2 c^2 - m_4^2 c^2),$$
$$E_4 = \frac{c}{2\sqrt{s}}(s + m_4^2 c^2 - m_3^2 c^2). \tag{12.29}$$

As in the previous section, using the total energy—three momentum relation, we can straightforwardly obtain the expressions for $|\mathbf{p}|$ and $|\mathbf{p}'|$

$$|\mathbf{p}| = \frac{1}{2\sqrt{s}}\lambda^{1/2}(s, m_1^2 c^2, m_2^2 c^2),$$
$$|\mathbf{p}'| = \frac{1}{2\sqrt{s}}\lambda^{1/2}(s, m_3^2 c^2, m_4^2 c^2). \tag{12.30}$$

Finally, let us calculate the scattering angle. From Fig. 12.2 we observe that

$$\mathbf{p} \cdot \mathbf{p}' = |\mathbf{p}||\mathbf{p}'| \cos \theta . \tag{12.31}$$

We can thus calculate the scalar product $p_1 \cdot p_3$ for example, and obtain

$$p_1 \cdot p_3 = \frac{E_1 E_3}{c^2} - \mathbf{p} \cdot \mathbf{p}' = \frac{E_1 E_3}{c^2} - |\mathbf{p}||\mathbf{p}'| \cos \theta . \tag{12.32}$$

On the other hand we have

$$t = (p_1 - p_3)^2 = p_1^2 + p_3^2 - 2p_1 \cdot p_3 = m_1^2 c^2 + m_3^2 c^2 - 2p_1 \cdot p_3 , \tag{12.33}$$

and therefore

$$p_1 \cdot p_3 = \frac{1}{2}(m_1^2 c^2 + m_3^2 c^2 - t) . \tag{12.34}$$

We thus obtain

$$\cos \theta = \frac{1}{|\mathbf{p}||\mathbf{p}'|} \left(\frac{E_1 E_3}{c^2} - \frac{1}{2}(m_1^2 c^2 + m_3^2 c^2 - t) \right) . \tag{12.35}$$

Introducing the explicit expressions for E_1, E_2, $|\mathbf{p}|$ and $|\mathbf{p}'|$ and simplifying, the previous equation reads

$$\cos \theta = \frac{(m_1^2 - m_2^2)(m_3^2 - m_4^2)c^4 + s(s + 2t - \sum_i m_i^2 c^2)}{\lambda^{1/2}(s, m_1^2 c^2, m_2^2 c^2)\lambda^{1/2}(s, m_3^2 c^2, m_4^2 c^2)} . \tag{12.36}$$

Using (12.22) it we can further simplify the previous expression and obtain the standard expression for $\cos \theta$

$$\cos \theta = \frac{(m_1^2 - m_2^2)(m_3^2 - m_4^2)c^4 + s(t - u)}{\lambda^{1/2}(s, m_1^2 c^2, m_2^2 c^2)\lambda^{1/2}(s, m_3^2 c^2, m_4^2 c^2)} . \tag{12.37}$$

Let us now calculate the same quantities in the Lab reference frame. We define this reference frame as the one where m_2 is at rest ($\mathbf{p}_2^L = \mathbf{0}$), as shown in Fig. 12.3. The four-momenta of the colliding particles will be given by

$$p_{1,L}^\mu = (E_1^L/c, \mathbf{p}_1^L),$$
$$p_{2,L}^\mu = (m_2 c, \mathbf{0}) . \tag{12.38}$$

For the outgoing particles we have

$$p_{3,L}^\mu = (E_3^L/c, \mathbf{p}_3^L),$$
$$p_{4,L}^\mu = (E_4^L/c, \mathbf{p}_4^L) . \tag{12.39}$$

Fig. 12.3 The
$1 + 2 \to 3 + 4$ process in
the Laboratory reference
frame where m_2 is at rest

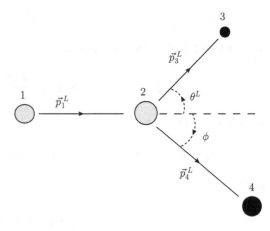

The total four-momentum of the system in the Lab. frame is thus

$$\mathcal{P}_L^\mu = p_{1,L}^\mu + p_{2,L}^\mu = p_{3,L}^\mu + p_{4,L}^\mu \, . \tag{12.40}$$

Its magnitude (as it is a Lorentz invariant quantity) will be given by $\mathcal{P}_L^2 = \mathcal{P}^2 = s$
(where $\mathcal{P}^\mu$ is the total four-momentum in the CM frame). This is on of the advantages
of working with Lorentz invariant quantities. One can calculate them in the most
convenient reference frame and use the results in any other reference frame. In order
to calculate the energies we can benefit from the fact that $\mathbf{p}_2^l = \mathbf{0}$. Thus

$$\begin{aligned}
s &= (p_{1,L} + p_{2,L})^2 \\
&= p_{1,L}^2 + p_{2,L}^2 + 2 p_{1,L} \cdot p_{2,L} \\
&= m_1^2 c^2 + m_2^2 c^2 + 2 E_1^L m_2 \, ,
\end{aligned} \tag{12.41}$$

and so, we obtain

$$E_1^L = \frac{1}{2m_2}(s - m_1^2 c^2 - m_2^2 c^2) \, . \tag{12.42}$$

In a similar way, using the definitions of t and u (the ones that involve p_2) we obtain

$$\begin{aligned}
E_3^L &= \frac{1}{2m_2}(m_2^2 c^2 + m_3^2 c^2 - u) \, , \\
E_4^L &= \frac{1}{2m_2}(m_2^2 c^2 + m_4^2 c^2 - t) \, .
\end{aligned} \tag{12.43}$$

The calculation of the magnitudes of the three-momenta are (just as previously)
straightforward. Using the relation $E^2 = m^2 c^4 + \mathbf{p}^2 c^2$, we obtain

$$|\mathbf{p}_1^L| = \frac{1}{2\,m_2\,c}\lambda^{1/2}(s, m_2^2 c^2, m_1^2 c^2)\,,$$

$$|\mathbf{p}_3^L| = \frac{1}{2\,m_2\,c}\lambda^{1/2}(u, m_2^2 c^2, m_3^2 c^2)\,,$$

$$|\mathbf{p}_4^L| = \frac{1}{2\,m_2\,c}\lambda^{1/2}(t, m_2^2 c^2, m_4^2 c^2)\,. \tag{12.44}$$

The angle θ^L is equally straightforward to calculate by using the Lorentz invariant product $p_{1,L}^\mu (p_{3,L})_\mu$ and relating it to the Mandelstam variable t. Hence, we can express $\cos\theta^L$ as

$$\cos\theta^L = \frac{1}{|\mathbf{p}_1^L||\mathbf{p}_3^L|}\left(\frac{E_1^L E_3^L}{c^2} - \frac{1}{2}(m_1^2 c^2 + m_3^2 c^2 - t)\right). \tag{12.45}$$

Inserting the explicit expression for the energies and momenta we obtain

$$\cos\theta^L = \frac{\mathcal{F}(m_i, s, t, u)}{\lambda^{1/2}(s, m_2^2 c^2, m_1^2 c^2)\lambda^{1/2}(u, m_2^2 c^2, m_3^2 c^2)}\,, \tag{12.46}$$

with

$$\mathcal{F}(m_i, s, t, u) = c^2(m_3^2 s + m_1^2 u + m_2^2(s + 2t + u)) - s\,u$$
$$- c^4(m_2^4 + 3m_2^2 m_3^2 + 3m_1^2 m_2^2 + m_1^2 m_3^2)\,, \tag{12.47}$$

which is a rather ugly expression, that can be somewhat further simplified using (12.22). Instead, one can use the relation of $\cos\theta^L$ with $\cos\theta$ (which is left as an exercise) given by the following expression

$$\tan\theta^L = \frac{m_2 c^2\,|\mathbf{p}'|\,\sin\theta}{|\mathbf{p}|\,E_3 + |\mathbf{p}'|\,E_2\cos\theta}\,. \tag{12.48}$$

All the quantities from the right side of the previous expression are the ones corresponding to the CM frame. As we shall see in the Proposed Exercises section this relation can be easily obtained by relating the CM frame with the Lab. frame through a Lorentz boost.

In this chapter we have briefly introduced the reader to the realm of relativistic collisions and decays. These processes, and more complex ones, are further studied in particle physics. There, it is usual to make use of natural units i.e., $c = 1 = \hbar$. Therefore one should not be surprised to find all the previous expressions with $c = 1$ all along the literature. As a conclusion, we must insist upon the fact that all the kinematics from particle accelerators (where the colliding particles travel at speeds greater than $0.99c$) are relativistic and, up till now there hasn't been observed any deviation from the predictions of Special Relativity. Far from intending to make the reader accept Special Relativity as a dogma, these comments are only made to clarify the fact that, even if this theory seems counter-intuitive, it has demonstrated to work flawlessly in all experiments.

12.4 Proposed Exercises

12.1 Consider an unstable particle at rest (in the CM frame) of mass m_a. In some other reference frame i.e., the Lab. frame, this particle moves along a straight path with constant velocity. This mother particle finally decays into two daughter particles. If in the CM frame the angle formed by the direction of a daughter particle and the direction of flight of the mother particle is θ, calculate this angle as seen from the Lab. reference frame. Use a Lorentz boost.

Solution: Let us consider without loss of generality that the mother particle travels along the $\mathbf{k}$ axis (we can always find a reference frame in which this holds true). In the CM frame the trajectories of the daughter particles belong to a straight line. This straight line and the direction of flight of the mother particle (before decaying) form a plane. Let us consider, again without loss of generality that this plane is (y, z) (we can always find a reference frame in which, besides the fact that the mother particle travels along the $\mathbf{k}$ axis, the decay takes place in the (y, z) plane).

With the previous considerations, the four momenta of the daughter particles in the CM frame can be written as

$$p_1^\mu = (E_1/c, 0, |\mathbf{p}| \sin \theta, |\mathbf{p}| \cos \theta)$$
$$p_2^\mu = (E_2/c, 0, -|\mathbf{p}| \sin \theta, -|\mathbf{p}| \cos \theta). \qquad (12.49)$$

Let us now consider that the mother particle moves with $\mathbf{p}_a = \gamma \, m_a v_a \, \mathbf{k}$, $v_a > 0$. If we identify the $\mathcal{O}$ reference frame with the CM frame, and $\mathcal{O}'$ with the Lab. frame, then the Lorentz transformation

$$\Lambda^\mu_\nu = \begin{pmatrix} \gamma_a & 0 & 0 & \gamma_a \, \beta_a \\ 0 & 1 & 0 & 0 \\ 0 & 0 & 1 & 0 \\ \gamma_a \, \beta_a & 0 & 0 & \gamma_a \end{pmatrix}, \qquad (12.50)$$

with $\gamma_a = \gamma(v_a)$ and $\beta_a = v_a/c$, corresponds to (11.4) with $\mathcal{V} = -v_a$ or $\beta = -\beta_a$ (as the particle moves with velocity $v_a \, \mathbf{k}$ with respect to $\mathcal{O}'$, this is equivalent to saying that the reference frame $\mathcal{O}'$ moves with velocity $-v_a \, \mathbf{k}$ with respect to the particle, or equivalently, with respect to the CM frame of the particle). One can also check the correctness (of the sign) of the previous transformation by applying the boost to the four momentum of the mother particle in the CM frame.[1]

Applying the previous Lorentz boost to p_1^μ we obtain

$$
\begin{pmatrix} E_1^L/c \\ 0 \\ |\mathbf{p}_1^L| \sin \theta^L \\ |\mathbf{p}_1^L| \cos \theta^L \end{pmatrix} = \begin{pmatrix} \gamma_a & 0 & 0 & \gamma_a \beta_a \\ 0 & 1 & 0 & 0 \\ 0 & 0 & 1 & 0 \\ \gamma_a \beta_a & 0 & 0 & \gamma_a \end{pmatrix} \begin{pmatrix} E_1/c \\ 0 \\ |\mathbf{p}| \sin \theta \\ |\mathbf{p}| \cos \theta \end{pmatrix}.
\tag{12.51}
$$

Therefore

$$
\begin{aligned}
|\mathbf{p}_1^L| \sin \theta^L &= |\mathbf{p}| \sin \theta, \\
|\mathbf{p}_1^L| \cos \theta^L &= \gamma_a \beta_a E_1/c + \gamma_a |\mathbf{p}| \cos \theta.
\end{aligned}
\tag{12.52}
$$

As $|\mathbf{p}| = \gamma_1 m_1 v_1$ (with $|\mathbf{v}_1| = v_1$) in the CM frame, one can easily check that

$$
\frac{E_1/c}{|\mathbf{p}|} = \frac{1}{\beta_1}, \qquad \beta_1 = \frac{v_1}{c}.
\tag{12.53}
$$

Thus, taking the quotient of the two expressions from (12.52) and simplifying, we obtain

$$
\tan \theta^L = \frac{\sin \theta}{\gamma_a (\beta_a/\beta_1 + \cos \theta)}.
\tag{12.54}
$$

12.2 Can a massive particle decay into a photon? How about to N photons ($N > 1$)? If the process for $N = 2$ is kinematically allowed, calculate the energy of the photons in the CM frame.

Solution: The simplest way to answer the first question is by analysing momentum conservation in the CM frame. Before decaying, the massive particle has four-momentum

$$
p^\mu = (mc, \mathbf{0}),
\tag{12.55}
$$

which also corresponds to the total momentum of the system. If the particle decays into one unique photon, the final four-momentum of the system will be given by the four-momentum of the photon

---

[1]In the CM frame $p_a^\mu = (m_a c, \mathbf{0})$ and after applying the boost we obtain $p_a^{\mu, L} = (\gamma_a m_a c, 0, 0, \gamma_a m_a v_a)$, which is the correct result.

$$p_\gamma^\mu = (E_\gamma, \mathbf{p}_\gamma). \tag{12.56}$$

As $\mathbf{p}_\gamma \neq \mathbf{0}$ (a photon will always have a non zero three-momentum as there is no reference frame where the photon is at rest), we conclude that the decay of a massive particle into a photon is forbidden. On the other hand, the decay into N ($N > 1$) photons is allowed. If $p_i^\mu = (E_i, \mathbf{p}_i)$ with $i = 1, \ldots, N$, are the four-momenta of the final state photons, momentum conservation imposes the following constraint

$$\sum_i \mathbf{p}_i = \mathbf{0}. \tag{12.57}$$

Let us now solve the two-photon problem in the CM. The final state momenta will be given by

$$p_1^\mu = (\hbar\omega_1/c, \mathbf{p}), \qquad p_1^\mu = (\hbar\omega_2/c, -\mathbf{p}). \tag{12.58}$$

Remember that the three-momentum of a photon of energy $\hbar\omega$ is given by

$$\mathbf{p} = \frac{\hbar\omega}{c}\mathbf{u}_k \tag{12.59}$$

where $\mathbf{u}_k$ (as explained in the previous chapter) is the unitary wave vector that corresponds to the direction and sense of the photon propagation. We conclude that momentum conservation (in the case of two final state photons in the CM) implies $\omega_1 = \omega_2 \equiv \omega$. Thus

$$mc = 2\hbar\omega/c, \tag{12.60}$$

and so, the energy of any of the two photons will be given by

$$\hbar\omega = mc^2/2. \tag{12.61}$$

12.3 Consider a massive particle of mass m that decays into another massive particle of mass m' ($m' < m$) and a photon. What are the kinematical constraints in the CM frame, on the three-momentum of the final state particle of mass m'? How about if the initial particles decays is into a massive particle of mass m' and two photons?

Solution: As in the previous exercise, the four momentum of the decaying particle in the CM frame is given by

$$p^\mu = (mc, \mathbf{0}). \tag{12.62}$$

After the decay we have

$$p_1^\mu = (\gamma m'c, \gamma m'\mathbf{v}) = \left(\frac{E_1}{c}, \mathbf{p}\right), \tag{12.63}$$

and

$$p_2^\mu = \frac{\hbar\omega}{c}(1, \mathbf{u}_k) = \left(\frac{E_\gamma}{c}, -\mathbf{p}\right). \tag{12.64}$$

As the photon can never be found at rest $\mathbf{p} \neq 0$, the massive particle cannot be at rest either. This restriction is no longer valid if in the final state, instead of one photon we have two photons. For the last case, if m' is at rest, the two photons will have the same energy and they will emerge back-to-back.

12.4 For a $2 \to 2$ scattering process, check that

$$s + t + u = (m_1^2 + m_2^2 + m_3^2 + m_4^2)c^2. \tag{12.65}$$

Solution: The solution to this exercise is rather straightforward. One only needs to explicitly write down both expressions for each term of the sum s, t and u, as follows

$$\begin{aligned} s + t + u &= \frac{1}{2}\big[(p_1 + p_2)^2 + (p_3 + p_4)^2 + (p_1 - p_3)^2 \\ &\quad + (p_2 - p_4)^2 + (p_1 - p_4)^2 + (p_2 - p_3)^2\big] \\ &= (m_1^2 + m_2^3 + m_3^2 + m_4^2)c^2. \end{aligned} \tag{12.66}$$

12.5 Obtain the relation between the CM scattering angle and the Lab. scattering angle for a $2 \to 2$ reaction,

$$\tan \theta^L = \frac{m_2 c^2 \, |\mathbf{p}'| \sin \theta}{|\mathbf{p}| \, E_3 + |\mathbf{p}'| \, E_2 \cos \theta}, \tag{12.67}$$

where all the quantities from the right side of the previous equation are expressed in the CM frame. Use a Lorentz boost.

Solution: Let us consider, without loss of generality, that in the CM frame, the particles before the collision move along the $\mathbf{k}$ axis.[2] Therefore, before the collision

---

[2]Before the frontal collision takes place, the particles move along a straight path. Thus we can always choose a reference frame in which this path is the $\mathbf{k}$ axis.

we have the following four-momenta

$$
\begin{aligned}
p_1^\mu &= (E_1/c, 0, 0, |\mathbf{p}|) \ , \\
p_2^\mu &= (E_2/c, 0, 0, -|\mathbf{p}|) \ .
\end{aligned}
\tag{12.68}
$$

We can also consider, without loss of generality, that the process takes place in the (y, z) plane.[3] Thus, the final state momenta read

$$
\begin{aligned}
p_3^\mu &= \left(E_3/c, 0, |\mathbf{p}'| \sin\theta, |\mathbf{p}'| \cos\theta\right) \ , \\
p_4^\mu &= \left(E_4/c, 0, -|\mathbf{p}'| \sin\theta, -|\mathbf{p}'| \cos\theta\right) \ .
\end{aligned}
\tag{12.69}
$$

Let us now use a Lorentz boost to obtain the four-momenta in the Lab. reference frame. In this frame m_2 is at rest. If we call the CM frame $\mathcal{O}$ and the Lab. frame $\mathcal{O}'$ then, $\mathcal{O}'$ moves with respect to $\mathcal{O}$ with $\mathcal{V} = V\mathbf{k}$ where $\mathcal{V} = -v_2 < 0$, $v_2 = |\mathbf{p}|/(\gamma_2 m_2)$ and $\gamma_2 = \gamma(v_2)$. This corresponds to a Lorentz boost (11.4) with $\mathcal{V} = -v_2$. Note that $\gamma_2 m_2 c = E_2/c$ therefore we can express the γ_2 function as

$$
\gamma_2 = \frac{E_2}{m_2 c^2} \ .
\tag{12.70}
$$

We can also express the product $\gamma_2 m_2$ as

$$
\gamma_2 m_2 = E_2/c^2 \quad \Rightarrow \quad v_2 = \frac{|\mathbf{p}|}{\gamma_2 m_2} = \frac{|\mathbf{p}| c^2}{E_2} \ ,
\tag{12.71}
$$

and so

$$
\beta_2 = \frac{v_2}{c} = \frac{|\mathbf{p}| c}{E_2} \ .
\tag{12.72}
$$

As a cross check one can verify that

$$
p_2'^\mu = \Lambda^\mu_\nu p_2^\nu = (m_2 c, 0) = p_2^{\mu, L} \ .
\tag{12.73}
$$

Let us now apply this Lorentz boost to p_3^μ. We obtain

$$
\begin{pmatrix} E_3^L/c \\ 0 \\ |\mathbf{p}_3^L| \sin\theta^L \\ |\mathbf{p}_3^L| \cos\theta^L \end{pmatrix}
=
\begin{pmatrix}
\frac{E_2}{m_2 c^2} & 0 & 0 & \frac{|\mathbf{p}|}{m_2 c} \\
0 & 1 & 0 & 0 \\
0 & 0 & 1 & 0 \\
\frac{|\mathbf{p}|}{m_2 c} & 0 & 0 & \frac{E_2}{m_2 c^2}
\end{pmatrix}
\begin{pmatrix} E_3/c \\ 0 \\ |\mathbf{p}'| \sin\theta \\ |\mathbf{p}'| \cos\theta \end{pmatrix} \ ,
\tag{12.74}
$$

---

[3] We know that the process takes place in a plane, thus we can choose the reference frame for which this plane is (y, z).

Fig. 12.4 Schematic representation of Compton scattering in the Lab. reference frame

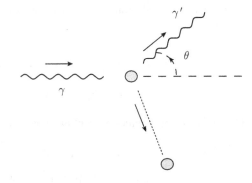

which translates into

$$|\mathbf{p}_3^L| \sin \theta^L = |\mathbf{p}'| \sin \theta \,,$$

$$|\mathbf{p}_3^L| \cos \theta^L = \frac{|\mathbf{p}| E_3}{m_2 c^2} + \frac{|\mathbf{p}'| E_2}{m_2 c^2} \cos \theta \,. \tag{12.75}$$

By taking the quotient of the two previous expressions, one obtains the relation we were looking for.

12.6 Compton scattering consists in a photon of energy E_γ that is scattered by an electron. The final state consists in a photon of energy E'_γ and the scattered electron. Prove that the Compton scattering angle in the Lab. frame (where the electron is at rest) can be written as

$$\cos \theta = 1 - m_e c^2 \left(\frac{1}{E_{\gamma'}} - \frac{1}{E_\gamma} \right) , \tag{12.76}$$

where m_e is the mass of the electron and $E_\gamma = \hbar\omega$, $E_{\gamma'} = \hbar\omega'$ are the energies of the incoming and the outgoing photons.

Solution: This collision is schematically shown in Fig. 12.4. The four-momentum configuration before the collision for the electron and the photon are given by

$$p_\gamma^\mu = \frac{\hbar\omega}{c}(1, \mathbf{u}_k) = (E_\gamma/c, \mathbf{p}_\gamma) \,,$$

$$p_e^\mu = (m_e c, \mathbf{0}) \,. \tag{12.77}$$

After the collision, the four momenta of the resulting photon and the scattered electron are given by

$$
\begin{aligned}
p_{\gamma'}^{\mu} &= \frac{\hbar \omega'}{c}(1, \mathbf{u}_k') = (E_{\gamma'}/c, \mathbf{p}_{\gamma'}), \\
p_e'^{\mu} &= (E_e'/c, \mathbf{p}_e').
\end{aligned}
\tag{12.78}
$$

On one hand, from three-momentum conservation we have

$$
\begin{aligned}
|\mathbf{p}_e'|^2 &= (\mathbf{p}_\gamma - \mathbf{p}_{\gamma'}) \cdot (\mathbf{p}_\gamma - \mathbf{p}_{\gamma'}) \\
&= \mathbf{p}_\gamma^2 + \mathbf{p}_{\gamma'}^2 - 2\mathbf{p}_\gamma \cdot \mathbf{p}_{\gamma'} \\
&= \frac{E_\gamma^2}{c^2} + \frac{E_{\gamma'}^2}{c^2} - \frac{2 E_\gamma E_{\gamma'}}{c^2} \cos \theta.
\end{aligned}
\tag{12.79}
$$

Multiplying by c^2 on both sides of the previous expression we obtain

$$
|\mathbf{p}_e'|^2 c^2 = E_\gamma^2 + E_{\gamma'}^2 - 2 E_\gamma E_{\gamma'} \cos \theta.
\tag{12.80}
$$

On the other hand, from energy conservation we have

$$
E_e' = E_\gamma - E_{\gamma'} + m_e c^2,
\tag{12.81}
$$

and so

$$
\begin{aligned}
E_e'^2 - m_e^2 c^4 &= |\mathbf{p}_e'|^2 c^2 \\
&= (E_\gamma - E_{\gamma'} + m_e c^2)^2 - m_e^2 c^4 \\
&= E_\gamma^2 + E_{\gamma'}^2 - 2 E_\gamma E_{\gamma'} + 2 m_e c^2 E_\gamma - 2 m_e c^2 E_{\gamma'}.
\end{aligned}
\tag{12.82}
$$

By combining the two expressions (12.80) and (12.82) we obtain the result we were looking for.

We shall not go deeper into the analysis of more complex processes, as they no longer belong to the realm of classical mechanics but to more advanced subjects such as particle physics. However, with our short list of exercises together with the theoretical introduction, we have managed to tackle a great deal of interesting and relevant results on the subject.

Further Reading

1. J.V. José, E.J. Saletan, *Classical Dynamics: A Contemporary Approach*. Cambridge University Press
2. S.T. Thornton, J.B. Marion, *Classical Dynamics of Particles and Systems*
3. H. Goldstein, C. Poole, J. Safko, *Classical Mechanics*, 3rd edn. Addison Wesley

4. J.R. Taylor, *Classical Mechanics*
5. D.T. Greenwood, *Classical Dynamics*. Prentice-Hall Inc.
6. D. Kleppner, R. Kolenkow, *An Introduction to Mechanics*
7. C. Lanczos, *The Variational Principles of Mechanics*. Dover Publications Inc.
8. W. Greiner, *Classical Mechanics: Systems of Particles and Hamiltonian Dynamics*. Springer
9. H.C. Corben, P. Stehle, *Classical Mechanics*, 2nd edn. Dover Publications Inc.
10. T.W.B. Kibble, F.H. Berkshire, *Classical Mechanics*. Imperial College Press
11. M.G. Calkin, *Lagrangian and Hamiltonian Mechanics*
12. A.J. French, M.G. Ebison, *Introduction to Classical Mechanics*

Chapter 13
Non-relativistic Lagrangian and Hamiltonian Mechanics

Abstract This chapter is intended to introduce the basic notions of the Lagrangian and Hamiltonian formalisms as well as Noether's theorem. Using this theorem we reveal the relation between symmetries and conserved quantities. The formalism will be built from scratch and the stationary action principle (widely known as the minimum action principle) will be also introduced and analysed for non-relativistic systems. Some illustrative solved exercises will be given at the end of the chapter.

13.1 Lagrangian Formalism

Consider Newton's law of motion for a point-like particle with constant mass.[1]

$$\mathbf{F} = m\,\mathbf{a}\,. \tag{13.1}$$

Let us now consider that $\mathbf{F}$ is conservative, thus $\mathbf{F} = -\nabla V$, and so the equations of motion can be written as

$$-\frac{\partial V}{\partial x_i} = m\,\ddot{x}_i\,, \tag{13.2}$$

where we have introduced the usual "dot" notation for the time derivative and where $i = 1, 2, 3$ with $x_1 = x$, $x_2 = y$ and $x_3 = z$. Let us now consider the following mathematical construction

$$\boxed{L = K - V}\,, \tag{13.3}$$

where L is called the Lagrangian of the system, K is the kinetic energy and V the potential energy. By calculating the following quantity

$$\frac{d}{dt}\left(\frac{\partial L}{\partial \dot{x}_i}\right) - \frac{\partial L}{\partial x_i}\,, \tag{13.4}$$

---

[1] In this chapter we will not treat Lagrangians corresponding to variable mass systems.

© Springer Nature Switzerland AG 2020
V. Ilisie, *Lectures in Classical Mechanics*, Undergraduate Lecture
Notes in Physics, https://doi.org/10.1007/978-3-030-38585-9_13

one obtains that it is equal to

$$\frac{\partial V}{\partial x_i} + m \ddot{x}_i \,. \tag{13.5}$$

We conclude that Newton's equations of motion for conservative forces are equivalent to

$$\boxed{\frac{d}{dt}\left(\frac{\partial L}{\partial \dot{x}_i}\right) - \frac{\partial L}{\partial x_i} = 0} \,. \tag{13.6}$$

These are called the Euler-Lagrange equations of motion. The previous expression can be generalized for the case when non-conservative forces are also present. We have

$$-\frac{\partial V}{\partial x_i} + F_i^{NC} = m \ddot{x}_i \,, \tag{13.7}$$

which is equivalent to

$$\boxed{\frac{d}{dt}\left(\frac{\partial L}{\partial \dot{x}_i}\right) - \frac{\partial L}{\partial x_i} = F_i^{NC}} \,. \tag{13.8}$$

It should be obvious that the previous equations are valid for an arbitrary system with a finite (N) number of particles i.e.,

$$\frac{d}{dt}\left(\frac{\partial L}{\partial \dot{x}_i^j}\right) - \frac{\partial L}{\partial x_i^j} = F_i^{j,NC} \,, \tag{13.9}$$

where j stands for the label of the particle $j = 1, \ldots, N$. In order to keep the notation more compact we can redefine the indices i and j in one unique label i, with $i = 1, \ldots, 3N$. Therefore, using this compact notation we recover the original expression

$$\frac{d}{dt}\left(\frac{\partial L}{\partial \dot{x}_i}\right) - \frac{\partial L}{\partial x_i} = F_i^{NC} \,. \tag{13.10}$$

In the following we shall demonstrate that the previous equations (with a slight modification of F_i^{NC}) is valid for any general coordinate system (not only in Cartesian coordinates).

Consider a set of general coordinates (that can be for example curvilinear[2]) $q_i(x_i)$ for an N-particle system with $i = 1, \ldots, 3N$ as previously, and where q_i are functions

---

[2]In any case we shall always refer to a regular invertible change of basis, whose determinant of the Jacobian matrix is different from zero.

of x_i. As this transformation is invertible we can also write the x_i coordinates as functions of q_i and so

$$\dot{x}_k = \sum_i \frac{\partial x_k}{\partial q_i} \dot{q}_i + \frac{\partial x_k}{\partial t}. \tag{13.11}$$

Normally (in all the cases we have treated in this book) the last term is zero, and we shall only keep it for completeness.

Let us now calculate $\partial \dot{x}_k / \partial q_j$. We obtain

$$\frac{\partial \dot{x}_k}{\partial q_j} = \sum_i \left(\frac{\partial^2 x_k}{\partial q_i \partial q_j} \dot{q}_i + \frac{\partial^2 x_k}{\partial t \partial q_j} + \frac{\partial x_k}{\partial q_i} \frac{\partial \dot{q}_i}{\partial q_j} \right). \tag{13.12}$$

As $\partial \dot{q}_i / \partial q_j = 0$ in general, the last term is zero. Therefore one easily obtains the following relation

$$\frac{\partial \dot{x}_k}{\partial q_j} = \frac{d}{dt} \left(\frac{\partial x_k}{\partial q_j} \right). \tag{13.13}$$

One more relation is still needed. Let us calculate $\partial \dot{x}_k / \partial \dot{q}_j$. We get

$$\frac{\partial \dot{x}_k}{\partial \dot{q}_j} = \sum_i \left(\frac{\partial^2 x_k}{\partial q_i \partial \dot{q}_j} + \frac{\partial^2 x_k}{\partial t \partial \dot{q}_j} + \frac{\partial x_k}{\partial q_i} \frac{\partial \dot{q}_i}{\partial \dot{q}_j} \right). \tag{13.14}$$

Again, as the coordinates do not depend on the velocities $\partial x_k / \partial \dot{q}_j = 0$, and as $\partial \dot{q}_i / \partial \dot{q}_j = \delta_{ij}$, we obtain the following simple relation

$$\frac{\partial \dot{x}_k}{\partial \dot{q}_j} = \frac{\partial x_k}{\partial q_j}. \tag{13.15}$$

Let us now go back to the expression for the kinetic energy. Using our compact notation with $k = 1, \ldots, 3N$ we have

$$K(\dot{x}_k) = \frac{1}{2} \sum_k m_k \dot{x}_k^2. \tag{13.16}$$

By inserting (13.11) into the previous equation we obtain

$$K(q_i, \dot{q}_i) = \frac{1}{2} \sum_k m_k \left(\sum_i \frac{\partial x_k}{\partial q_i} \dot{q}_i + \frac{\partial x_k}{\partial t} \right)^2. \tag{13.17}$$

One should not be confused by the previous dependency of K on both q_i and $\dot{q}_i$. In Cartesian coordinates the kinetic energy only depends on the velocities $\dot{x}_k$, however, one can easily check that in curvilinear coordinates, the velocities depend both on q_i

and $\dot{q}_i$ (where $q_i = r, \theta, \phi$ in spherical coordinates, for example). Having made the previous clarification, the associated momentum p_k in Cartesian coordinates can be expressed as

$$p_j = \frac{\partial K(\dot{x}_k)}{\partial \dot{x}_j}. \tag{13.18}$$

Similarly, the *canonical conjugated momentum* (or the *generalized momentum*) associated to the Lagrangian L is defined as

$$\boxed{\pi_j \equiv \frac{\partial L}{\partial \dot{q}_j} = \frac{\partial K}{\partial \dot{q}_j}}, \tag{13.19}$$

where the last equality is valid because we will only consider velocity-independent potentials i.e., $V = V(x_i) = V(q_i)$. Let us now take the time derivative of the previously introduced quantity. On one hand we have

$$\dot{\pi}_j = \frac{d}{dt}\left(\frac{\partial L}{\partial \dot{q}_j}\right). \tag{13.20}$$

On the other hand

$$\begin{aligned}
\dot{\pi}_j &= \frac{d}{dt}\left(\frac{\partial K}{\partial \dot{q}_j}\right) \\
&= \frac{d}{dt}\left(\sum_k \frac{\partial K}{\partial \dot{x}_k}\frac{\partial \dot{x}_k}{\partial \dot{q}_j}\right) \\
&= \frac{d}{dt}\left(\sum_k p_k \frac{\partial \dot{x}_k}{\partial \dot{q}_j}\right) \\
&= \sum_k \dot{p}_k \frac{\partial \dot{x}_k}{\partial \dot{q}_j} + \sum_k p_k \frac{d}{dt}\left(\frac{\partial \dot{x}_k}{\partial \dot{q}_j}\right) \\
&= \sum_k \dot{p}_k \frac{\partial x_k}{\partial q_j} + \sum_k p_k \frac{\partial \dot{x}_k}{\partial q_j},
\end{aligned} \tag{13.21}$$

where, in order to get to the last line we have used (13.13) and (13.15). As $\dot{p}_k$ includes the sum of both conservative and non-conservative contributions, we can write

$$\dot{p}_k = -\frac{\partial V}{\partial x_k} + F_k^{NC}. \tag{13.22}$$

Defining the generic non-conservative force (in arbitrary coordinates) as

$$Q_j^{NC} \equiv \sum_k F_k^{NC} \frac{\partial x_k}{\partial q_j}, \tag{13.23}$$

and using

$$-\sum_k \frac{\partial V}{\partial x_k} \frac{\partial x_k}{\partial q_j} = -\frac{\partial V}{\partial q_j}, \tag{13.24}$$

and

$$\sum_k p_k \frac{\partial \dot{x}_k}{\partial q_j} = \sum_k \frac{\partial K}{\partial \dot{x}_k} \frac{\partial \dot{x}_k}{\partial q_j} = \frac{\partial K}{\partial q_j}, \tag{13.25}$$

we finally obtain

$$\dot{\pi}_j = \frac{\partial K}{\partial q_j} - \frac{\partial V}{\partial q_j} + Q_j^{NC} = \frac{\partial L}{\partial q_j} + Q_j^{NC}. \tag{13.26}$$

By introducing (13.20) we obtain

$$\boxed{\frac{d}{dt}\left(\frac{\partial L}{\partial \dot{q}_j}\right) - \frac{\partial L}{\partial q_j} = Q_j^{NC}}, \tag{13.27}$$

which proves our previous statement, that the Euler-Lagrange equations of motion are valid in any arbitrary system of coordinates.

13.1.1 Cyclic Coordinates and Constants of Motion

From the previous results we can immediately deduce the following important theorem:
If the quantity

$$\dot{\pi}_j = \frac{\partial L}{\partial q_j} + Q_j^{NC} = 0, \tag{13.28}$$

for some j, then the canonical momentum associated to the variable q_j is conserved

$$\pi_j = \frac{\partial L}{\partial \dot{q}_j} = constant. \tag{13.29}$$

In the literature, one normally finds a less general formulation of this theorem. If one restricts the analysis only to conservative forces i.e., $Q_j^{NC} = 0$, $\forall j$, then the theorem simply takes the form:

If the Lagrangian does not exhibit an explicit dependence on a variable q_j i.e., the quantity

$$\dot{\pi}_j = \frac{\partial L}{\partial q_j} = 0 \,, \tag{13.30}$$

then the canonical momentum associated to the variable q_j is conserved

$$\pi_j = \frac{\partial L}{\partial \dot{q}_j} = constant \,. \tag{13.31}$$

In both cases the coordinate q_j is called *cyclic*.

13.1.2   Lagrangian with Boundary Conditions

So far we have considered coordinates and velocities with no boundaries. Consider however the motion of a particle on a surface, a simple or a double pendulum. In such cases there are restrictions imposed on the coordinates. In more general cases there can be restrictions on both the coordinates and the velocities. Depending on the case, sometimes one can introduce the constraints directly in the equations of motion. Another method is by introducing Lagrange multipliers, as presented in the following.

Consider a given set of M boundary conditions of the type

$$f_\alpha(q_i, t) + c_\alpha = 0 \,, \tag{13.32}$$

where $\alpha = 1, \ldots, M$, c_α constants and, f_α differentiable functions. The equations of motion will be given by the system

$$\frac{d}{dt}\left(\frac{\partial L}{\partial \dot{q}_j}\right) - \frac{\partial L}{\partial q_j} = Q_j^{NC} + \sum_\alpha \lambda_\alpha \frac{\partial f_\alpha}{\partial q_j} \,,$$
$$f_\alpha(q_i, t) + c_\alpha = 0 \,, \tag{13.33}$$

with λ_α constants. These equations of motion can be directly obtained by defining and extended Lagrangian that accounts for the constraints i.e.,

$$\bar{L} = K - V + \sum_\alpha \lambda_\alpha(f_\alpha + c_\alpha) \,, \tag{13.34}$$

and applying the Euler-Lagrange equations (considering each λ_α as a generalized coordinate)

$$\frac{d}{dt}\left(\frac{\partial \bar{L}}{\partial \dot{q}_j}\right) - \frac{\partial \bar{L}}{\partial q_j} = Q_j^{NC} \,. \tag{13.35}$$

The Lagrange multiplier approach is also valid for conditions of the type $g_\alpha(q_i, \dot{q}_i, t)$ that can be integrated, to obtain functions in the form $f_\alpha(q_i, t)$. Some illustrative exercises regarding systems with boundary conditions will be solved at the end of the chapter.

In the following we are going to introduce Hamilton's equations of motion (which are tightly related to the Euler-Lagrange equations).

13.2 Hamiltonian Formalism

In this section we will introduce the Hamiltonian of a system of particles and deduce Hamilton's equations of motion. The Hamiltonian of a system of N particles ($i = 1, \ldots, 3N$) is defined as

$$\boxed{H = \sum_i \pi_i \dot{q}_i - L}. \tag{13.36}$$

In order to obtain the equations of motion we first have to calculate dH. On one hand we have

$$dH = \sum_i \left(\frac{\partial H}{\partial \pi_i} d\pi_i + \frac{\partial H}{\partial q_i} dq_i + \frac{\partial H}{\partial \dot{q}_i} d\dot{q}_i + \frac{\partial H}{\partial t} dt \right), \tag{13.37}$$

where we have supposed a generic case, where the Hamiltonian also exhibits an explicit time dependence. On the other hand we have

$$dH = \sum_i \left(\dot{q}_i d\pi_i + \pi d\dot{q}_i - \frac{\partial L}{\partial q_i} dq_i - \frac{\partial L}{\partial \dot{q}_i} d\dot{q}_i - \frac{\partial L}{\partial t} dt \right), \tag{13.38}$$

where we have also supposed a generic Lagrangian with an explicit time dependence. Introducing the defining expression of π_i in the previous equation, the second and the fourth term cancel each other and so, by comparing the last two expressions we obtain

$$\frac{\partial H}{\partial \pi_i} = \dot{q}_i, \quad \frac{\partial H}{\partial q_i} = -\frac{\partial L}{\partial q_i}, \quad \frac{\partial H}{\partial t} = -\frac{\partial L}{\partial t}, \tag{13.39}$$

and

$$\frac{\partial H}{\partial \dot{q}_i} = 0. \tag{13.40}$$

We conclude that the Hamiltonian exhibits no dependence on $\dot{q}_i$. Let us now introduce $\dot{\pi}_i$. In the previous section we have deduced

$$\dot{\pi}_i = \frac{d}{dt}\left(\frac{\partial L}{\partial \dot{q}_i}\right) = \frac{\partial L}{\partial q_i} + Q_i^{NC} = -\frac{\partial H}{\partial q_i} + Q_i^{NC} . \qquad (13.41)$$

Hence, the Hamiltonian equations of motion simply read

$$\boxed{\frac{\partial H}{\partial \pi_i} = \dot{q}_i, \qquad \dot{\pi}_i = -\frac{\partial H}{\partial q_i} + Q_i^{NC}, \qquad \frac{\partial H}{\partial t} = -\frac{\partial L}{\partial t}} . \qquad (13.42)$$

One normally finds the in the literature the particular case for the previous expressions with $Q_j^{NC} = 0$. Here we preferred presenting the general case.

There is however, a much easier way to directly deduce the previous expressions for the Euler-Lagrange and the Hamiltonian equations of motion. It consists in using the stationary action principle. It does, however, have a drawback, and that is, it cannot be applied to systems with non-conservative forces. In the following, in order to emphasize this feature, we shall refer to these systems as conservative systems.

13.3 Stationary Action Principle

We define the action associated to a discrete conservative system ($Q_j^{NC} = 0$) formed by N point-like particles, thus with $3N$ degrees of freedom ($i = 1, \ldots, 3N$), as

$$S(q_i) = \int_{t_1}^{t_2} dt \, L(q_i, \dot{q}_i, t) , \qquad (13.43)$$

where t_1 and t_2 are two consecutive time instants of the system. In order to obtain the Euler-Lagrange equations of motion we must consider small variations of the generalized coordinates q_i keeping the extremes fixed, as schematically shown in Fig. 13.1, i.e., by considering the transformations

$$q_i \rightarrow q_i' = q_i + \delta q_i , \qquad \delta q_i(t_1) = \delta q_i(t_2) = 0 . \qquad (13.44)$$

The first order Taylor expansion of L then gives

$$L(q_i + \delta q_i, \dot{q}_i + \delta \dot{q}_i, t) = L(q_i, \dot{q}_i, t) + \sum_i \frac{\partial L}{\partial q_i}\delta q_i + \sum_i \frac{\partial L}{\partial \dot{q}_i}\delta \dot{q}_i$$

$$\equiv L(q_i, \dot{q}_i, t) + \delta L . \qquad (13.45)$$

It is straightforward to demonstrate that the *variation* and the *differentiation* operators commute

$$\delta q_i(t) = q_i'(t) - q_i(t) \Rightarrow \frac{d}{dt}(\delta q_i) = \dot{q}_i'(t) - \dot{q}_i(t) = \delta \dot{q}_i(t) . \qquad (13.46)$$

Fig. 13.1 Schematic representation of the variation of the physical path $q_i(t)$, keeping the extremes fixed

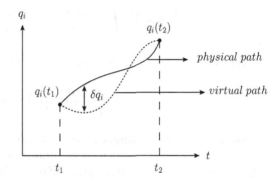

We thus obtain the following expression for δL

$$\delta L = \sum_i \frac{\partial L}{\partial q_i} \delta q_i + \sum_i \frac{\partial L}{\partial \dot{q}_i} \delta \dot{q}_i = \sum_i \frac{\partial L}{\partial q_i} \delta q_i + \sum_i \frac{\partial L}{\partial \dot{q}_i} \frac{d}{dt}(\delta q_i) , \quad (13.47)$$

which can be written as

$$\delta L = \sum_i \left(\frac{\partial L}{\partial q_i} - \frac{d}{dt}\left(\frac{\partial L}{\partial \dot{q}_i}\right)\right)\delta q_i + \sum_i \frac{d}{dt}\left(\frac{\partial L}{\partial \dot{q}_i}\delta q_i\right) . \quad (13.48)$$

In order to obtain the equations of motion we will now apply the **Stationary Action Principle**: *For the physical path, the action must be a maximum, a minimum or an inflexion point.* This translates mathematically into

$$\boxed{\delta S = \delta \int_{t_1}^{t_2} dt\, L = \int_{t_1}^{t_2} dt\, \delta L = 0} . \quad (13.49)$$

Explicitly introducing δL we get

$$\delta S = \sum_i \int_{t_1}^{t_2} dt\left(\frac{\partial L}{\partial q_i} - \frac{d}{dt}\left(\frac{\partial L}{\partial \dot{q}_i}\right)\right)\delta q_i + \sum_i \int_{t_1}^{t_2} dt \frac{d}{dt}\left(\frac{\partial L}{\partial \dot{q}_i}\delta q_i\right) = 0 . \quad (13.50)$$

Because $\delta q_i(t_1) = \delta q_i(t_2) = 0$ the second integral vanishes $\forall\, q_i$,

$$\int_{t_1}^{t_2} dt \frac{d}{dt}\left(\frac{\partial L}{\partial \dot{q}_i}\delta q_i\right) = \int_{t_1}^{t_2} d\left(\frac{\partial L}{\partial \dot{q}_i}\delta q_i\right) = \left[\frac{\partial L}{\partial \dot{q}_i}\delta q_i\right]_{t_1}^{t_2} = 0 . \quad (13.51)$$

Therefore, we are left with

$$\delta S = \sum_i \int_{t_1}^{t_2} dt \left(\frac{\partial L}{\partial q_i} - \frac{d}{dt}\left(\frac{\partial L}{\partial \dot{q}_i} \right) \right) \delta q_i = 0 , \tag{13.52}$$

for arbitrary δq_i. Thus the following equations must hold

$$\frac{\partial L}{\partial q_i} - \frac{d}{dt}\left(\frac{\partial L}{\partial \dot{q}_i} \right) = 0 , \tag{13.53}$$

$\forall q_i$. We have therefore obtained once again, the Euler-Lagrange equations of motion. As we have already mentioned, there is a drawback associated to this principle, that is, it only works for conservative systems. If one includes non-zero Q_j^{NC} the previous principle no longer works and one would have to look for some generalized principle. We shall however, not be concerned with this problem here, as it is way off-topic.

From (13.51) we can also deduce an important aspect of Lagrangians, that they are not uniquely defined

$$\boxed{L(q_i, \dot{q}_i, t) \text{ and } \widetilde{L}(q_i, \dot{q}_i, t) = L(q_i, \dot{q}_i, t) + \frac{dF(q_i, t)}{dt}} \tag{13.54}$$

generate the same equations of motion. We have an alternative way to directly check that adding a function of the form $dF(q_i, t)/dt$ to the Lagrangian, doesn't alter the equations of motion. If we apply the Euler-Lagrange equations of motion to $dF(q_i, t)/dt$ we obtain:

$$\frac{\partial}{\partial q_i}\left(\frac{dF(q_i, t)}{dt} \right) - \frac{d}{dt}\left(\frac{\partial}{\partial \dot{q}_i}\left(\frac{dF(q_i, t)}{dt} \right) \right) \equiv D . \tag{13.55}$$

By explicitly expanding the expressions of the derivatives, one obtains $D = 0$.

The Hamiltonian equations of motion can be obtained just as previously by applying the principle of the stationary action (this time considering both variations of q_i and π_i)

$$\delta S = \int_{t_1}^{t_2} dt \delta L$$

$$= \sum_i \int_{t_1}^{t_2} dt \delta(\pi_i \dot{q}_i - H)$$

$$= \sum_i \int_{t_1}^{t_2} dt \left(\delta \pi_i \dot{q}_i + \pi_i \delta \dot{q}_i - \frac{\partial H}{\partial q_i}\delta q_i - \frac{\partial H}{\partial \pi_i}\delta \pi_i \right)$$

$$= \sum_i \int_{t_1}^{t_2} dt \left(\delta \pi_i \dot{q}_i + \frac{d}{dt}(\pi_i \delta q_i) - \dot{\pi}_i \delta q_i - \frac{\partial H}{\partial q_i}\delta q_i - \frac{\partial H}{\partial \pi_i}\delta \pi_i \right). \tag{13.56}$$

Grouping terms we finally obtain

$$\delta S = \sum_i \int_{t_1}^{t_2} dt \left(\delta \pi_i \left[\dot{q}_i - \frac{\partial H}{\partial \pi_i} \right] + \delta q_i \left[-\dot{\pi}_i - \frac{\partial H}{\partial q_i} \right] \right) + \sum_i \int_{t_1}^{t_2} d(\pi_i \, \delta q_i)$$

$$= \sum_i \int_{t_1}^{t_2} dt \left(\delta \pi_i \left[\dot{q}_i - \frac{\partial H}{\partial \pi_i} \right] + \delta q_i \left[-\dot{\pi}_i - \frac{\partial H}{\partial q_i} \right] \right) = 0 . \tag{13.57}$$

This must hold for arbitrary $\delta \pi_i$ y δq_i, therefore, we again obtain the Hamiltonian equations of motion

$$\dot{q}_i = \frac{\partial H}{\partial \pi_i} , \qquad \dot{\pi}_i = -\frac{\partial H}{\partial q_i} . \tag{13.58}$$

Just as in the previous section, if the Hamiltonian exhibits an explicit time dependence, it can be easily related to the time dependence of the Lagrangian i.e.,

$$\frac{\partial H}{\partial t} = -\frac{\partial L}{\partial t} . \tag{13.59}$$

Next we will present one of the most important theorems of analytical mechanics, a powerful tool that allows us to relate the symmetries of a system with conserved quantities. We shall restrict our analysis only to conservative systems. In Chaps. 4 and 5 we talked about the energy, momentum, angular momentum etc., conservations and we mentioned that they are direct consequences of the symmetries of our Universe, however, without getting into details. By making use of Noether's theorem, the relations between symmetry and conservation will become transparent.

13.4 Noether's Theorem

There is a conserved quantity associated to every symmetry of the Lagrangian.
Let's consider a transformation of the type

$$q_i \to q_i' = q_i + \delta q_i , \tag{13.60}$$

so that the variation of the Lagrangian can be written as the exact differential of some function F:

$$L(q_i', \dot{q}_i', t) = L(q_i, \dot{q}_i, t) + \frac{dF(q_i, \dot{q}_i, t)}{dt} \Rightarrow \delta L = \frac{dF(q_i, \dot{q}_i, t)}{dt} . \tag{13.61}$$

Note that here we allow F to also depend on $\dot{q}_i$ (that was not the case for (13.54)). On the other hand, we know that we can write δL as:

$$\delta L = \sum_i \left(\frac{\partial L}{\partial q_i} - \sum_i \frac{d}{dt}\left(\frac{\partial L}{\partial \dot{q}_i} \right) \right) \delta q_i + \sum_i \frac{d}{dt}\left(\frac{\partial L}{\partial \dot{q}_i} \delta q_i \right)$$

$$= \sum_i \frac{d}{dt}\left(\frac{\partial L}{\partial \dot{q}_i} \delta q_i \right) . \tag{13.62}$$

To get to the last equality we used the equations of motion. Let's now write δq_i as an infinitesimal variation of the form

$$q_i' = q_i + \delta q_i = q_i + \epsilon f_i , \tag{13.63}$$

with $|\epsilon| \ll 1$ a constant, and f a smooth, well behaved function. In the limit $\epsilon \to 0$ we obtain

$$\lim_{\epsilon \to 0} q_i' = q_i \Rightarrow \lim_{\epsilon \to 0} \delta L = 0 . \tag{13.64}$$

Thus, necessarily F must be of the form $F = \epsilon \widetilde{F} + \mathcal{O}(\epsilon^2)$, and so

$$\sum_i \frac{d}{dt}\left(\frac{\partial L}{\partial \dot{q}_i}(\epsilon f_i) \right) = \epsilon \frac{d\widetilde{F}(q_i, \dot{q}_i, t)}{dt}. \tag{13.65}$$

Integrating in t we obtain

$$\sum_i \frac{\partial L}{\partial \dot{q}_i} f_i = \widetilde{F}(q_i, \dot{q}_i, t) + C , \tag{13.66}$$

with C an integration constant. We therefore conclude, that the conserved quantity associated to our infinitesimal symmetry is:

$$\boxed{ C = \sum_i \frac{\partial L}{\partial \dot{q}_i} f_i - \widetilde{F}(q_i, \dot{q}_i, t) } . \tag{13.67}$$

13.5 Symmetries and Conservation

Next, we are going to apply this simple formula to a few interesting cases and reproduce some typical results such as energy and momentum conservation, angular momentum conservation, etc., and show that they are nothing but consequences of the symmetries of our Universe.

13.5.1 Time Translations

Let's consider an infinitesimal time shift: $t \to t + \epsilon$. By using a first order Taylor expansion we obtain

$$\delta q_i = q_i(t + \epsilon) - q_i(t) = \epsilon \dot{q}_i(t) + O(\epsilon^2) ,$$

$$\delta \dot{q}_i = \dot{q}_i(t + \epsilon) - \dot{q}_i(t) = \epsilon \ddot{q}_i(t) + O(\epsilon^2) = \frac{d}{dt}(\delta q_i) . \qquad (13.68)$$

If the Lagrangian does not exhibit an explicit time dependence ($\partial L/\partial t = 0$) then

$$\delta L = \epsilon \sum_i \frac{\partial L}{\partial q_i} \dot{q}_i + \epsilon \sum_i \frac{\partial L}{\partial \dot{q}_i} \ddot{q}_i = \epsilon \frac{dL}{dt} \Rightarrow \tilde{F} = L . \qquad (13.69)$$

Thus, the conserved quantity is given by the following expression

$$\sum_i \frac{\partial L}{\partial \dot{q}_i} \dot{q}_i - L = E , \qquad (13.70)$$

where E is the associated energy of the system. We conclude that the energy conservation is nothing but a consequence of the time-shift invariance of a system.

In the **three remaining subsections** we shall only consider **Cartesian coordinates**. This is done to ensure that the canonical momentum π_j corresponds to the physical momentum i.e., $\pi_j = p_j$. This way we can relate momentum and angular momentum conservation with the corresponding symmetries (one can check that in curvilinear coordinates one cannot ensure that π_j corresponds to the physical momentum).

13.5.2 Spatial Translations

Let's consider a Lagrangian of the form $L = K - V$, where K is the kinetic energy of the system and V a central potential energy. In this case the momentum p_i defined as

$$p_i \equiv \frac{\partial L}{\partial \dot{x}_i} , \qquad (13.71)$$

obeys $p_i = \partial K/\partial \dot{x}_i$. Due to the fact that the potential is central and $K \neq K(x_i)$ the Lagrangian obeys

$$L(\mathbf{r}_\alpha + \epsilon \mathbf{n}, \mathbf{v}_\alpha) = L(\mathbf{r}_\alpha, \mathbf{v}_\alpha) , \qquad (13.72)$$

with $\mathbf{r}_\alpha$ the coordinates of the particle α and $\mathbf{n}$ a vector corresponding to an arbitrary spatial direction and sense (with $|\mathbf{n}| = 1$). We conclude that $\delta L = 0$. Under this spatial translation the coordinates of the particle α transform the following way:

$$\mathbf{r}_\alpha \to \mathbf{r}'_\alpha = \mathbf{r}_\alpha + \epsilon \mathbf{n} \quad \Rightarrow \quad r_{\alpha j} \to r'_{\alpha j} = r_{\alpha j} + \epsilon n_j , \tag{13.73}$$

with $j = 1, 2, 3$. Therefore $f_j = n_j$. The conserved quantity is straightforwardly obtained

$$C = \sum_{\alpha,j} \frac{\partial L}{\partial \dot{q}_{\alpha j}} n_j = \sum_{\alpha,j} p_{\alpha j} n_j = \sum_\alpha \mathbf{p}_\alpha \cdot \mathbf{n} = \mathbf{P} \cdot \mathbf{n} , \tag{13.74}$$

for an arbitrary $\mathbf{n}$. Thus, the constant associated to this transformations is the total momentum $\mathbf{P}$ of the system.

13.5.3 Rotations

Again, let's consider a Lagrangian with the same properties as in the previous example. Under an infinitesimal rotation we have

$$\mathbf{r}'_\alpha = \mathbf{r}_\alpha - \epsilon \mathbf{n} \times \mathbf{r}_\alpha, \qquad r'_{\alpha j} = r_{\alpha j} + \epsilon \epsilon_{jkm} n_m r_{\alpha k} , \tag{13.75}$$

and just as previously $\delta L = 0$. It is straightforward to observe that $f_j = \epsilon_{jkm} n_m r_{\alpha k}$ (where ϵ_{jkm} is the totally antisymmetric three-dimensional Levi-Civita tensor). The conserved quantity is therefore:

$$\begin{aligned} C &= \sum_{\alpha,j,k,m} \frac{\partial L}{\partial \dot{q}_{\alpha j}} \epsilon_{jkm} n_m r_{\alpha k} \\ &= \sum_{\alpha,j,k,m} p_{\alpha j} \epsilon_{jkm} n_m r_{\alpha k} \\ &= \sum_\alpha (\mathbf{p}_\alpha \times \mathbf{r}_\alpha) \cdot \mathbf{n} = -\mathbf{L} \cdot \mathbf{n} . \end{aligned} \tag{13.76}$$

Again, this holds for an arbitrary $\mathbf{n}$, thus, the conserved quantity is the total angular momentum $\mathbf{L}$ of the system.

13.5.4 Galilean Transformations

For this last example we shall again consider the same type of Lagrangian as in the previous cases. A Galileo transformation reads

$$\mathbf{r}_\alpha \to \mathbf{r}'_\alpha = \mathbf{r}_\alpha + \mathbf{v}t , \tag{13.77}$$

with $\mathbf{v}$ a constant velocity vector, therefore:

$$\dot{\mathbf{r}}_\alpha \to \dot{\mathbf{r}}'_\alpha = \dot{\mathbf{r}}_\alpha + \mathbf{v} . \tag{13.78}$$

Under these transformations $\delta L = \delta K$. Let's calculate K' explicitly:

$$\begin{aligned}
K' &= \frac{1}{2} \sum_\alpha m_\alpha (\dot{\mathbf{r}}_\alpha + \mathbf{v})^2 \\
&= K + \sum_\alpha m_\alpha \dot{\mathbf{r}}_\alpha \mathbf{v} + \sum_\alpha \frac{1}{2} m_\alpha \mathbf{v}^2 \\
&= K + \frac{1}{2} M \mathbf{v}^2 + \frac{d}{dt} \sum_\alpha (m_\alpha \mathbf{r}_\alpha \mathbf{v}) \\
&= K + \frac{1}{2} M \mathbf{v}^2 + \frac{d}{dt} (M\mathbf{R} \cdot \mathbf{v}) .
\end{aligned} \tag{13.79}$$

Considering an infinitesimal transformation $\mathbf{v} = \epsilon \mathbf{n}$ with $|\epsilon| \ll 1$ and ignoring terms of $O(\epsilon^2)$ we have

$$\delta L = \delta K = \epsilon \frac{d}{dt} (M\mathbf{R} \cdot \mathbf{n}) . \tag{13.80}$$

The conserved quantity is then given by:

$$C = \sum_{\alpha, j} p_{\alpha j} n_j \, t - M\mathbf{R} \cdot \mathbf{n} = (\mathbf{P}t - M\mathbf{R}) \cdot \mathbf{n} , \tag{13.81}$$

for an arbitrary $\mathbf{n}$. The conserved quantity associated to this transformation is then $\mathbf{P}t - M\mathbf{R}$.

Having introduced all the previous theoretical background, we shall now move on to some practical exercises by working out some simple examples.

13.6 Proposed Exercises

13.1 Write down the Lagrangian of a particle of mass m moving under the influence of a potential with spherical symmetry, in Cartesian and in spherical coordinates. Find in each case the cyclic coordinates and the corresponding constants of motion. Comment upon the results.

Solution: The corresponding Lagrangian in Cartesian coordinates simply reads

$$L = \frac{1}{2}m(\dot{x}^2 + \dot{y}^2 + \dot{z}^2) - V(x^2 + y^2 + z^2). \tag{13.82}$$

We immediately realize that there is no cyclic coordinate as they all appear as parameters of the potential V. Let us therefore continue and switch to spherical coordinates. The Lagrangian in this case is given by

$$L = \frac{1}{2}m\left(\dot{r}^2 + r^2\dot{\theta}^2 + r^2\sin^2\theta\,\dot{\phi}^2\right) - V(r). \tag{13.83}$$

We observe that there is one cyclic coordinate, namely ϕ, that does not appear explicitly in the Lagrangian. Hence, the canonical momentum π_ϕ is constant

$$\pi_\phi = \frac{\partial L}{\partial\dot{\phi}} = mr^2\sin^2\theta\,\dot{\phi}. \tag{13.84}$$

Notice that the Lagrangian can be re-written in terms of the previously calculated constant i.e.,

$$L = \frac{1}{2}m\left(\dot{r}^2 + r^2\dot{\theta}^2 + \frac{\pi_\phi}{m}\dot{\phi}\right) - V(r), \tag{13.85}$$

and so, we conclude that there is a second cyclic coordinate θ. Thus, the second constant of motion is

$$\pi_\theta = \frac{\partial L}{\partial\dot{\theta}} = mr^2\dot{\theta}. \tag{13.86}$$

Let us now calculate the angular momentum in spherical coordinates. We obtain

$$\begin{aligned}
\mathbf{L} = \mathbf{r} \times \mathbf{p} &= m\left(r\,\mathbf{u}_r \times (\dot{r}\,\mathbf{u}_r + r\dot{\theta}\,\mathbf{u}_\theta + r\sin\theta\dot{\phi}\,\mathbf{u}_\phi)\right) \\
&= mr^2\dot{\theta}(\mathbf{u}_r \times \mathbf{u}_\theta) + mr^2\sin\theta\,\dot{\phi}(\mathbf{u}_r \times \mathbf{u}_\phi) \\
&= mr^2\dot{\theta}\,\mathbf{u}_\phi - mr^2\sin\theta\,\dot{\phi}\,\mathbf{u}_\theta \\
&= \pi_\theta\,\mathbf{u}_\phi - \frac{\pi_\phi}{\sin\theta}\,\mathbf{u}_\theta.
\end{aligned} \tag{13.87}$$

If we chose a reference system where $\pi_\phi = 0$ (this is always possible as the system has spherical symmetry), then $\dot{\phi} = 0$ and so ϕ is constant. In this case the angular momentum reads

$$\mathbf{L} = \pi_\theta\,\mathbf{u}_\phi, \tag{13.88}$$

with $\mathbf{u}_\phi$ given by

$$\mathbf{u}_\phi = -\sin\phi\,\mathbf{i} + \cos\phi\,\mathbf{j}. \tag{13.89}$$

We therefore conclude that the angular momentum is conserved.

13.2 Obtain the Euler-Lagrange equations of motion of the previous system in spherical coordinates. Can this initial three-dimensional problem be reduced to a one-dimensional problem using the conserved quantities associated to the system i.e., π_θ, π_ϕ and the total energy E? Simplify the final result by choosing the reference frame where $\pi_\phi = 0$.

Solution: By applying the Euler-Lagrange equations of motion for θ and ϕ one simply obtains the constants of motion

$$\pi_\phi = mr^2\sin^2\theta\,\dot\phi,$$
$$\pi_\theta = mr^2\dot\theta. \tag{13.90}$$

For r we have

$$\frac{\partial L}{\partial r} = mr\dot\theta^2 + mr\sin^2\theta\dot\phi - \frac{\partial V}{\partial r}, \tag{13.91}$$

with $V = V(r)$, and

$$\frac{\partial L}{\partial \dot r} = m\dot r \quad\Rightarrow\quad \frac{d}{dt}\left(\frac{\partial L}{\partial \dot r}\right) = m\ddot r. \tag{13.92}$$

The final equation reads

$$m\ddot r + \frac{\partial V}{\partial r} - \frac{\pi_\phi^2}{mr^3\sin^2\theta} - \frac{\pi_\theta^2}{mr^3} = 0. \tag{13.93}$$

Note that we are not able to reduce the previous differential equation to a equation with one variable only, as we still have one term that contains $\cos\theta$. Introducing the expression for total energy E, which is also a constant of motion

$$E = K + V = \frac{1}{2}m\left(\dot r^2 + r^2\dot\theta^2 + \frac{\pi_\phi}{m}\dot\phi\right) + V(r)$$
$$= \frac{1}{2}m\left(\dot r^2 + \frac{\pi_\theta^2}{m^2r^2} + \frac{\pi_\phi^2}{m^2r^2\sin^2\theta}\right) + V(r), \tag{13.94}$$

one can express $\cos\theta$ as a function of r and $\dot{r}$ and reduce (13.93) to a (very complicated) one-variable differential equation. By choosing the initial conditions as $\pi_\phi = 0$ the problem simplifies to

$$m\ddot{r} + \frac{\partial V}{\partial r} - \frac{\pi_\theta^2}{mr^3} = 0, \tag{13.95}$$

which should already result familiar to the reader i.e., particularizing for $V = -k/r$ we obtain Newton's equations corresponding to the motion of a particle under the influence of Kepler's potential.

Next, we are going to re-analyse the previous results in terms of the Hamiltonian equations of motion.

13.3 Write down the Hamiltonian corresponding to the previous system, obtain Hamilton's equations of motion and particularize the results for $\pi_\phi = 0$ and $V = -k/r$.

Solution: The Hamiltonian is this case is given by

$$H = \dot{r}\pi_r + \dot{\theta}\pi_\theta + \dot{\phi}\pi_\phi - L. \tag{13.96}$$

The canonical momenta π_ϕ and π_θ were calculated previously. The expression for π_r is simply

$$\pi_r = \frac{\partial L}{\partial \dot{r}} = m\dot{r}. \tag{13.97}$$

We thus obtain

$$\dot{r} = \frac{\pi_r}{m}, \qquad \dot{\theta} = \frac{\pi_\theta}{mr^2}, \qquad \dot{\phi} = \frac{\pi_\phi}{mr^2 \sin^2\theta}, \tag{13.98}$$

and so the Hamiltonian takes the form

$$
\begin{aligned}
H &= \dot{r}\pi_r + \dot{\theta}\pi_\theta + \dot{\phi}\pi_\phi - \frac{1}{2}(\dot{r}\pi_r + \dot{\theta}\pi_\theta + \dot{\phi}\pi_\phi) + V(r) \\
&= \frac{1}{2}(\dot{r}\pi_r + \dot{\theta}\pi_\theta + \dot{\phi}\pi_\phi) + V(r).
\end{aligned} \tag{13.99}
$$

Remember however that H is a function of π_i and q_i and not $\dot{q}_i$. We can thus re-write the previous result as

$$H = \frac{1}{2m}\left(\pi_r^2 + \frac{\pi_\theta^2}{r^2} + \frac{\pi_\phi^2}{r^2\sin^2\theta}\right) + V(r) = K(q_i, \pi_i) + V(r), \tag{13.100}$$

where $K(q_i, \pi_i)$ is the kinetic energy written in terms of the generalized coordinates and the canonical momenta. The equations of motion $\dot{q}_i = \partial H/\partial \pi_i$, in this case read

$$\dot{r} = \frac{\pi_r}{m}, \quad \dot{\theta} = \frac{\pi_\theta}{mr^2}, \quad \dot{\phi} = \frac{\pi_\phi}{mr^2 \sin^2 \theta}, \tag{13.101}$$

which are the already known results and, $\dot{\pi}_i = -\partial H/\partial q_i$, which correspond

$$\dot{\pi}_r = -\frac{\partial H}{\partial r} = \frac{\pi_\theta^2}{mr^3} + \frac{\pi_\phi^2}{mr^3 \sin^2 \theta} - \frac{\partial V}{\partial r},$$

$$\dot{\pi}_\theta = -\frac{\partial H}{\partial \theta} = \frac{\pi_\phi^2}{mr^2 \sin^3 \theta}, \tag{13.102}$$

$$\dot{\pi}_\phi = -\frac{\partial H}{\partial \phi} = 0.$$

If we choose the reference frame where $\pi_\phi = 0$ and $V = -k/r$ (Kepler's potential energy) we are left with

$$\dot{r} = \frac{\pi_r}{m}, \quad \dot{\pi}_r = \frac{\pi_\theta^2}{mr^3} - \frac{k}{r^2}. \tag{13.103}$$

Combining the two equations we finally obtain

$$m\ddot{r} - \frac{\pi_\theta^2}{mr^3} + \frac{k}{r^2} = 0. \tag{13.104}$$

13.4 Write down the Lagrangian corresponding to the *heavy symmetric spinning top* and obtain the associated constants of motion.

Solution: In general, analytical mechanics is a highly powerful tool that allows us to straightforwardly obtain symmetries and conserved quantities for a given system, that otherwise, would require a huge amount of effort and this is one perfect example.

The reader should remember that we have previously obtained both the expression for the kinetic energy (10.77) and the potential energy (10.76). The Lagrangian is thus simply given by

$$L = \frac{1}{2}I_1'(\dot{\phi}^2 \sin^2 \theta + \dot{\theta}^2) + \frac{1}{2}I_3'(\dot{\psi} + \dot{\phi} \cos \theta)^2 + Mgl \cos \theta. \tag{13.105}$$

In this case, our generalized coordinates are Euler's angles θ, ϕ and ψ. We can observe that neither ϕ nor ψ appear in the Lagrangian, and so, they are cyclic coordinates. The canonical momenta π_ϕ and π_ψ will thus be two constants of motion. Their expressions read

$$\pi_\phi = \frac{\partial L}{\partial \dot\phi} = I_1' \sin^2\theta\, \dot\phi + I_3' \cos\theta(\dot\psi + \dot\phi\cos\theta)\,,$$

$$\pi_\psi = \frac{\partial L}{\partial \dot\psi} = I_3'(\dot\psi + \dot\phi\cos\theta)\,. \tag{13.106}$$

The third constant of motion is just the total energy of the system.

Note that these expressions correspond to the ones obtained previously in Chap. 10, however, this time with a very little amount of effort employed, and without having to perform any basis transformation (for the angular momentum). Thus, as we just mentioned, this formalism, even if it is mathematically equivalent, many times can significantly reduce our working time.

13.5 Consider once again the simple pendulum introduced in Chap. 4, shown in Fig. 13.2. Write down the Lagrangian in polar coordinates and obtain the equations of motion. Do the same for Cartesian coordinates using Lagrangian multipliers. Why wasn't it necessary in polar coordinates to introduce Lagrangian multipliers?

Solution: The potential energy is given as usual by $U = mgh$ with $g, h > 0$. In our case we have $h = l - y = l(1 - \cos\theta)$. Therefore, the Lagrangian in polar coordinates reads

$$L = \frac{1}{2}ml^2\dot\theta^2 - mgl(1 - \cos\theta)\,. \tag{13.107}$$

The constant term mgl does not modify the equations of motion, we can therefore write down the simplified equivalent Lagrangian

$$L = \frac{1}{2}ml^2\dot\theta^2 + mgl\cos\theta\,. \tag{13.108}$$

We obtain

$$\frac{\partial L}{\partial \theta} = -mgl\sin\theta\,, \qquad \frac{\partial L}{\partial \dot\theta} = ml^2\dot\theta\,, \tag{13.109}$$

and so, the equation of motion is simply

$$ml^2\ddot\theta + mgl\sin\theta = 0\,, \tag{13.110}$$

which is an already familiar expression. Let us now turn to the Cartesian case. The corresponding constraint is given by

Fig. 13.2 Simple pendulum with motion in the (x, y) plane

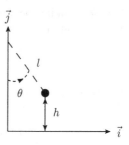

$$x^2 + y^2 - l^2 = 0. \tag{13.111}$$

The extended Lagrangian will be given by

$$L = \frac{1}{2}m(\dot{x}^2 + \dot{y}^2) + mgy + \lambda(x^2 + y^2 - l^2), \tag{13.112}$$

where again, we have ignored the mgl constant term.

The equations of motion can be easily obtained

$$m\ddot{x} = 2\lambda x \qquad m\ddot{y} = mg + 2\lambda y, \qquad x^2 + y^2 - l^2 = 0. \tag{13.113}$$

These equations can be obtained directly by applying Newton's laws in Cartesian coordinates. One can identify the 2λ term with $2\lambda = -T/l$ where T is the tension of the string.

To answer the last question, in polar coordinates one does not need to include Lagrange multipliers, as the constraints are already included in the expression for the kinetic energy!

13.6 Given the Lagrangian of the one-dimensional harmonic oscillator

$$L = \frac{1}{2}m\dot{x}^2 - \frac{1}{2}kx^2, \tag{13.114}$$

check that it is invariant under the infinitesimal transformation

$$x \rightarrow x' = x + \epsilon \sin(\omega t) \qquad \text{with} \qquad \omega = \sqrt{\frac{k}{m}}, \tag{13.115}$$

and obtain the correspondent conserved quantity. Write down the correspondent Hamiltonian and obtain the equations of motion.

Solution: Taking the time derivative of the previous transformation of the coordinate x, we obtain the transformation of $\dot{x}$

$$\dot{x} \rightarrow \dot{x}' = \dot{x} + \epsilon\omega\cos(\omega t). \tag{13.116}$$

Let us now write down the Lagrangian in terms of the new variables. We obtain

$$\begin{aligned} L &= \frac{1}{2}m\left(\dot{x} + \epsilon\omega\cos(\omega t)\right)^2 - \frac{1}{2}k\left(x + \epsilon\sin(\omega t)\right)^2 \\ &= \frac{1}{2}m\dot{x}^2 - \frac{1}{2}kx^2 + \epsilon\dot{x}\,m\,\omega\cos(\omega t) - \epsilon k\,x\sin(\omega t) + \mathcal{O}(\epsilon^2) \\ &= \frac{1}{2}m\dot{x}^2 - \frac{1}{2}kx^2 + \epsilon\frac{d}{dt}\left(x\,m\,\omega\cos(\omega t)\right) + \mathcal{O}(\epsilon^2). \end{aligned} \tag{13.117}$$

We can thus identify the following functions

$$f = \sin(\omega t), \qquad \tilde{F} = x\,m\,\omega\cos(\omega t), \tag{13.118}$$

and so, the conserved quantity is given by

$$C = \frac{\partial L}{\partial \dot{x}}f - \tilde{F} = m\dot{x}\sin(\omega t) - m\,x\,\omega\cos(\omega t). \tag{13.119}$$

The canonical momentum for this simple system reads

$$\pi_x = p_x = \frac{\partial L}{\partial \dot{x}} = m\dot{x}, \tag{13.120}$$

and so, the Hamiltonian is simply

$$H = \frac{p_x^2}{2m} + \frac{1}{2}kx^2. \tag{13.121}$$

The equations of motion can be straightforwardly obtained

$$\dot{x} = \frac{\partial H}{\partial p_x} = \frac{p_x}{m}, \qquad \dot{p}_x = -\frac{\partial H}{\partial x} = -kx. \tag{13.122}$$

This implies

$$m\ddot{x} + kx = 0. \tag{13.123}$$

13.7 Check explicitly, in Cartesian coordinates, that for a particle moving under the influence of a central potential, the previously calculated *conserved energy* in (13.70)

$$\sum_i \frac{\partial L}{\partial \dot{q}_i} \dot{q}_i - L = E\,, \qquad (13.124)$$

indeed corresponds to the total energy of the system. How is it related to the Hamiltonian?

Solution: Without presenting the explicit calculation, which should be already straightforward by now, one obtains the following expression for E

$$E = \frac{1}{2}m(\dot{x}^2 + \dot{y}^2 + \dot{z}^2) + V(r)\,, \qquad (13.125)$$

which indeed corresponds to the total energy of the system. One should note that the expression (13.124) corresponds to the formal definition of the Hamiltonian, which is a result that we have already encountered in some of the previously solved exercises. The only difference is that the energy is expressed in terms of q_i and $\dot{q}_i$ and the Hamiltonian is a function of π_i and q_i, i.e, in this case it would correspond to

$$H = \frac{p_x^2}{2m} + \frac{p_y^2}{2m} + \frac{p_z^2}{2m} + V(r)\,. \qquad (13.126)$$

One should also be aware of the following. If the energy of a system is conserved it does not mean that, by taking the partial derivatives of the Hamiltonian, the result will be zero. In order to obtain the equations of motion, one first has to *ignore* the fact that the energy is conserved, and only after obtaining the equations of motion, introduce the corresponding constants associated to the system.

13.8 Consider a particle of mass m moving over the surface of an inverted cone under the influence of gravity, as shown in Fig. 13.3. Write down the Lagrangian of the system (accounting for the corresponding constraints implicitly, without making use of the Lagrange multipliers) and find the constants of motion. Construct the Hamiltonian and obtain the equations of motion.

Solution: In polar coordinates the Kinetic energy of the particle is

$$K = \frac{1}{2}m(\dot{\rho}^2 + \dot{z}^2 + \rho^2 \dot{\phi}^2)\,. \qquad (13.127)$$

Fig. 13.3 A point-like
particle moving over the
surface of an inverted cone,
under the influence of gravity

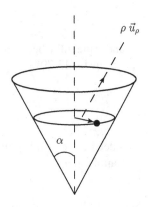

As the mass never leaves the surface of the cone, we have the following boundary
condition

$$\tan \alpha = \frac{\rho}{z} \quad \Rightarrow \quad \dot{z} = \frac{\dot{\rho}}{\tan \alpha} \,, \tag{13.128}$$

hence we can eliminate $\dot{z}$ from the kinetic energy and obtain

$$K = \frac{1}{2} m \left(\frac{\dot{\rho}^2}{\sin^2 \alpha} + \rho^2 \dot{\phi}^2 \right) . \tag{13.129}$$

The potential energy is given by $V = mgz$, which can be re-written as
$V = mg\rho / \tan \alpha$. The Lagrangian finally reads

$$L = \frac{1}{2} m \left(\frac{\dot{\rho}^2}{\sin^2 \alpha} + \rho^2 \dot{\phi}^2 \right) - \frac{mg\rho}{\tan \alpha} \,, \tag{13.130}$$

which apparently corresponds to a two-dimensional problem. As ϕ does not appear
explicitly in the Lagrangian, its corresponding canonical momentum is a conserved
quantity

$$\pi_\phi = \frac{\partial L}{\partial \dot{\phi}} = m\rho^2 \dot{\phi}. \tag{13.131}$$

As the system is conservative, the total energy of the system is also conserved. As
for π_ρ, we have

$$\pi_\rho = \frac{\partial L}{\partial \dot{\rho}} = \frac{m\dot{\rho}}{\sin^2 \alpha} \,. \tag{13.132}$$

The corresponding Hamiltonian can be easily deduced. Its expression reads

$$H = \frac{\pi_\phi^2}{2m\rho^2} + \frac{\pi_\rho^2}{2m} \sin^2 \alpha + \frac{mg\rho}{\tan \alpha} \,. \tag{13.133}$$

Note that, as π_ϕ is constant, the problem is further reduced to a one-dimensional problem. Hence, the equations of motion are given by

$$\pi_\rho = \frac{m\dot\rho}{\sin^2 \alpha} \,, \qquad \dot\pi_\rho = \frac{\pi_\phi^2}{m\rho^3} - \frac{mg}{\tan \alpha} \,. \tag{13.134}$$

This implies

$$\frac{m\ddot\rho}{\sin^2 \alpha} - \frac{\pi_\phi^2}{m\rho^3} + \frac{mg}{\tan \alpha} = 0 \,. \tag{13.135}$$

As a final comment, in this book we only presented some basic notions and some representative exercises on analytical mechanics however, one should be aware that there is an infinite amount of additional material and associated problems and exercises on this topic. Here, we roughly covered a very small percentage of the all the possible aspects of Lagrangian and Hamiltonian mechanics. As it is not the goal of this book to go further on this topic, we shall end here.

Further Reading

1. J.V. José, E.J. Saletan, *Classical Dynamics: A Contemporary Approach*. Cambridge University Press
2. S.T. Thornton, J.B. Marion, *Classical Dynamics of Particles and Systems*
3. H. Goldstein, C. Poole, J. Safko, *Classical Mechanics*, 3rd edn. Addison Wesley
4. J.R. Taylor, *Classical Mechanics*
5. D.T. Greenwood, *Classical Dynamics*. Prentice-Hall Inc.
6. D. Kleppner, R. Kolenkow, *An Introduction to Mechanics*
7. C. Lanczos, *The Variational Principles of Mechanics*. Dover Publications Inc.
8. W. Greiner, *Classical Mechanics: Systems of Particles and Hamiltonian Dynamics*. Springer
9. H.C. Corben, P. Stehle, *Classical Mechanics*, 2nd edn. Dover Publications Inc.
10. T.W.B. Kibble, F.H. Berkshire, *Classical Mechanics*. Imperial College Press
11. M.G. Calkin, *Lagrangian and Hamiltonian Mechanics*
12. A.J. French, M.G. Ebison, *Introduction to Classical Mechanics*

Appendix A
Tensor Formalism in Cartesian Coordinates

This appendix is meant to clarify the concept of tensors in Cartesian coordinates and introduce the concept of *metric*, which will be necessary to better understand the formalism of Special Relativity. Thus, it is intended to simplify the understanding of Chap. 11 and offer the reader a more complete vision of the problem. To simplify things we only introduce the minimum needed amount of information about tensors. However, with the basic notions introduced here, the interested reader will be able to read more advanced courses on tensors.

In order to properly work with tensors we needed to slightly change the vector notation we have been using in the previous chapters. We are used to write a vector, in a given basis, in Cartesian coordinates as

$$\mathbf{v} = v_x \mathbf{i} + v_y \mathbf{j} + v_z \mathbf{k} = \sum_{i=1}^{3} v_i \mathbf{e}_i. \tag{A.1}$$

Here, we shall use an upper index for the vector components and a lower index for the basis, thus one can write a vector in a generic basis as

$$\mathbf{v} = \sum_{i=1}^{3} v^i \mathbf{e}_i \equiv v^i \mathbf{e}_i, \tag{A.2}$$

where, for the right side of the previous expression we suppressed the summation symbol and adopted the standard Einstein summation convention for repeated indices (one is simply supposed to sum over repeated indices). In matrix form, the upper indices will stand for the row and the lower indices will stand for the column. Thus, a vector will be written as

$$\mathbf{v} = (\mathbf{e}_1, \mathbf{e}_2, \mathbf{e}_3) \begin{pmatrix} v^1 \\ v^2 \\ v^3 \end{pmatrix}, \tag{A.3}$$

© Springer Nature Switzerland AG 2020
V. Ilisie, *Lectures in Classical Mechanics*, Undergraduate Lecture
Notes in Physics, https://doi.org/10.1007/978-3-030-38585-9

as usual, as in Chap. 1. Consider now, a transformation of the reference system so that the components of a vector change according to

$$v^i \;\; \rightarrow \;\; v'^i = \Lambda^i_j \, v^j \,, \tag{A.4}$$

where Λ can be any invertible matrix in this case, not only a proper orthogonal matrix as we considered in Chap. 1. Again, the upper index of Λ stands for the row and the lower one for the column. Also remember that, if not stated otherwise, summation is always performed over repeated indices.

We can thus obtain the transformation law for the basis

$$\mathbf{v} = v'^i \, \mathbf{e}'_i = \Lambda^i_j \, v^j \, \mathbf{e}'_i = v^j \, \mathbf{e}_j \,, \tag{A.5}$$

and so

$$\mathbf{e}_j = \Lambda^i_j \, \mathbf{e}'_i \,. \tag{A.6}$$

Equivalently, the inverse relation reads

$$\mathbf{e}'_j = (\Lambda^{-1})^i_j \, \mathbf{e}_i \,. \tag{A.7}$$

Expressing the previous equation in matrix notation, we obtain

$$(\mathbf{e}'_1, \, \mathbf{e}'_2, \, \mathbf{e}'_3) = (\mathbf{e}_1, \, \mathbf{e}_2, \, \mathbf{e}_3) \, \Lambda^{-1} \,. \tag{A.8}$$

Transposing we finally get

$$\begin{pmatrix} \mathbf{e}'_1 \\ \mathbf{e}'_2 \\ \mathbf{e}'_3 \end{pmatrix} = (\Lambda^{-1})^T \begin{pmatrix} \mathbf{e}_1 \\ \mathbf{e}_2 \\ \mathbf{e}_3 \end{pmatrix} \,. \tag{A.9}$$

Writing down (A.4) in matrix form we also obtain

$$\begin{pmatrix} v'^1 \\ v'^2 \\ v'^3 \end{pmatrix} = \Lambda \begin{pmatrix} v^1 \\ v^2 \\ v^3 \end{pmatrix} \,. \tag{A.10}$$

If we consider proper orthogonal transformations, by defining

$$R \equiv \Lambda = (\Lambda^{-1})^T \,, \tag{A.11}$$

we recover the expressions (1.28) and (1.31) from Chap. 1.

So far we have done nothing new, but a slight update of our notation and introduced a more general transformation Λ. Next we shall introduce a new concept, the dual space.

A.1 Dual Space

The dual space of $\mathbb{R}^3$ is defined as the space of all the linear maps (applications) β from $\mathbb{R}^3$ to $\mathbb{R}$:

$$\beta : \mathbb{R}^3 \to \mathbb{R}$$
$$\beta : \mathbf{v} \mapsto \beta(\mathbf{v}) , \tag{A.12}$$

with the following linear property

$$\beta(\lambda_1 \mathbf{v}_1 + \lambda_2 \mathbf{v}_2) = \lambda_1 \beta(\mathbf{v}_1) + \lambda_2 \beta(\mathbf{v}_2) , \tag{A.13}$$

$\forall \ \mathbf{v}_1, \mathbf{v}_2 \in \mathbb{R}^3$ and $\forall \ \lambda_1, \lambda_2 \in \mathbb{R}$. The dual space of $\mathbb{R}^3$ is also a vector space of dimension 3, that we can also identify with $\mathbb{R}^3$. Its elements are called covectors and we shall denote its basis components as $\mathbf{e}^i$, in order to distinguish between the two spaces. Thus, β can be written as

$$\beta = \beta_i \mathbf{e}^i . \tag{A.14}$$

Given a vector $\mathbf{v} \in \mathbb{R}^3$ the linear map $\beta(\mathbf{v}) \in \mathbb{R}$ can be explicitly written as

$$\beta(v) = \beta_i \mathbf{e}^i (v^j \mathbf{e}_j) = \beta_i v^j \mathbf{e}^i (\mathbf{e}_j) . \tag{A.15}$$

The basis of the dual space can be arbitrary, however we shall work with the most convenient one, the dual basis, that has the following property

$$\mathbf{e}^i (\mathbf{e}_j) = \delta^i_j , \tag{A.16}$$

where δ^i_j is the Kronecker-delta defined the usual way ($\delta^i_j = 0$ if $i \neq j$ and $\delta^i_j = 1$ if $i = j$). Thus $\beta(\mathbf{v})$ will be given by the sum

$$\beta(\mathbf{v}) = \beta_i v^i . \tag{A.17}$$

Let us suppose that we perform a change of basis (A.7), and that under this transformation, the transformation of the dual basis is given a matrix $\bar{\Lambda}$:

$$\mathbf{e}'^i = \bar{\Lambda}^i_l \mathbf{e}^l . \tag{A.18}$$

Using the property (A.16) for the new bases, we easily obtain the relation between the matrices Λ and $\bar{\Lambda}$.

$$\mathbf{e}'^i(\mathbf{e}'_j) = \bar{\Lambda}^i_l \mathbf{e}^l \left((\Lambda^{-1})^k_j \mathbf{e}_k\right) = \bar{\Lambda}^i_l (\Lambda^{-1})^k_j \mathbf{e}^l(\mathbf{e}_k) = \bar{\Lambda}^i_l (\Lambda^{-1})^k_j \delta^l_k = \delta^i_j . \quad \text{(A.19)}$$

Therefore, we obtain the following

$$\bar{\Lambda}^i_k (\Lambda^{-1})^k_j = \delta^i_j \quad \Rightarrow \quad \bar{\Lambda}\,\Lambda^{-1} = I \quad \Rightarrow \quad \bar{\Lambda} = \Lambda , \quad \text{(A.20)}$$

where I is the 3×3 identity matrix. In conclusion β transforms as

$$\beta'_i = (\Lambda^{-1})^j_i \beta_j , \qquad \mathbf{e}'^i = \Lambda^i_j \mathbf{e}^j . \quad \text{(A.21)}$$

A.2   Covariant and Contravariant Laws of Transformation

Summing up, given $\mathbf{v}$ a vector ($v = v^i \mathbf{e}_i$) and β a covector ($\beta = \beta_i \mathbf{e}^i$), we have the following law of transformation for $\mathbf{e}_i$ and β_i

$$\boxed{\mathbf{e}'_i = (\Lambda^{-1})^j_i \mathbf{e}_j , \qquad \beta'_i = (\Lambda^{-1})^j_i \beta_j} . \quad \text{(A.22)}$$

We shall call this, the **covariant** law of transformation. For $\mathbf{e}^i$ and v^i we have

$$\boxed{\mathbf{e}'^i = \Lambda^i_j \mathbf{e}^j , \qquad v'^i = \Lambda^i_j v^j} . \quad \text{(A.23)}$$

We shall call this, the **contravariant** law of transformation. This is the reason why we have introduced this new notation with upper and lower indices, to be able to make the difference between covariant and contravariant quantities. Normally, the coordinates are also used to be written with an upper index as x^i with $i = 1, 2$ and 3. However, as we have insisted many times, strictly speaking, coordinates are not vectors, so even though they will carry an upper index we shall not consider them contravariant quantities.

As we have already seen in Chap. 1, if v^i are the components of a vector in Cartesian coordinates (due to the fact that Λ is a constant matrix), the time derivatives of v^i will also behave as the components of a vector i.e.,

$$v'^i = \Lambda^i_j v^j , \qquad \dot{v}'^i = \Lambda^i_j \dot{v}^j , \qquad \ddot{v}'^i = \Lambda^i_j \ddot{v}^j , \quad \text{(A.24)}$$

and so on. Thus, as the velocity $\mathbf{v}$ is a vector, the acceleration and higher derivatives are also vectors, and their components in a given basis will follow the previous transformations. In curvilinear coordinates things will be a little bit different, as we shall see in Appendix B.

A.3   Rank-Two Covariant Tensor

In the same fashion we have defined β as linear map over a vector space $\mathbb{R}^3$ one can define a bi-linear map over $\mathbb{R}^3 \times \mathbb{R}^3$ as

$$\mathbf{g} : \mathbb{R}^3 \times \mathbb{R}^3 \rightarrow \mathbb{R}$$
$$\mathbf{g} : (\mathbf{v}_1, \mathbf{v}_2) \mapsto \mathbf{g}(\mathbf{v}_1, \mathbf{v}_2) \,, \tag{A.25}$$

with the following two properties

1. $\mathbf{g}(\lambda_1 \mathbf{v}_1 + \lambda_2 \mathbf{v}_2, \mathbf{v}) = \lambda_1 \mathbf{g}(\mathbf{v}_1, \mathbf{v}) + \lambda_2 \mathbf{g}(\mathbf{v}_2, \mathbf{v})$
2. $\mathbf{g}(\mathbf{v}, \lambda_1 \mathbf{v}_1 + \lambda_2 \mathbf{v}_2) = \lambda_1 \mathbf{g}(\mathbf{v}, \mathbf{v}_1) + \lambda_2 \mathbf{g}(\mathbf{v}, \mathbf{v}_2) \,, \tag{A.26}$

$\forall \mathbf{v}, \mathbf{v}_1, \mathbf{v}_2 \in \mathbb{R}^3$ and $\forall \lambda_1, \lambda_2 \in \mathbb{R}$. It is straightforward to deduce that the following property also holds

$$\mathbf{g}(\lambda_1 \mathbf{v}_1, \lambda_2 \mathbf{v}_2) = \lambda_1 \lambda_2 \, \mathbf{g}(\mathbf{v}_1, \mathbf{v}_2) \,. \tag{A.27}$$

The linear map $\mathbf{g}$ is called a rank-two covariant tensor (rank-two because it acts on two copies of $\mathbb{R}^3$ and covariant because its components transform, just as β, following the covariant law, as we shall see next).[1]

Without getting into unnecessary details we shall also introduce the tensor product of $\mathbf{e}^i$ and $\mathbf{e}^j$, as the map

$$\otimes \, : \, (\mathbf{e}^i, \mathbf{e}^j) \mapsto \mathbf{e}^i \otimes \mathbf{e}^j \,, \tag{A.28}$$

with the following properties

1. $\mathbf{e}^i \otimes (\lambda \mathbf{e}^j) = (\lambda \mathbf{e}^i) \otimes \mathbf{e}^j \equiv \lambda \, \mathbf{e}^i \otimes \mathbf{e}^j$ $\tag{A.29}$
2. $\mathbf{e}^i \otimes \mathbf{e}^j \neq \mathbf{e}^j \otimes \mathbf{e}^i \,, \tag{A.30}$

with $\lambda \in \mathbb{R}$. The tensor product $\mathbf{e}^i \otimes \mathbf{e}^j$ is itself a map (application) that acts over covariant bases of $\mathbb{R}^3 \times \mathbb{R}^3$ as follows

$$\mathbf{e}^i \otimes \mathbf{e}^j \, (\mathbf{e}_k, \mathbf{e}_l) = \mathbf{e}^i(\mathbf{e}_k) \, \mathbf{e}^j(\mathbf{e}_l) = \delta_k^i \, \delta_l^j \,. \tag{A.31}$$

Thus, one can write $\mathbf{g}$ using the tensor product as

$$\mathbf{g} = g_{ij} \, \mathbf{e}^i \otimes \mathbf{e}^j \,, \tag{A.32}$$

---

[1] In a similar way one can define higher rank covariant or contravariant tensors. However, general tensor algebra is a topic which lays far beyond the goal of this book and that shall not be treated here.

and $\mathbf{g}(\mathbf{v}_1, \mathbf{v}_2)$ will take the form

$$
\begin{aligned}
\mathbf{g}(\mathbf{v}_1, \mathbf{v}_2) &= g_{ij}\, \mathbf{e}^i \otimes \mathbf{e}^j\, (\mathbf{v}_1, \mathbf{v}_2) \\
&= g_{ij}\, \mathbf{e}^i \otimes \mathbf{e}^j\, (v_1^k \mathbf{e}_k,\, v_2^l \mathbf{e}_l) \\
&= g_{ij}\, v_1^k v_2^l\, \mathbf{e}^i \otimes \mathbf{e}^j\, (\mathbf{e}_k,\, \mathbf{e}_l) \\
&= g_{ij}\, v_1^k v_2^l\, \mathbf{e}^i\, (\mathbf{e}_k)\, \mathbf{e}^j\, (\mathbf{e}_l) \\
&= g_{ij}\, v_1^k v_2^l\, \delta_k^i\, \delta_l^j \\
&= g_{ij}\, v_1^i v_2^j\, .
\end{aligned}
\tag{A.33}
$$

Let us now deduce how the components of $\mathbf{g}$ must change under a transformation $\mathbf{e}^i \to \mathbf{e}'^i = \Lambda^i_j\, \mathbf{e}^j$. We have

$$
\mathbf{g} = g_{kl}\, \mathbf{e}^k \otimes \mathbf{e}^l = g_{kl}\, (\Lambda^{-1})_i^k\, (\Lambda^{-1})_j^l\, \mathbf{e}'^i \otimes \mathbf{e}'^j = g'_{ij}\, \mathbf{e}'^i \otimes \mathbf{e}'^j\, .
\tag{A.34}
$$

We therefore obtain the law of transformation for the components g_{ij} to be

$$
\boxed{g'_{ij} = g_{kl}\, (\Lambda^{-1})_i^k\, (\Lambda^{-1})_j^l}\, ,
\tag{A.35}
$$

as expected (for a rank two covariant tensor). The previous expression is the intrinsic definition of a tensor and it is what one normally finds in many physics books, skipping this way the introduction and usage of the tensor product.

It is trivial to check that given two vectors $\mathbf{v}_1$ and $\mathbf{v}_2$ the quantity we have previously calculated in (A.33)

$$
\mathbf{g}(\mathbf{v}_1, \mathbf{v}_2) = g_{ij}\, v_1^i v_2^j\, ,
\tag{A.36}
$$

is basis invariant (or a scalar). It is also trivial to check that, given a vector $\mathbf{v} = v^i\, \mathbf{e}_i$, the quantities defined as

$$
v_i \equiv g_{ij}\, v^j\, ,
\tag{A.37}
$$

transform as the components of a covector.

A.4 Metric Tensor in $\mathbb{R}^3$

A metric tensor is just a rank-two covariant tensor with the additional two properties

$$
\begin{aligned}
&1.\, \text{symmetric}\, :\, g_{ij} = g_{ji}\, , \\
&2.\, \text{non-singular}\, :\, \det(\mathbf{g}) \neq 0\, ,
\end{aligned}
\tag{A.38}
$$

where $\det(\mathbf{g})$ is the determinant of $\mathbf{g}$ written in matrix form i.e., with components g_{ij} with i standing for the row and j for the column. The second property is just the requirement for $\mathbf{g}$ to be a well behaved invertible matrix and the first one ensures that the result of the *inner product* $\mathbf{g}(\mathbf{v}_1, \mathbf{v}_2)$ is not altered by the order in which it is performed i.e., $\mathbf{g}(\mathbf{v}_1, \mathbf{v}_2) = \mathbf{g}(\mathbf{v}_2, \mathbf{v}_1)$.

The quantity $\mathbf{g}(\mathbf{v}_1, \mathbf{v}_2)$ is nothing but the inner product defined in (1.6) with the Euclidean metric with the components g_{ij} given by

$$g_{ij} = \begin{pmatrix} 1 & 0 & 0 \\ 0 & 1 & 0 \\ 0 & 0 & 1 \end{pmatrix}. \tag{A.39}$$

Summing up, with this formalism, the inner product of two vectors simply becomes

$$\mathbf{v}_1 \cdot \mathbf{v}_2 = \mathbf{g}(\mathbf{v}_1, \mathbf{v}_2) = v_1^i \, v_2^j \, \mathbf{g}(\mathbf{e}_i, \mathbf{e}_j) = g_{ij} \, v_1^i \, v_2^j = v_1^i \, v_{2,i}, \tag{A.40}$$

and the squared modulus of a vector reads

$$\mathbf{v} \cdot \mathbf{v} = \mathbf{v}^2 = v^2 = \mathbf{g}(\mathbf{v}, \mathbf{v}) = g_{ij} \, v^i \, v^j = v^i \, v_i, \tag{A.41}$$

where we have made use of the definition (A.37). As one can imagine, under proper orthogonal rotations of the reference system, the components of the metric tensor remain unaltered. Under a spatial translation the the metric will also remain invariant. Notice that in the previous expression, for the inner product we have the components of a covariant and the components of a contravariant quantity. However, in Cartesian coordinates with the Euclidean metric $v^i = v_i$ so, it does not really make any difference. This will no longer hold true for Special Relativity where the metric is not Euclidean any more.

One could conclude from (A.40) that the components of the metric tensor are given by

$$g_{ij} = \mathbf{g}(\mathbf{e}_i, \mathbf{e}_j) = \mathbf{e}_i \cdot \mathbf{e}_j. \tag{A.42}$$

We should be really careful about this conclusion. The expression $\mathbf{g}(\mathbf{v}_1, \mathbf{v}_2)$ should be regarded as a generalization of the scalar product defined en 3D Euclidean space. If one uses the tensor formalism with a given metric, one should no longer use the *traditional* definition of the scalar product that was given in Chap. 1. Having said this, there is no danger in using (A.42) in Cartesian coordinates or in spherical, polar and cylindrical coordinates. The problem arises in Special Relativity for example, where some components of the metric are negative (as we shall see next). In that case we would have to define all the previous formalism over complex numbers, which supposes further complications and is not at all necessary. In more generic cases we could also find the problem of non-separability of the metric tensor i.e., the components of the metric g_{ij} cannot be expressed as a product, as in the previous equation. This is however a very advanced topic and we shall not get into further details.

A.5   Metric Tensor in Special Relativity

This subsection is just a short overview of the tensor analysis needed for Special Relativity which will be useful in Chap. 11, where we shall make use of the previously defined concepts.

In Minkowski space we have four coordinates (three spatial coordinates and a fourth one corresponding to time) that can be mapped into $\mathbb{R}^4$. The Minkowski metric tensor is given by

$$g_{\mu\nu} = \begin{pmatrix} 1 & 0 & 0 & 0 \\ 0 & -1 & 0 & 0 \\ 0 & 0 & -1 & 0 \\ 0 & 0 & 0 & -1 \end{pmatrix}. \tag{A.43}$$

The short-hand notation for this is $g_{\mu\nu} = \text{diag}\{1, -1, -1, -1\}$. Our coordinates in $\mathbb{R}^4$ are the space-time coordinates x^μ that can be written in matrix form as

$$\begin{pmatrix} x^0 \\ x^1 \\ x^2 \\ x^3 \end{pmatrix} \equiv \begin{pmatrix} ct \\ x \\ y \\ z \end{pmatrix}, \tag{A.44}$$

where c is the speed of light in the vacuum. We shall call these space-time coordinates, events (that take place at x^i at instant t). Even though an upper index corresponds to a column matrix (as in the previous expression), many times we shall make an **abuse of notation** and write

$$x^\mu = (x^0, x^1, x^2, x^3) \equiv (ct, x^i) \equiv (ct, \mathbf{x}). \tag{A.45}$$

Note that we use Greek letters $\mu = 0, \ldots, 4$ for space-time coordinates and Roman letters $i = 1, \ldots, 3$ only for the spatial coordinates. Some authors define x_μ as

$$x_\mu \equiv g_{\mu\nu} x^\nu = (x_0, x_1, x_2, x_3) = (ct, x_i) \equiv (ct, -x^i) = (ct, -\mathbf{x}), \tag{A.46}$$

but we have to be really careful about that. Similar to what we saw in $\mathbb{R}^3$, x^μ **must be considered coordinates, not vector components!** Thus x_μ must nor be regarded as the components of a co-vector either!

Under Lorentz transformations Λ^μ_ν and space-time translations (with a^μ constants) i.e.,

$$x^\mu \;\rightarrow\; x'^\mu = \Lambda^\mu_\nu x^\nu + a^\mu \,, \tag{A.47}$$

the metric tensor in Special Relativity (in Cartesian coordinates) remains invariant. In matrix form the previous expression reads

$$\begin{pmatrix} x'^0 \\ x'^1 \\ x'^2 \\ x'^3 \end{pmatrix} = \Lambda \begin{pmatrix} x^0 \\ x^1 \\ x^2 \\ x^3 \end{pmatrix} + \begin{pmatrix} a^0 \\ a^1 \\ a^2 \\ a^3 \end{pmatrix}. \tag{A.48}$$

The Lorentz transformations corresponding to Λ^μ_ν can be rotations or boosts (see Chap. 11). Again, just as in $\mathbb{R}^3$ for rotations, the matrix elements Λ^μ_ν are constants, thus everything we have previously learned can be directly applied to Minkowski space. Given a vector with components v^μ, under a Lorentz transformation its components will change as

$$v'^\mu = \Lambda^\mu_\nu v^\nu \,. \tag{A.49}$$

In particular v^μ can be the four velocity

$$v'^\mu = \frac{dx'^\mu}{d\tau} = \Lambda^\mu_\nu v^\nu = \Lambda^\mu_\nu \frac{dx^\mu}{d\tau} \,, \tag{A.50}$$

where $d\tau$ is the differential Lorentz-invariant proper time (given by $ds = g_{\mu\nu} dx^\mu dx^\nu = c\, d\tau$) that is discussed in detail in Chap. 11. Therefore, higher derivatives, such as the acceleration, will also transform as vectors

$$\frac{dv'^\mu}{d\tau} = a'^\mu = \Lambda^\mu_\nu a^\nu = \Lambda^\mu_\nu \frac{dv^\nu}{d\tau} \,, \tag{A.51}$$

and so on. The metric tensor will transform as (A.35)

$$g'_{\mu\nu} = g_{\alpha\beta} (\Lambda^{-1})^\alpha_\mu (\Lambda^{-1})^\beta_\nu \,. \tag{A.52}$$

However, the metric remains invariant under these transformations i.e., $g'_{\mu\nu} = g_{\alpha\beta} = \mathrm{diag}\{1, -1, -1, -1\}$. Given a vector v with components

$$v^\mu = (v^0, v^1, v^2, v^3) = (v^0, \mathbf{v}) \,, \tag{A.53}$$

one can define the quantity

$$v_\mu = g_{\mu\nu} v^\nu = (v_0, v_1, v_2, v_3) = (v^0, -v^1, -v^2, -v^3) = (v^0, -\mathbf{v}) \,, \tag{A.54}$$

which will transform as a covector. One can also define the squared modulus

$$v^2 \;=\; v \cdot v \;=\; g_{\mu\nu}\, v^\mu v^\nu \;=\; v^\mu v_\mu \;=\; (v^0)^2 - \mathbf{v}^2$$
$$= (v^0)^2 - (v^1)^2 - (v^2)^2 - (v^3)^2\,, \tag{A.55}$$

which is Lorentz-invariant. Finally, given another vector u, one can define the inner product

$$u \cdot v \;=\; g_{\mu\nu}\, u^\mu v^\nu \;=\; u^\mu v_\mu \;=\; u_\mu v^\mu$$
$$= u^0 v^0 - \mathbf{u} \cdot \mathbf{v} \;=\; u^0 v^0 - u^1 v^1 - u^2 v^2 - u^3 v^3$$
$$= u^0 v_0 + u^1 v_1 + u^2 v_2 + u^3 v_3$$
$$= u_0 v^0 + u_1 v^1 + u_2 v^2 + u_3 v^3\,, \tag{A.56}$$

which is also a Lorentz-invariant quantity.

Appendix B
Tensors in Curvilinear Coordinates

As we have already mentioned, the lecture of the previous section is somewhat needed in order to better understand the tensor formalism of Special Relativity. On the other hand, the following section is presented for completeness for the interested reader. Its lecture is not at all necessary in order to be able to properly follow any of the chapters of this book. However, the lecture of the previous Appendix is mandatory, in order to follow the calculations presented in the following.

In Chap. 2 we have deduced the transformation that relates the Euclidean basis to the curvilinear basis, in terms of the matrix given in (2.6). There, we have introduced the normalization factors $h_{\alpha,\beta,\gamma}$ in order to ensure that the curvilinear basis was unitary. Consider now that we don't need to ensure the unitarity of the curvilinear basis.[2] The transformation matrix would then take the form

$$
\begin{pmatrix} \mathbf{u}_\alpha \\ \\ \mathbf{u}_\beta \\ \\ \mathbf{u}_\gamma \end{pmatrix} = \begin{pmatrix} \dfrac{\partial x}{\partial \alpha} & \dfrac{\partial y}{\partial \alpha} & \dfrac{\partial z}{\partial \alpha} \\ \dfrac{\partial x}{\partial \beta} & \dfrac{\partial y}{\partial \beta} & \dfrac{\partial z}{\partial \beta} \\ \dfrac{\partial x}{\partial \gamma} & \dfrac{\partial y}{\partial \gamma} & \dfrac{\partial z}{\partial \gamma} \end{pmatrix} \begin{pmatrix} \mathbf{i} \\ \\ \mathbf{j} \\ \\ \mathbf{k} \end{pmatrix} . \tag{B.1}
$$

Let's now introduce the following notations

$$ \{x^1, x^2, x^3\} \equiv \{x, y, z\}, \qquad \{\bar{x}^1, \bar{x}^2, \bar{x}^3\} \equiv \{\alpha, \beta, \gamma\}, \tag{B.2} $$

for the coordinates, and

$$ \{\mathbf{e}_1, \mathbf{e}_2, \mathbf{e}_3\} \equiv \{\mathbf{i}, \mathbf{j}, \mathbf{k}\}, \qquad \{\mathbf{u}_1, \mathbf{u}_2, \mathbf{u}_3\} \equiv \{\mathbf{u}_\alpha, \mathbf{u}_\beta, \mathbf{u}_\gamma\}, \tag{B.3} $$

---

[2]When using tensor formalism we have to give up on this condition, however, with no terrible consequences as we shall see later on.

© Springer Nature Switzerland AG 2020

V. Ilisie, *Lectures in Classical Mechanics*, Undergraduate Lecture Notes in Physics, https://doi.org/10.1007/978-3-030-38585-9

for the components of the two bases. We will also use the following notation for the components of a vector in Cartesian and curvilinear coordinates

$$\{v^1, v^2, v^3\} \equiv \{v^x, v^y, v^z\}, \qquad \{\bar{v}^1, \bar{v}^2, \bar{v}^3\} \equiv \{v^\alpha, v^\beta, v^\gamma\}. \tag{B.4}$$

We can thus express the previous transformation (B.1) as

$$\boxed{\mathbf{u}_j = \frac{\partial x^i}{\partial \bar{x}^j} \mathbf{e}_i}, \tag{B.5}$$

which is the **covariant** law of transformation, and where again, we have adopted the Einstein summation convention. By comparing this expression to (A.7) we conclude that

$$\frac{\partial x^i}{\partial \bar{x}^j} = (\Lambda^{-1})^i_j. \tag{B.6}$$

Note that the matrix introduced in (B.1) is in general not orthogonal thus $\Lambda^T \neq \Lambda^{-1}$, however this is not at all an issue. As we have already mentioned, with the tensor formalism we relaxed the unitarity condition for the curvilinear basis, which translates into non-orthogonal transformation matrices. Note that this does not imply that the components of the new basis are not orthogonal. When using orthogonal curvilinear coordinates (cylindrical, polar and spherical) they still are, but they are just not unitary any more.

Consider now a vector in Cartesian coordinates, i.e., the velocity vector $\mathbf{v} = d\mathbf{r}(x^j)/dt$. If the coordinates x^j are functions of some parameters $\bar{x}^i$ i.e., $x^j = x^j(\bar{x}^i)$ then

$$\mathbf{v} = \frac{d\mathbf{r}}{dt} = v^j \mathbf{e}_j = \frac{dx^j}{dt} \mathbf{e}_j = \frac{\partial x^j}{\partial \bar{x}^i} \frac{d\bar{x}^i}{dt} \mathbf{e}_j = \frac{d\bar{x}^i}{dt} \mathbf{u}_i = \bar{v}^i \mathbf{u}_i. \tag{B.7}$$

We can conclude that

$$\mathbf{v} = \bar{v}^i \mathbf{u}_i = \bar{v}^i \frac{\partial x^j}{\partial \bar{x}^i} \mathbf{e}_j = v^j \mathbf{e}_j, \tag{B.8}$$

and so, the components of a vector transform as

$$\boxed{\bar{v}^i = \frac{\partial \bar{x}^i}{\partial x^j} v^j}, \tag{B.9}$$

which is the **contravariant** law of transformation, and where we have used the property

$$\frac{\partial x^j}{\partial \bar{x}^i} \frac{\partial \bar{x}^i}{\partial x^k} = \delta_k^j .$$ (B.10)

Similarly the following property also holds

$$\frac{\partial \bar{x}^j}{\partial x^i} \frac{\partial x^i}{\partial \bar{x}^k} = \delta_k^j .$$ (B.11)

By comparing (B.9) to (A.4) we conclude that

$$\frac{\partial \bar{x}^i}{\partial x^j} = \Lambda_j^i .$$ (B.12)

The difference however, with the case of Cartesian coordinates is that, in our present case Λ is not a constant matrix, but depends on the coordinates.

By taking the time derivative of (B.9) we obtain

$$\frac{d\bar{v}^i}{dt} = \frac{\partial \bar{x}^i}{\partial x^j} \frac{dv^j}{dt} + v^j \frac{\partial^2 \bar{x}^i}{\partial x^j \partial x^k} \frac{dx^k}{dt} .$$ (B.13)

Obviously $d\bar{v}^i/dt$ does not transform as vector. This is a result that we already found in Chap. 2 for the acceleration i.e., if $\bar{v}^j$ are the components of the velocity in curvilinear coordinates, the components of the acceleration vector are not equal to the time derivatives of $\bar{v}^j$ i.e., $\bar{a}^j \neq d\bar{v}^j/dt$. This was due to the fact that the basis is local (depends on the point P) hence, we also have to take into account the derivatives of the basis components. We shall deduce the expression of the acceleration in the following. First we must introduce the **covariant derivative**. Consider a vector in Cartesian coordinates

$$\mathbf{v} = v^i \, \mathbf{e}_i .$$ (B.14)

The partial derivative of $\mathbf{v}$ with respect to the coordinate x^j will take the form

$$\frac{\partial \mathbf{v}}{\partial x^j} = \frac{\partial v^i}{\partial x^j} \mathbf{e}_i \equiv (\partial_j v^i) \, \mathbf{e}_i .$$ (B.15)

In curvilinear coordinates $\mathbf{v} = \bar{v}^i \, \mathbf{u}_i$, and the previous equation does not hold as the basis $\mathbf{u}_i$ also depends on the coordinates. The partial derivative must therefore, also act on the basis:

$$\frac{\partial \mathbf{v}}{\partial \bar{x}^j} \equiv \frac{\partial \bar{v}^i}{\partial \bar{x}_j} \mathbf{u}_i + \bar{v}^i \frac{\partial \mathbf{u}_i}{\partial \bar{x}_j} = (\bar{\partial}_j \, \bar{v}^i) \mathbf{u}_i + \bar{v}^i (\bar{\partial}_j \mathbf{u}_i) .$$ (B.16)

Here we have introduced the short-hand notation

$$\bar{\partial}_j \equiv \frac{\partial}{\partial \bar{x}^j} \, . \tag{B.17}$$

As the curvilinear basis is a complete basis, we can write $\bar{\partial}_j \mathbf{u}_i$ as a linear combination of the components of the basis (as we argued in (2.15) in Chap. 2)

$$\boxed{\bar{\partial}_j \mathbf{u}_i \ = \ \Gamma_{ji}^m \, \mathbf{u}_m} \, . \tag{B.18}$$

The quantities Γ_{ji}^m are called the Christoffel symbols (which are all equally zero in Cartesian coordinates). Note that in the previous equation an implicit sum is performed over the repeated index m. We can thus write down $\bar{\partial}_j \, \mathbf{v}$ as

$$\bar{\partial}_j \, \mathbf{v} \ = \ (\bar{\partial}_j \, \bar{v}^i) \mathbf{u}_i + \bar{v}^i (\Gamma_{ji}^m \, \mathbf{u}_m) \ = \ (\bar{\partial}_j \, \bar{v}^m + \bar{v}^i \, \Gamma_{ji}^m) \, \mathbf{u}_m \, , \tag{B.19}$$

where we have changed one mute index from an implicit sum i.e., we have used $\bar{v}^i \, \mathbf{u}_i \ = \ \bar{v}^m \, \mathbf{u}_m$.

We define the covariant derivative ∇_j as

$$\boxed{\bar{\partial}_j \, \mathbf{v} \equiv (\nabla_j \, \bar{v}^m) \, \mathbf{u}_m \ = \ (\bar{\partial}_j \, \bar{v}^m + \bar{v}^i \, \Gamma_{ji}^m) \, \mathbf{u}_m} \, , \tag{B.20}$$

where $\nabla_j \, \bar{v}^m$ is called the covariant derivative of $\bar{v}^m$ (do not confuse ∇_j with ∇ which is the ordinary gradient/divergence).

Having introduced the previous concepts we can express the time derivative of a vector in curvilinear coordinates as

$$\begin{aligned}
\frac{d\mathbf{v}}{dt} &= \frac{d\bar{v}^i}{dt} \mathbf{u}_i + \bar{v}^i \frac{d\mathbf{u}_i}{dt} \\
&= \dot{\bar{v}}^i \mathbf{u}_i + \bar{v}^i \frac{\partial \mathbf{u}_i}{\partial \bar{x}^j} \frac{d\bar{x}^j}{dt} \\
&= \dot{\bar{v}}^i \mathbf{u}_i + \bar{v}^i \bar{v}^j \bar{\partial}_j \mathbf{u}_i ,
\end{aligned} \tag{B.21}$$

where we have used the fact that the components of the basis $\mathbf{u}_i$ only depend on the coordinates $\bar{x}^j$. Introducing the Christoffel symbols the previous expression turns into

$$\frac{d\mathbf{v}}{dt} \ = \ \left(\dot{\bar{v}}^m + \bar{v}^j \bar{v}^i \, \Gamma_{ji}^m \right) \mathbf{u}_m \, . \tag{B.22}$$

If we consider that $\mathbf{v}$ is the velocity vector, then the acceleration in curvilinear coordinates will be written as

$$\mathbf{a} \ = \ \bar{a}^m \, \mathbf{u}_m \, , \qquad \bar{a}^m \ = \ \dot{\bar{v}}^m + \bar{v}^j \bar{v}^i \, \Gamma_{ji}^m \, . \tag{B.23}$$

We now have to prove that the components of the acceleration transform according to the contravariant law (B.9) i.e., if the acceleration in Cartesian coordinates is written as $\mathbf{a} = a^j \, \mathbf{e}_j$ and in curvilinear coordinates as $\mathbf{a} = \bar{a}^m \, \mathbf{u}_m$, then

$$\bar{a}^m = \frac{\partial \bar{x}^m}{\partial x^j} \, a^j , \tag{B.24}$$

or equivalently

$$a^j = \frac{\partial x^j}{\partial \bar{x}^m} \, \bar{a}^m . \tag{B.25}$$

In order obtain the desired proof we have to explicitly find the expressions for the Christoffel symbols. Equation (B.18) can be written in terms of partial derivatives of $\mathbf{r}$ as

$$\frac{\partial}{\partial \bar{x}^j} \frac{\partial \mathbf{r}}{\partial \bar{x}^i} = \Gamma_{ji}^m \frac{\partial \mathbf{r}}{\partial \bar{x}^m} . \tag{B.26}$$

Expressing $\mathbf{r} = x^k \, \mathbf{e}_k$ in Cartesian coordinates, the previous expression turns into

$$\frac{\partial^2 x^k}{\partial \bar{x}^j \partial \bar{x}^i} = \Gamma_{ji}^m \frac{\partial x^k}{\partial \bar{x}^m} , \tag{B.27}$$

$\forall k$. Multiplying both sides of the previous expression by $\partial \bar{x}^l / \partial x^k$ and using the property (B.11) we finally obtain

$$\Gamma_{ji}^l = \frac{\partial \bar{x}^l}{\partial x^k} \frac{\partial^2 x^k}{\partial \bar{x}^j \partial \bar{x}^i} = \Gamma_{ij}^l . \tag{B.28}$$

Note that we have deduced an important property of the Christoffel symbols, that they are symmetric in the lower indices.[3]

Now remember that we have found that the velocity vector transforms as

$$v^l = \frac{\partial x^l}{\partial \bar{x}^m} \bar{v}^m . \tag{B.29}$$

Taking the time derivative on both sides of the previous equation and keeping in mind that in Cartesian coordinates the components of the acceleration are given by $a^l = \dot{v}^l$ we obtain

---

[3] This property does not hold in general. There are space-time geometries that present what is called *torsion* and for which, the previously introduced expression is not correct. This is however a topic that we will not get into.

$$a^l = \frac{\partial x^l}{\partial \bar{x}^m} \dot{\bar{v}}^m + \bar{v}^m \frac{d}{dt}\left(\frac{\partial x^l}{\partial \bar{x}^m}\right)$$

$$= \frac{\partial x^l}{\partial \bar{x}^m} \dot{\bar{v}}^m + \bar{v}^m \bar{v}^n \frac{\partial^2 x^l}{\partial \bar{x}^m \partial \bar{x}^n}. \tag{B.30}$$

Introducing the expression of $\dot{\bar{v}}^m$ from (B.23) we obtain

$$a^l = \left(\bar{a}^m - \bar{v}^i \bar{v}^j \frac{\partial \bar{x}^m}{\partial x^k} \frac{\partial^2 x^k}{\partial \bar{x}^j \partial \bar{x}^i}\right) \frac{\partial x^l}{\partial \bar{x}^m} + \bar{v}^m \bar{v}^n \frac{\partial^2 x^l}{\partial \bar{x}^m \partial \bar{x}^n}$$

$$= \frac{\partial x^l}{\partial \bar{x}^m} \bar{a}^m - \bar{v}^i \bar{v}^j \frac{\partial^2 x^l}{\partial \bar{x}^j \partial \bar{x}^i} + \bar{v}^m \bar{v}^n \frac{\partial^2 x^l}{\partial \bar{x}^m \partial \bar{x}^n}$$

$$= \frac{\partial x^l}{\partial \bar{x}^m} \bar{a}^m \tag{B.31}$$

which is the expression we were looking for.

The formalism we have just introduced is valid for any general (regular[4]) type of curvilinear coordinates. There are many other types of curvilinear coordinates besides the ones studied here and the previous formalism is completely generic. However, we have not studied transformations between two sets of curvilinear coordinates (only between Cartesian and curvilinear). This is would further push us towards the realm of differential geometry and it is not what we intend here. This topic is rather advanced and requires the reading of dedicated literature, thus it shall not be treated further.

In the following we shall apply some of the previously acquired knowledge to cylindrical and spherical coordinates.

B.1 Metric Tensor and Scalar Products

Similar to what we saw in the case of Cartesian coordinates, it is straightforward to deduce that the transformation law for the metric tensor from Cartesian to curvilinear coordinates is given by

$$\bar{g}_{ij} = \frac{\partial x^k}{\partial \bar{x}^i} \frac{\partial x^l}{\partial \bar{x}^j} g_{kl}, \tag{B.32}$$

where $\bar{g}_{ij}$ is the metric tensor in curvilinear coordinates. Thus one can check that, as expected, the scalar product of two vectors defined as in (A.40) is basis invariant i.e.,

---

[4]Whose Jacobian matrix corresponding to the coordinate transformation has non-zero determinant.

$$\mathbf{u} \cdot \mathbf{v} = \bar{g}_{ij} \bar{u}^i \bar{v}^j = \frac{\partial x^k}{\partial \bar{x}^i} \frac{\partial x^l}{\partial \bar{x}^j} g_{kl} \frac{\partial \bar{x}^i}{\partial x^m} \frac{\partial \bar{x}^j}{\partial x^n} u^m v^n = g_{mn} u^m v^n . \tag{B.33}$$

There is one easy way to obtain the components of the metric by using (A.42). Consider the scalar product of of two vectors ω and $\mathbf{v}$. If we denote the unitary basis in orthogonal curvilinear coordinates by $\mathbf{u}'_i$ and, the components of the vectors in this basis by ω'^i and v'^i then

$$\omega \cdot \mathbf{v} = \omega'^i v'^j \, \mathbf{u}'_i \cdot \mathbf{u}'_j = \omega'^i v'^j \delta_{ij} , \tag{B.34}$$

where δ_{ij} is the Kronecker delta. Let's now write the previous product in terms of the non-unitary basis denoted here by $\mathbf{u}_i$

$$\omega \cdot \mathbf{v} = \omega'^i v'^j \frac{1}{h_i h_j} \mathbf{u}_i \cdot \mathbf{u}_j = \frac{\omega'^i}{h_i} \frac{v'^j}{h_j} \bar{g}_{ij} \equiv \bar{\omega}^i \bar{v}^j \bar{g}_{ij} , \tag{B.35}$$

and where $\bar{\omega}^i$ and $\bar{v}^j$ are the components of ω and $\mathbf{v}$ in the non-unitary basis. By comparing the previous expression to (B.34) we deduce that the components of the metric tensor in curvilinear coordinates are nothing but the product of the normalization factors h_i and h_j i.e.,

$$\bar{g}_{ij} = h_i h_j . \tag{B.36}$$

B.2 Vector Operators

We shall shortly extend the previously introduced notions to vector operators. Consider the operator

$$\frac{\partial}{\partial x^i} , \tag{B.37}$$

in Cartesian coordinates. Consider now a transformation $x^j \to x'^j$, where the primed coordinates can be Cartesian or curvilinear. By using the chain rule the previous operator will transform as

$$\frac{\partial}{\partial x^i} = \frac{\partial x'^j}{\partial x^i} \frac{\partial}{\partial x'^j} . \tag{B.38}$$

Using the short-hand notation

$$\partial_i = \frac{\partial x'^j}{\partial x^i} \partial'_j , \tag{B.39}$$

which is the covariant law of transformation. If we now consider that the operator ∂_i acts on a scalar field $\phi(x) = \phi(x')$ then

$$\partial_i \phi(x) = \frac{\partial x'^j}{\partial x^i} \partial_j' \phi(x'), \tag{B.40}$$

and therefore, $\partial_i \phi(x)$ transform as the components of a covector i.e., following the covariant law of transformation. This seems at first sight to be in conflict to what we deduced in Chaps. 1 and 2, when we stated that the components of $\nabla \phi$ transform as the components of a vector. This is true. However, as we have always worked with transformations that were given by orthogonal matrices (i.e., $R^T = R^{-1}$), when written by columns, both vectors and covectors transform the same way. Thus, in those particular cases everything is simpler and we don't have to make a clear distinction between vectors and covectors.[5]

Consider now the divergence of a vector field in Cartesian coordinates

$$\nabla \cdot \mathbf{A} = \partial_i A^i. \tag{B.41}$$

It should be obvious from the transformation rules of ∂_i and A^i that the previous result is an invariant quantity under a transformation from Cartesian coordinates to other Cartesian coordinates. However, this is not true for curvilinear coordinates. For curvilinear coordinates we have to substitute the ordinary partial derivative with the covariant derivative. Thus, if $\bar{A}^i$ are the components of $\mathbf{A}$ in curvilinear coordinates, then

$$\nabla \cdot \mathbf{A} = \partial_i A^i = \nabla_i \bar{A}^i. \tag{B.42}$$

Similar considerations can be made for the remaining operations over vectors, however as the expressions will be slightly more complicated we will not go into further details.

B.3   Practical Exercise

As a simple exercise to put in practice the previously introduced formalism and check its consistency, one can for example, calculate the Christoffel symbols in spherical coordinates, calculate the expression for the acceleration, and finally check that the result is consistent with the expression found in Chap. 2. As a second part one could obtain the expression for the divergence of a vector in curvilinear coordinates by using the covariant derivative i.e., $\nabla \cdot \mathbf{A} = \nabla_k \bar{A}^k$ (where $\bar{A}^k$ just as previously,

---

[5]In fact we can go through most physics subjects without even realising that these two entities are, in general, different.

denotes the components of **A** in curvilinear coordinates) and again check that the result is identical to the one obtained previously in Chap. 2.

Solution: We can calculate in a simple way the components of Γ_{ij}^l by using the expression (B.28). In spherical coordinates we have the following non-null Cristoffel symbols $\Gamma_{\theta\theta}^r$, $\Gamma_{\phi\phi}^r$, $\Gamma_{r\theta}^\theta = \Gamma_{\theta r}^\theta$, $\Gamma_{\phi\phi}^\theta$, $\Gamma_{r\phi}^\phi = \Gamma_{\phi r}^\phi$ and $\Gamma_{\theta\phi}^\phi = \Gamma_{\theta\theta}^\phi$. In order to calculate them we need the following relations

$$
\begin{aligned}
x &= r \sin\theta \cos\phi, \\
y &= r \sin\theta \sin\phi, \\
z &= r \cos\theta,
\end{aligned}
\tag{B.43}
$$

and the inverse relations we have also deduced in Chap. 2

$$
\begin{aligned}
r &= \sqrt{x^2 + y^2 + z^2}, \\
\theta &= \arccos\left(\frac{z}{r}\right), \\
\phi &= \arctan\left(\frac{y}{x}\right).
\end{aligned}
$$

We shall explicitly calculate the first coefficient $\Gamma_{\theta\theta}^r$ and directly present the results for the remaining ones. We obtain

$$
\Gamma_{\theta\theta}^r = \frac{\partial r}{\partial x^k}\frac{\partial^2 x^k}{\partial \theta^2} = \frac{\partial r}{\partial x}\frac{\partial^2 x}{\partial \theta^2} + \frac{\partial r}{\partial y}\frac{\partial^2 y}{\partial \theta^2} + \frac{\partial r}{\partial z}\frac{\partial^2 z}{\partial \theta^2}.
\tag{B.44}
$$

Introducing the explicit expressions we obtain

$$
\begin{aligned}
\Gamma_{\theta\theta}^r &= \frac{x}{r}r(-\sin\theta)\cos\phi + \frac{y}{r}r(-\sin\theta)\sin\phi + \frac{z}{r}r(-\cos\theta) \\
&= -r\sin^2\theta\cos^2\phi - r\sin^2\theta\sin^2\phi - r\cos^2\theta \\
&= -r.
\end{aligned}
\tag{B.45}
$$

The remaining coefficients are given by

$$
\Gamma_{\phi\phi}^r = -r\sin^2\theta, \quad \Gamma_{r\theta}^\theta = \frac{1}{r}, \quad \Gamma_{\phi\phi}^\theta = -\sin\theta\cos\theta,
$$

$$
\Gamma_{r\phi}^\phi = \frac{1}{r}, \quad \Gamma_{\theta\phi}^\phi = \frac{\cos\theta}{\sin\theta}.
\tag{B.46}
$$

Let's now move on and calculate the expression for the acceleration using (B.23). First we need to obtain the components of the velocity. We easily obtain them from their expressions in Cartesian coordinates, by applying the contravariant transformation law (B.9) i.e.,

$$v^r = \frac{\partial r}{\partial x}v^x + \frac{\partial r}{\partial y}v^y + \frac{\partial r}{\partial z}v^z = \frac{\partial r}{\partial x}\dot{x} + \frac{\partial r}{\partial y}\dot{y} + \frac{\partial r}{\partial z}\dot{z} = \dot{r}, \tag{B.47}$$

and

$$v^\theta = \dot{\theta}, \qquad v^\phi = \dot{\phi}, \tag{B.48}$$

which is simply the result of the expressions from (2.41) divided by the corresponding normalization factors h_r, h_θ and h_ϕ, as found in (B.35). Thus. the expressions for the components of the acceleration are simply given by

$$
\begin{aligned}
a^r &= \ddot{r} + \dot{\phi}^2\, \Gamma^r_{\phi\phi} + \dot{\theta}^2\, \Gamma^r_{\theta\theta} = \ddot{r} - r\dot{\phi}^2 \sin^2\theta - r\dot{\theta}^2, \\
a^\theta &= \ddot{\theta} + 2\dot{r}\dot{\theta}\, \Gamma^\theta_{r\theta} + \dot{\phi}^2\, \Gamma^\theta_{\phi\phi} = \ddot{\theta} + \frac{2}{r}\dot{r}\dot{\theta} - \dot{\phi}^2 \sin\theta \cos\theta, \\
a^\phi &= \ddot{\phi} + 2\dot{r}\dot{\phi}\, \Gamma^\phi_{r\phi} + 2\dot{\theta}\dot{\phi}\, \Gamma^\phi_{\theta\phi} = \ddot{\phi} + \frac{2}{r}\dot{r}\dot{\phi} + 2\frac{\cos\theta}{\sin\theta}\dot{\theta}\dot{\phi},
\end{aligned}
\tag{B.49}
$$

which again, are nothing but the expressions obtained in (2.43), divided by the corresponding normalization factors.

It is worth mentioning that having defined the scalar product using the metric tensor, and having defined the covariant derivative and the covariant and contravariant laws of transformation, the explicit use of the basis is no longer necessary. All the previous formalism is completely generic and, in Special or General Relativity for example, one can completely forget about the existence of the bases.

Let's now move on to the divergence of $\mathbf{A}$. Its expression reads

$$
\begin{aligned}
\nabla_k \bar{A}^k &= \bar{\partial}_k \bar{A}^k + \bar{A}^i \Gamma^k_{ki} \\
&= \partial_r A^r + \partial_\theta A^\theta + \partial_\phi A^\phi + A^\theta \Gamma^\phi_{\phi\theta} + A^r(\Gamma^\phi_{\phi r} + \Gamma^\theta_{\theta r}) \\
&= \partial_r A^r + \partial_\theta A^\theta + \partial_\phi A^\phi + A^\theta \frac{\cos\theta}{\sin\theta} + \frac{2}{r}A^r.
\end{aligned}
\tag{B.50}
$$

The previous result might seem different from the one obtained in (2.59). This is not true however. Again, the components of $\mathbf{V} \cdot \mathbf{A}$ that we have used in Chap. 2 are written in the unitary basis. Thus, if we call A'^r, A'^θ and A'^ϕ these components in the unitary basis, the relation between them and the components from the non-unitary basis that we are using here is, just as in (B.35)

$$A^r = A'^r, \qquad A^\theta = \frac{A'^\theta}{r}, \qquad A^\phi = \frac{A'^\phi}{r \sin\theta}, \tag{B.51}$$

which corresponds indeed to the same result for the divergence of $\mathbf{A}$.

Appendix C
Passive and Active Transformations

Passive transformations is a notion that we are very used to in physics. It has to do with some observer describing some physical phenomena from different reference frames. This is why the transformations are called passive. On the other hand, for active transformations, the observer does not change the reference frame at any time, but modifies the coordinates of the studied object itself. Both transformations are related as we shall see in the following. Here we will analyse two representative examples: translations and two-dimensional rotations. Any other transformation can be extrapolated or deduced from the main ideas presented here.

Let's start with a translation along the $\mathbf{i}$ axis with respect to an $\{\mathcal{O}, \mathbf{i}, \mathbf{j}, \mathbf{k}\}$ reference system. Consider a point-like particle with coordinates ($x > 0$)

$$\mathbf{r} = x\,\mathbf{i}, \tag{C.1}$$

in the previously defined reference system. If we make a translation

$$\mathbf{c} = -c_x\,\mathbf{i}, \qquad c_x > 0, \tag{C.2}$$

of the origin $\mathcal{O}$ of the reference system (as shown in Fig. C.1, left), then the coordinates of the particle in the $\{\mathcal{O}', \mathbf{i}, \mathbf{j}, \mathbf{k}\}$ frame will be given by

$$\mathbf{r}' = (x + c_x)\,\mathbf{i}. \tag{C.3}$$

This transformation of the coordinates, which is passive, is equivalent to a translation of the object

$$-\mathbf{c} = c_x\,\mathbf{i}, \qquad c_x > 0, \tag{C.4}$$

as shown in Fig. C.1, right. $\mathbf{r}'$ would have the same expression as in (C.3) in this case. This is an active translation, because we haven't changed the reference frame but directly displaced the object.

© Springer Nature Switzerland AG 2020

V. Ilisie, *Lectures in Classical Mechanics*, Undergraduate Lecture Notes in Physics, https://doi.org/10.1007/978-3-030-38585-9

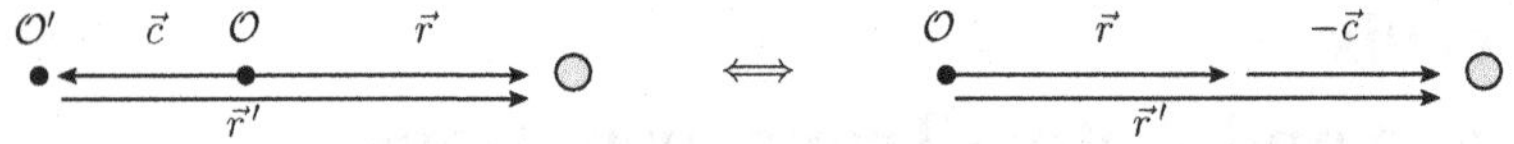

Fig. C.1 Passive (left) versus active (right) translation

We observe that a passive translation **c** can be expressed as an active translation in the opposite sense i.e., −**c**. This will always be the case. By changing the sign of some magnitude involved in the transformation (some coordinates, angles, velocities, etc.) we can switch from an active to a passive transformation. Although the result turns out to be the same in both cases, there is one subtle difference. In the case of a passive transformation the new coordinates of the object are expressed in terms of a new reference system, obtained from the initial one by means of some transformation of the basis. For an active transformation the new coordinates are given in terms of the old basis, but for the translated object.

Let's now turn our attention to a 2D rotation. Consider a rotation about the **k** axis of the $\{\mathcal{O}, \mathbf{i}, \mathbf{j}, \mathbf{k}\}$ reference system, as shown in Fig. C.2 (left), with $\theta > 0$ (counterclockwise). We have seen in Chap. 1 that the new basis $\{\mathbf{i}', \mathbf{j}', \mathbf{k}'\}$ (with the same origin) can be written as

$$\begin{pmatrix} \mathbf{i}' \\ \mathbf{j}' \end{pmatrix} = \begin{pmatrix} \cos\theta & \sin\theta \\ -\sin\theta & \cos\theta \end{pmatrix} \begin{pmatrix} \mathbf{i} \\ \mathbf{j} \end{pmatrix}, \tag{C.5}$$

and $\mathbf{k} = \mathbf{k}'$. Thus the coordinates $\mathbf{r} = x\,\mathbf{i} + y\,\mathbf{j}$ of a point-like particle (the gray blob) will transform as

$$\begin{pmatrix} x' \\ y' \end{pmatrix} = \begin{pmatrix} \cos\theta & \sin\theta \\ -\sin\theta & \cos\theta \end{pmatrix} \begin{pmatrix} x \\ y \end{pmatrix}. \tag{C.6}$$

This transformation of coordinates is equivalent to an active transformation of angle $-\theta$ as shown in Fig. C.2 (right). Again notice, that in the case of the active transformation, the coordinates will transform as in (C.6), but there is no corresponding transformation of the basis (C.5) as there is no new basis.

Even if it may seem redundant we shall also make the following clarification due to the fact that some authors use active transformations. An active transformation of angle $\theta > 0$ (counter-clockwise) will be equivalent to a passive transformation of $-\theta$. In both cases the coordinate transformation reads

$$\begin{pmatrix} x' \\ y' \end{pmatrix} = \begin{pmatrix} \cos\theta & -\sin\theta \\ \sin\theta & \cos\theta \end{pmatrix} \begin{pmatrix} x \\ y \end{pmatrix}. \tag{C.7}$$

This is why we find in the literature the previous transformation, where the "−" sign of $\sin\theta$ is present in the first row of the transformation matrix and not in the second one. This is schematically shown in Fig. C.3.

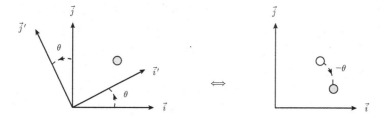

Fig. C.2 Passive (left) versus active (right) rotation about the **z** axis

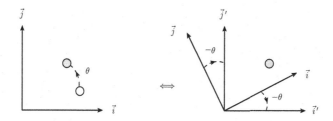

Fig. C.3 Active (left) versus passive (right) rotation about the **z** axis

The results presented here can be further generalized. An active translation with an arbitrary direction and sense **c** is equivalent to a passive translation −**c**. A generic rotation (about an arbitrary axis) can be expressed in terms of the three Euler angles as we shall see in Appendix F. We can therefore switch from active to passive rotations by changing the sign of the there Euler angles. This is of course valid also for Galilean transformations an Lorentz boosts. By changing the sign of the velocity we go from passive to active transformations and vice-versa.

Passive rotations about the **i, j** and the **k** axes of angle ϕ are given by the following matrices[6]

$$R_\phi^{\mathbf{i}} = \begin{pmatrix} \cos\phi & \sin\phi & 0 \\ -\sin\phi & \cos\phi & 0 \\ 0 & 0 & 1 \end{pmatrix}, \qquad R_\phi^{\mathbf{j}} = \begin{pmatrix} \cos\phi & 0 & -\sin\phi \\ 0 & 1 & 0 \\ \sin\phi & 0 & \cos\phi \end{pmatrix},$$

$$R_\phi^{\mathbf{k}} = \begin{pmatrix} 0 & 0 & 1 \\ 0 & \cos\phi & \sin\phi \\ 0 & -\sin\phi & \cos\phi \end{pmatrix}. \tag{C.8}$$

The corresponding active rotations are given by changing the sign of $\sin\phi$.

---

[6]The sign of $\sin\phi$ in $R_\phi^{\mathbf{j}}$ is correct, it is not a typo.

Appendix D
Vector Operators in Curvilinear Coordinates

For this first part we shall make use of Gauss's theorem. For a volume V enclosed by a surface ∂V, given a vector field $\mathbf{A}$, the theorem reads

$$\iiint_V \nabla \cdot \mathbf{A}\, dV = \oiint_{\partial V} \mathbf{A} \cdot d\mathbf{s}. \tag{D.1}$$

If the volume is differential, then the previous expression simplifies, and it reduces to

$$\nabla \cdot \mathbf{A} = \frac{1}{dV} \sum_i \mathbf{A} \cdot d\mathbf{s}^{(i)}, \tag{D.2}$$

where we have passed dV from the left side to the right side of the equation, and where the sum runs over the differential surfaces that enclose the differential volume dV. This means that the divergence of a field $\mathbf{A}$ evaluated at a point P, is equal to the flux of the field through the surface that encloses the differential volume that contains the point P (times a normalization factor $1/dV$). As we are working with infinitesimal volumes and surfaces i.e., in the limit s, $V \rightarrow 0$, it does not matter where we place the point P inside the volume. For simplicity, for the following calculations we shall assume that it is placed in one of the corners, as it is shown in Fig. D.1.

Expanding $\mathbf{A}$ in terms if the curvilinear basis we get

$$\mathbf{A} = A_\alpha\, \mathbf{u}_\alpha + A_\beta\, \mathbf{u}_\beta + A_\gamma\, \mathbf{u}_\gamma. \tag{D.3}$$

We shall only write down explicitly the contribution to (D.2) of the two surfaces whose normal vectors are $\mathbf{u}_\gamma$ and $-\mathbf{u}_\gamma$ (denoted as (2) and (1) in Fig. D.1), as the rest of the contributions can be derived trivially from this one. The minus sign, in front of $\mathbf{u}_\gamma$ for the surface (1), is due to the fact that the normal vector always points towards the exterior region of the cube. We have

© Springer Nature Switzerland AG 2020
V. Ilisie, *Lectures in Classical Mechanics*, Undergraduate Lecture Notes in Physics, https://doi.org/10.1007/978-3-030-38585-9

Fig. D.1 The infinitesimal
volume dV that contains the
point $P(\alpha, \beta, \gamma)$

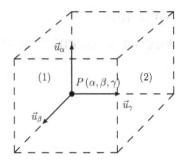

$$\nabla \cdot \mathbf{A} = \frac{1}{dV} \left(A_\gamma^{(2)} h_\alpha^{(2)} h_\beta^{(2)} - A_\gamma^{(1)} h_\alpha^{(1)} h_\beta^{(1)} \right) d\alpha \, d\beta + \cdots$$

$$= \frac{1}{dV} \Big[A_\gamma(\gamma + d\gamma) \, h_\alpha(\gamma + d\gamma) \, h_\beta(\gamma + d\gamma)$$

$$- A_\gamma(\gamma) \, h_\alpha(\gamma) \, h_\beta(\gamma) \Big] d\alpha \, d\beta + \cdots . \tag{D.4}$$

Remembering that the definition of the partial derivative is given by the following
expression

$$\frac{\partial f}{\partial x} = \lim_{\delta x \to 0} \frac{f(x + \delta x, y, z) - f(x, y, z)}{\delta x}, \tag{D.5}$$

we simply obtain

$$\nabla \cdot \mathbf{A} = \frac{d\alpha \, d\beta \, d\gamma}{dV} \frac{\partial}{\partial \gamma} \left(A_\gamma \, h_\alpha \, h_\beta \right) + \cdots . \tag{D.6}$$

Putting together all the three contributions we obtain the expression we were interested in deducing

$$\nabla \cdot \mathbf{A} = \frac{1}{h_\alpha \, h_\beta \, h_\gamma} \left[\frac{\partial}{\partial \alpha} \left(A_\alpha \, h_\beta \, h_\gamma \right) + \frac{\partial}{\partial \beta} \left(A_\beta \, h_\alpha \, h_\gamma \right) + \frac{\partial}{\partial \gamma} \left(A_\gamma \, h_\alpha \, h_\beta \right) \right] . \tag{D.7}$$

We will now calculate the expression for the curl $\nabla \times \mathbf{A}$ in a similar way. Consider
Stokes' theorem. For a given arbitrary surface s (simply connected i.e., with no holes)
enclosed by a curve ∂s, given a vector field $\mathbf{A}$, the theorem states

$$\iint_s (\nabla \times \mathbf{A}) \cdot ds = \oint_{\partial s} \mathbf{A} \cdot d\mathbf{r} . \tag{D.8}$$

For a differential surface the previous expression turns into

Fig. D.2 The infinitesimal
surface $ds_{\beta\gamma}$ that contains
the point $P(\alpha, \beta, \gamma)$

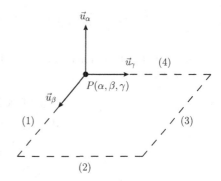

$$(\nabla \times \mathbf{A}) \cdot d\mathbf{s} = \sum_i \mathbf{A} \cdot d\mathbf{r}^{(i)}, \qquad (D.9)$$

where $d\mathbf{r}^{(i)}$ are the differential segments that enclose the differential surface $d\mathbf{s}$. Defining the pseudo-vector

$$\mathbf{W} \equiv \nabla \times \mathbf{A}, \qquad (D.10)$$

and expanding it in terms of the curvilinear basis we have

$$\mathbf{W} = W_\alpha \, \mathbf{u}_\alpha + W_\beta \, \mathbf{u}_\beta + W_\gamma \, \mathbf{u}_\gamma. \qquad (D.11)$$

Keeping in mind that (D.9) is valid for any arbitrary surface, let us choose it conveniently. By choosing it $d\mathbf{s} = h_\beta \, h_\gamma \, d\beta \, d\gamma \, \mathbf{u}_\alpha = ds_{\beta\gamma} \, \mathbf{u}_\alpha$ we obtain

$$W_\alpha = \frac{1}{ds_{\beta\gamma}} \sum_i \mathbf{A} \cdot d\mathbf{r}^{(i)}, \qquad (D.12)$$

where $d\mathbf{r}^{(i)}$ are the segments that enclose our chosen surface. This is shown in Fig. D.2 and it must be interpreted the following way. Given a point P, the component of the curl of a vector field $\mathbf{A}$, along $\mathbf{u}_\alpha$ evaluated in P, is given by the contour integral (which in this case is a discrete sum) over the curve that encloses differential surface whose normal vector is $\mathbf{u}_\alpha$ and that contains P (times a normalization factor $1/ds_{\beta\gamma}$).

The result of the contour integral along the path $1 \to 2 \to 3 \to 4$ is

$$
\begin{aligned}
W_\alpha &= \frac{1}{ds_{\beta\gamma}} \left(A_\beta^{(1)} h_\beta^{(1)} d\beta + A_\gamma^{(2)} h_\gamma^{(2)} d\gamma - A_\beta^{(3)} h_\beta^{(3)} d\beta - A_\gamma^{(4)} h_\gamma^{(4)} d\gamma \right) \\
&= \frac{d\beta}{ds_{\beta\gamma}} \left(A_\beta^{(1)} h_\beta^{(1)} - A_\beta^{(3)} h_\beta^{(3)} \right) + \frac{d\gamma}{ds_{\beta\gamma}} \left(A_\gamma^{(2)} h_\gamma^{(2)} - A_\gamma^{(4)} h_\gamma^{(4)} \right) \\
&= \frac{d\beta}{ds_{\beta\gamma}} \left[A_\beta(\gamma) h_\beta(\gamma) - A_\beta(\gamma + d\gamma) h_\beta(\gamma + d\gamma) \right]
\end{aligned}
$$

$$+ \frac{d\gamma}{ds_{\beta\gamma}}\left[A_\gamma(\beta+d\beta)h_\gamma(\beta+d\beta) - A_\gamma(\beta)h_\gamma(\beta)\right]. \qquad \text{(D.13)}$$

Again, using the definition of the partial derivative (D.5), we obtain the final expression for W_α

$$\begin{aligned}
W_\alpha &= \frac{d\gamma\, d\beta}{ds_{\beta\gamma}}\left[\frac{\partial(A_\gamma h_\gamma)}{\partial\beta} - \frac{\partial(A_\beta h_\beta)}{\partial\gamma}\right] \\
&= \frac{1}{h_\beta h_\gamma}\left[\frac{\partial(A_\gamma h_\gamma)}{\partial\beta} - \frac{\partial(A_\beta h_\beta)}{\partial\gamma}\right].
\end{aligned} \qquad \text{(D.14)}$$

In a similar way we can deduce the rest of the expression. The final result can be easily written in matrix form as the following determinant

$$\nabla \times \mathbf{A} = \frac{1}{h_\alpha h_\beta h_\gamma}\begin{vmatrix} h_\alpha\,\mathbf{u}_\alpha & h_\beta\,\mathbf{u}_\beta & h_\gamma\,\mathbf{u}_\gamma \\ \dfrac{\partial}{\partial\alpha} & \dfrac{\partial}{\partial\beta} & \dfrac{\partial}{\partial\gamma} \\ h_\alpha\,A_\alpha & h_\beta\,A_\beta & h_\gamma\,A_\gamma \end{vmatrix}. \qquad \text{(D.15)}$$

Appendix E
Variable Mass Equation

Here we present an alternative deduction of the variable mass problem, from the point of view of momentum conservation. Consider an infinitesimal mass Δm moving at velocity $\mathbf{u}$ that collides with an object of mass m moving at velocity $\mathbf{v}$ as shown below.

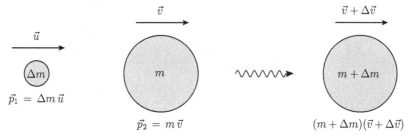

The center-of-mass momentum before the collision is

$$\mathcal{P}_i = \mathbf{p}_1 + \mathbf{p}_2 = \Delta m \mathbf{u} + m \mathbf{v}. \tag{E.1}$$

After the collision it will be given by

$$\mathcal{P}_f = (m + \Delta m)(\mathbf{v} + \Delta \mathbf{v}) = m\mathbf{v} + m\Delta\mathbf{v} + \Delta m \mathbf{v} + \mathcal{O}(\Delta^2). \tag{E.2}$$

Therefore the center-of-mass momentum variation simply reads

$$\begin{aligned}
\Delta \mathcal{P} &= \mathcal{P}_f - \mathcal{P}_f \\
&= m\Delta\mathbf{v} + \Delta m(\mathbf{v} - \mathbf{u}) + \mathcal{O}(\Delta^2) \\
&= m\Delta\mathbf{v} + \Delta m \mathbf{v}_r + \mathcal{O}(\Delta^2),
\end{aligned} \tag{E.3}$$

where we have defined the relative velocity $\mathbf{v}_r \equiv \mathbf{v} - \mathbf{u}$. Dividing the previous expression by Δt and taking the limit $\Delta \to 0$, neglecting internal forces, we obtain $\mathcal{O}(\Delta^2) \to 0$ and

V. Ilisie, *Lectures in Classical Mechanics*, Undergraduate Lecture Notes in Physics, https://doi.org/10.1007/978-3-030-38585-9

$$\mathbf{F}^{ext} = \frac{d\mathbf{P}}{dt} = m\frac{d\mathbf{v}}{dt} + \mathbf{v}_r\frac{dm}{dt}, \qquad (E.4)$$

which is the expected result. For a mass m that expels an infinitesimal dm one obtains the same expression.

Appendix F
Rotations, Euler Angles and Angular Velocity

Consider a reference system given by $\{\mathbf{i}, \mathbf{j}, \mathbf{k}\}$ in Cartesian coordinates. Consider a second reference frame given by $\{\mathbf{i'}, \mathbf{j'}, \mathbf{k'}\}$ with the same origin, that is obtained from the original basis through an arbitrary proper orthogonal rotation as in (1.28)

$$\begin{pmatrix} \mathbf{i'} \\ \mathbf{j'} \\ \mathbf{k'} \end{pmatrix} = R \begin{pmatrix} \mathbf{i} \\ \mathbf{j} \\ \mathbf{k} \end{pmatrix} . \tag{F.1}$$

So far we haven't explicitly treated the components of the rotation matrix R. Here we shall use Euler's angles in order to describe the matrix components R_{ij}. These angles will turn out to be extremely useful in describing general rigid solid rotations and thus, they can also be used to describe motion in non-inertial reference frames.

Consider a rotation about the $\mathbf{k}$ axis (of angle ϕ) of the initial basis. We obtain the intermediate basis $\{\mathbf{i}_1, \mathbf{j}_1, \mathbf{k}_1\}$ given by

$$\begin{pmatrix} \mathbf{i}_1 \\ \mathbf{j}_1 \\ \mathbf{k}_1 \end{pmatrix} = \begin{pmatrix} \cos\phi & \sin\phi & 0 \\ -\sin\phi & \cos\phi & 0 \\ 0 & 0 & 1 \end{pmatrix} \begin{pmatrix} \mathbf{i} \\ \mathbf{j} \\ \mathbf{k} \end{pmatrix} \equiv R_\phi \begin{pmatrix} \mathbf{i} \\ \mathbf{j} \\ \mathbf{k} \end{pmatrix} , \tag{F.2}$$

with $\mathbf{k}_1 = \mathbf{k}$. Now let us perform a second rotation about the new $\mathbf{i}_1$ axis (of angle θ). We obtain a second intermediate basis $\{\mathbf{i}_2, \mathbf{j}_2, \mathbf{k}_2\}$

$$\begin{pmatrix} \mathbf{i}_2 \\ \mathbf{j}_2 \\ \mathbf{k}_2 \end{pmatrix} = \begin{pmatrix} 1 & 0 & 0 \\ 0 & \cos\theta & \sin\theta \\ 0 & -\sin\theta & \cos\theta \end{pmatrix} \begin{pmatrix} \mathbf{i}_1 \\ \mathbf{j}_1 \\ \mathbf{k}_1 \end{pmatrix} \equiv R_\theta \begin{pmatrix} \mathbf{i}_1 \\ \mathbf{j}_1 \\ \mathbf{k}_1 \end{pmatrix} , \tag{F.3}$$

with $\mathbf{i}_1 = \mathbf{i}_2$. Finally let us perform a last rotation about the $\mathbf{k}_2$ axis (of angle ψ). We obtain the basis $\{\mathbf{i'}, \mathbf{j'}, \mathbf{k'}\}$

© Springer Nature Switzerland AG 2020
V. Ilisie, *Lectures in Classical Mechanics*, Undergraduate Lecture
Notes in Physics, https://doi.org/10.1007/978-3-030-38585-9

Fig. F.1 Schematic
representation of Euler's
angles

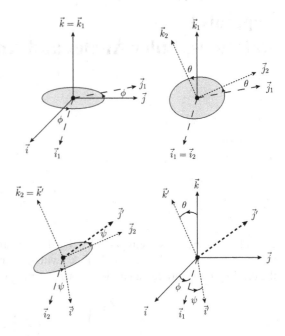

$$\begin{pmatrix} \mathbf{i}' \\ \mathbf{j}' \\ \mathbf{k}' \end{pmatrix} = \begin{pmatrix} \cos\psi & \sin\psi & 0 \\ -\sin\psi & \cos\psi & 0 \\ 0 & 0 & 1 \end{pmatrix} \begin{pmatrix} \mathbf{i}_2 \\ \mathbf{j}_2 \\ \mathbf{k}_2 \end{pmatrix} \equiv R_\psi \begin{pmatrix} \mathbf{i}_2 \\ \mathbf{j}_2 \\ \mathbf{k}_2 \end{pmatrix}, \qquad (\text{F.4})$$

with $\mathbf{k}_2 = \mathbf{k}'$.

With $\phi \in [0, 2\pi)$, $\theta \in [0, \pi]$, $\psi \in [0, 2\pi)$ one can describe any arbitrary proper
orthogonal rotation of a basis. Note that we have used twice the $\mathbf{k}$ axis and never
the $\mathbf{j}$ axis. This is just a matter of convention. One can use other set of angles,
however, we shall see in Chap. 10 that the three angles described here have direct
standard interpretation in physics i.e. *precession, nutation* and *spin*.[7] Schematically
the rotations and the corresponding angles described previously are shown in Fig. F.1.

We can use the matrix product to combine all three rotations into a single matrix

$$\begin{pmatrix} \mathbf{i}' \\ \mathbf{j}' \\ \mathbf{k}' \end{pmatrix} = R_\psi R_\theta R_\phi \begin{pmatrix} \mathbf{i} \\ \mathbf{j} \\ \mathbf{k} \end{pmatrix} = R_{\psi\theta\phi} \begin{pmatrix} \mathbf{i} \\ \mathbf{j} \\ \mathbf{k} \end{pmatrix} \equiv R \begin{pmatrix} \mathbf{i} \\ \mathbf{j} \\ \mathbf{k} \end{pmatrix}. \qquad (\text{F.5})$$

One can check as an exercise that the explicit expression of $R \equiv R_{\psi\theta\phi}$ is given by

---

[7]In aeronautical engineering one normally uses a rotation about $\mathbf{i}$, another one about $\mathbf{j}$ and finally
one about the $\mathbf{k}$ axis. These angles are associated to plane motion such as *pitch, yaw* and *roll*.

$$R_{\psi\theta\phi} = \begin{pmatrix} c\phi\,c\psi - s\phi\,c\theta\,s\psi & s\phi\,c\psi + c\phi\,c\theta\,s\psi & s\theta\,s\psi \\ -c\phi\,s\psi - s\phi\,c\theta\,c\psi & -s\phi\,s\psi + c\phi\,c\theta\,c\psi & s\theta\,c\psi \\ s\phi\,s\theta & -c\phi\,s\theta & c\theta \end{pmatrix}, \qquad (F.6)$$

where we have introduced the short-hand notation $c \equiv \cos$ and $s \equiv \sin$.

Let us now consider that these three angles vary with time and so $R = R(t)$.[8] Thus we can naturally define an angular velocity $\boldsymbol{\omega}$ as follows

$$\boldsymbol{\omega} = \dot{\phi}\mathbf{k} + \dot{\theta}\mathbf{i}_1 + \dot{\psi}\mathbf{k}'. \qquad (F.7)$$

Note that in the previous expression each term is expressed in a different basis. However, it is more useful to express the previous quantity with all its components in one basis. We shall choose this basis to be the primed one. Thus, rotating $\dot{\phi}\mathbf{k}$ and $\dot{\theta}\mathbf{i}_1$ into the primed basis we obtain

$$\boldsymbol{\omega} = \omega_1'\mathbf{i}' + \omega_2'\mathbf{j}' + \omega_3'\mathbf{k}', \qquad (F.8)$$

with its components given by

$$\begin{aligned} \omega_1' &= \dot{\phi}\sin\theta\sin\psi + \dot{\theta}\cos\psi, \\ \omega_2' &= \dot{\phi}\sin\theta\cos\psi - \dot{\theta}\sin\psi, \\ \omega_3' &= \dot{\phi}\cos\theta + \dot{\psi}. \end{aligned} \qquad (F.9)$$

We shall see in the following that this angular velocity its is actually related to $dR_{\psi\theta\phi}/dt$. As the angles depend on time one can straightforwardly calculate the variation of $R_{\psi\theta\phi}$. We can explicitly check as an exercise that $dR_{\psi\theta\phi}/dt$ is equal to the matrix product $\Omega\,R_{\psi\theta\phi}$

$$\frac{dR_{\psi\theta\phi}}{dt} = \Omega\,R_{\psi\theta\phi}, \qquad (F.10)$$

where Ω is given by

$$\Omega = \begin{pmatrix} 0 & \omega_3' & -\omega_2' \\ -\omega_3' & 0 & \omega_1' \\ \omega_2' & -\omega_1' & 0 \end{pmatrix}. \qquad (F.11)$$

This result will turn out to be useful for describing motion in non-inertial reference systems and motion of rigid bodies.

---

[8]Therefore $\mathbf{i}' = \mathbf{i}'(t)$, $\mathbf{j}' = \mathbf{j}'(t)$ and $\mathbf{k}' = \mathbf{k}'(t)$. However, in order to keep the notation simple we shall not write down explicitly this time dependence.

Appendix G
Three-Body Particle Decays

We shall only give the relevant results in the CM reference frame. The corresponding configuration is schematically shown in Fig. G.1. The four-momenta are given by

$$p_a^\mu = (m_a c, \mathbf{0}) \,,$$
$$p_i^\mu = (E_i/c, \mathbf{p}_i) \,, \tag{G.1}$$

satisfying the following relations

$$\sum_i \mathbf{p}_i = \mathbf{0} \,, \qquad m_a c = \sum_i E_i/c \,. \tag{G.2}$$

Due to the fact that $\mathbf{p}_1 + \mathbf{p}_2 + \mathbf{p}_3 = \mathbf{0}$ the process takes place in a same plane. Introducing the angles from Fig. G.1 we obtain

$$|\mathbf{p}_1| + |\mathbf{p}_2| \cos\theta_{12} + |\mathbf{p}_3| \cos\theta_{13} = 0 \,,$$
$$|\mathbf{p}_2| \sin\theta_{12} + |\mathbf{p}_3| \sin\theta_{13} = 0 \,. \tag{G.3}$$

We can now introduce the following Lorentz invariant kinematical variables

$$t_1 \equiv s_{23} \equiv (p_a - p_1)^2 = (p_2 + p_3)^2 \,,$$
$$t_2 \equiv s_{13} \equiv (p_a - p_2)^2 = (p_1 + p_3)^2 \,,$$
$$t_3 \equiv s_{12} \equiv (p_a - p_3)^2 = (p_1 + p_2)^2 \,. \tag{G.4}$$

One can check that the quantities t_i satisfy the following sum

$$\sum_i t_i = m_a^2 c^2 + \sum_i m_i^2 c^2 \,. \tag{G.5}$$

It can be easily shown that the next relation also holds

© Springer Nature Switzerland AG 2020
V. Ilisie, *Lectures in Classical Mechanics*, Undergraduate Lecture Notes in Physics, https://doi.org/10.1007/978-3-030-38585-9

Fig. G.1 Three-body decay
as seen from the center of
mass reference frame

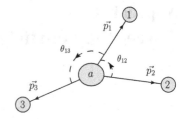

$$t_i = (p_a - p_i)^2 = p_a^2 + p_i^2 - 2p_a \cdot p_i = m_a^2 c^2 + m_i^2 c^2 - 2m_a\, E_i \ . \qquad \text{(G.6)}$$

Therefore we can express E_i and $|\mathbf{p}_i|$ in terms of Lorentz invariant quantities such as masses and the previously introduced t_i variables

$$E_i = \frac{1}{2m_a}(m_a^2 c^2 + m_i^2 c^2 - t_i) \ .$$

$$|\mathbf{p}_i| = \frac{1}{2m_a\, c}\lambda^{1/2}(m_a^2 c^2, m_i^2 c^2, t_i) \ . \qquad \text{(G.7)}$$

The given threshold values, for the decay to take place, for s and s_{ij} are denoted as s_{th} and s_{ij}^{th} and they satisfy

$$s \equiv p_a^2 = m_a^2 c^2 \geqslant s_{th} = (m_1 + m_2 + m_3)^2 c^2 \ ,$$

$$s_{ij} = t_k \geqslant s_{ij}^{th} = (m_i + m_j)^2 c^2 = t_k^{th} \ , \qquad \text{(G.8)}$$

with $(i \neq j \neq k)$. From (G.7) and (G.8) is easy to deduce that the maximum value of the energy of the particle i in the CM frame is

$$E_i^{\max} = \frac{c^2}{2m_a}(m_a^2 + m_i^2 - (m_j + m_k)^2) \ , \qquad \text{(G.9)}$$

with $(i \neq j \neq k)$.

Further Reading

1. J.V. José, E.J. Saletan, *Classical Dynamics: A Contemporary Approach*. Cambridge University Press
2. S.T. Thornton, J.B. Marion, *Classical Dynamics of Particles and Systems*
3. B.C. Consuelo, G.F. Antonio, R.D. Marcelo, *Campos electromagnéticos*. Editorial Universidad de Sevilla
4. H. Goldstein, C. Poole, J. Safko, *Classical Mechanics*, 3rd edn. Addison Wesley
5. J.R. Taylor, *Classical Mechanics*
6. D.T. Greenwood, *Classical Dynamics*. Prentice-Hall Inc.
7. D. Kleppner, R. Kolenkow, *An Introduction to Mechanics*

8. C. Lanczos, *The Variational Principles of Mechanics*. Dover Publications Inc.
9. W. Greiner, *Classical Mechanics: Systems of Particles and Hamiltonian Dynamics*. Springer
10. H.C. Corben, P. Stehle, *Classical Mechanics*, 2nd edn. Dover Publications Inc.
11. T.W.B. Kibble, F.H. Berkshire, *Classical Mechanics*. Imperial College Press
12. M.G. Calkin, *Lagrangian and Hamiltonian Mechanics*
13. A.J. French, M.G. Ebison, *Introduction to Classical Mechanics*
14. V. Ilisie, *Concepts in Quantum Field Theory: A Practitioner's Toolkit*. Springer
15. C.W. Misner, K.S. Thorne, J.A. Wheeler, *Gravitation*

Index

© Springer Nature Switzerland AG 2020
V. Ilisie, *Lectures in Classical Mechanics*, Undergraduate Lecture
Notes in Physics, https://doi.org/10.1007/978-3-030-38585-9

Printed in the United States
By Bookmasters